Co-crystals
Preparation, Characterization and Applications

Monographs in Supramolecular Chemistry

Series editors:
Philip Gale, *The University of Sydney, Australia*
Jonathan Steed, *Durham University, UK*

Titles in this series:
1: Cyclophanes
2: Calixarenes
3: Crown Ethers and Cryptands
4: Container Molecules and Their Guests
5: Membranes and Molecular Assemblies: The Synkinetic Approach
6: Calixarenes Revisited
7: Self-assembly in Supramolecular Systems
8: Anion Receptor Chemistry
9: Boronic Acids in Saccharide Recognition
10: Calixarenes: An Introduction, 2nd Edition
11: Polymeric and Self Assembled Hydrogels: From Fundamental Understanding to Applications
12: Molecular Logic-based Computation
13: Supramolecular Systems in Biomedical Fields
14: Synthetic Receptors for Biomolecules: Design Principles and Applications
15: Polyrotaxane and Slide-Ring Materials
16: Boron: Sensing, Synthesis and Supramolecular Self-Assembly
17: Porous Polymers: Design, Synthesis and Applications
18: Pillararenes
19: Supramolecular Chemistry at Surfaces
20: Aromatic Interactions: Frontiers in Knowledge and Application
21: Naphthalenediimide and its Congeners: From Molecules to Materials
22: Functional Supramolecular Materials: From Surfaces to MOFs
23: Supramolecular Amphiphiles
24: Co-crystals: Preparation, Characterization and Applications

How to obtain future titles on publication:
A standing order plan is available for this series. A standing order will bring delivery of each new volume immediately on publication.

For further information please contact:
Book Sales Department, Royal Society of Chemistry, Thomas Graham House, Science Park, Milton Road, Cambridge, CB4 0WF, UK
Telephone: +44 (0)1223 420066, Fax: +44 (0)1223 420247
Email: booksales@rsc.org
Visit our website at http://www.rsc.org/Shop/Books/

can provide an element of predictable structural periodicity. This, in turn, offers unique opportunities for synthesis of materials with bulk properties that can be 'dialled-in' simply by changing a molecular building block within a stable three-dimensional structural framework. The inherent modularity in co-crystal synthesis can act as a conduit between molecular structure, crystal structure, and solid-state properties that may present new opportunities for design, manufacture, and performance of a wide range of specialty chemicals and solid-state devices.

This book covers a wide range of topics related to co-crystallization, *e.g.* hydrogen- and halogen-bond based hierarchical assembly of co-crystals, mechanochemistry as a versatile route to co-crystal synthesis, pharmaceutical co-crystallization, co-crystals of energetic materials, magnetic and semiconducting co-crystals, and co-crystals in chemical separations. Although the list of topics is by no means exhaustive, we hope that this book will provide readers with a greater understanding of 'co-crystals', and also inspire new approaches to targeted and application-driven assembly of complex supramolecular architectures.

By now it is hopefully clear that the term 'co-crystals' represents a highly diverse scientific field of significant complexity and considerable promise for materials chemistry. Therefore, it would be quite a challenge to put together a multi-author book on co-crystals that provides a balanced, insightful, and up-to-date view of the field that is of equal use and importance to co-crystal novices and aficionados alike. We are absolutely delighted that we were able to gather an exceptionally strong field of co-authors, World-class scientists each and every one, and we would like to thank all of them for their outstanding contributions to this volume. All the credit for the high-points of this book should go to them; any mistakes and weaknesses are our responsibility alone.

Christer B. Aakeröy
Abhijeet S. Sinha

Co-crystals
Preparation, Characterization and Applications

Edited by

Christer B. Aakeröy
Kansas State University, USA
Email: aakeroy@ksu.edu

and

Abhijeet S. Sinha
Kansas State University, USA
Email: sinha@ksu.edu

ROYAL SOCIETY OF CHEMISTRY

Monographs in Supramolecular Chemistry No. 24

Print ISBN: 978-1-78801-115-0
PDF ISBN: 978-1-78801-287-4
EPUB ISBN: 978-1-78801-489-2
ISSN: 1368-8642

A catalogue record for this book is available from the British Library

The Royal Society of Chemistry is a charity, registered in England and Wales, Number 207890, and a company incorporated in England by Royal Charter (Registered No. RC000524), registered office: Burlington House, Piccadilly, London W1J 0BA, UK, Telephone: +44 (0) 207 4378 6556.

Visit our website at www.rsc.org/books

Printed in the United Kingdom by CPI Group (UK) Ltd, Croydon, CR0 4YY, UK

Preface

Interest in the design, synthesis, characterization, and applications of co-crystals has grown almost at an exponential rate in recent years, and the genuinely interdisciplinary nature of 'co-crystals' as a distinct research topic has made it one of the most prolific areas in current organic solid-state chemistry. The study of co-crystals is still considered to be a relatively new field (even though its history goes back almost two hundred years) and, within this research space, we have had opportunities to explore and improve our understanding of many fundamental issues related to intermolecular forces, molecular recognition events, nucleation, and crystallization.

It is also true that co-crystal synthesis represents an important aspect of chemical synthesis, and advances in this field have already positively impacted many different communities in the applied sciences. In practice, most protocols for effective and deliberate co-crystal synthesis will utilize a bottom-up approach for the assembly of discrete chemical entities (such as neutral organic molecules) into extended architectures with specific chemical compositions, topologies, and dimensionalities. The tool-box for co-crystal synthesis is filled with a wide variety of non-covalent forces including (but not limited to) $\pi \cdots \pi$ interactions, hydrogen- and halogen-bonds, and phenyl embraces. It is also clear that more 'exotic' representatives of non-covalent interactions such a chalcogen, pnictogen, and tetrel bonds (which have yet to be formally defined and officially named), can have important roles to play in future ventures in co-crystal synthesis.

No wonder then that this field cuts across many of the traditional areas of organic, physical, and solid-state chemistry, which, in turn has produced a vibrant scientific community that is supplying valuable insight into chemical bonding, new experimental techniques, and versatile strategies for non-covalent synthesis and practical crystal engineering. In some ways it feels as if we could be on the cusp of a minor scientific revolution because co-crystals

Contents

CHAPTER 1

Co-crystals: Introduction and Scope

CHRISTER B. AAKERÖY* AND ABHIJEET S. SINHA

Department of Chemistry, Kansas State University, Manhattan, Kansas 66506, USA
*Email: aakeroy@ksu.edu

1.1 Rationale and Scope

The structure and chemical composition of a molecule are responsible for a wide range of properties, especially in solution and in the gas-phase. However, the three-dimensional orientation and organization of molecules in a crystalline lattice determine many bulk properties of a substance, *e.g.* density, solubility, thermal stability, optical properties, hygroscopicity, and mechanical strength.[1] Consequently, if we are seeking to fine-tune or 'dial-in' physical properties through a bottom-up approach, then we need to acquire control over the metrics and topologies of the structural landscape that is inhabited by a particular target molecule. Unfortunately, this presents considerable challenges because molecular recognition events,[2] which provide the driving force behind any supramolecular assembly,[3] are the result of a delicate balancing act between relatively weak and reversible non-covalent interactions. Thankfully, supramolecular and structural chemistry have given us invaluable information about the inherent directionality and selectivity of many non-covalent interactions, and through the use of co-crystals we may be able to harness this knowledge in the preparation of materials with tailored physical properties.

One of the reasons why co-crystals offer unique opportunities for deliberate adjustments of bulk properties is that within a series of co-crystals of a target molecule, it may be possible to make modular changes to the crystalline framework that contains the 'active' molecules. This, in turn, may allow us to make incremental (and predictable!) changes to the physical properties and performance of a substance without having to alter any of the molecular properties of the target. The outcome from many systematic studies on synthesis and characterization of co-crystals are already having a positive impact on the deliberate design of new functional solids and materials. In addition, co-crystallization offers unique opportunities for probing intermolecular competitions, which in turn may help improve our central understanding of how fundamental laws of physics manifest themselves in crystalline materials. The inexorable (but not always obvious) connections between structure, crystal morphology and bulk properties means that with the ability to change specific aspects of the solid form of a compound comes new and unique prospects for materials design that could undoubtedly be highly beneficial to manufacturers and consumers alike. Finally, in order to continue to make progress in this area, we will need expertise from organic, inorganic, physical, materials, and theoretical chemists. This will require and facilitate unique interfaces of experimental and theoretical tools, and more extensive and open collaborations between academia and industry.

1.2 Covalent *Versus* Non-covalent Synthesis

Chemical transformations that proceed through bond-breaking and bond-making events involving covalent bonds have been studied and refined for well over 150 years.[4] A wealth of information has been acquired and as a result we now have a thorough understanding of the interplay between molecular structure, chemical reactivity, and reaction pathways. Consequently, organic synthesis has had an unparalleled impact on all aspects of modern society.

The design of synthetic pathways for co-crystal synthesis, and the practical implementation and optimization of such protocols, are undoubtedly part of chemical synthesis, and it is therefore inevitable that we occasionally try to draw comparisons between covalent and non-covalent synthesis. It is well known that organic synthetic chemists can guide site-specific reactivity *via* substituent-guided modulation of the electronic environment of a molecule. For example, the methyl group of toluene activates the benzene ring towards electrophilic aromatic substitution at *ortho* and *para* positions, whereas the nitro group deactivates the benzene ring in nitrobenzene and substitution occurs *meta* to the nitro group. However, in controlling and predicting the supramolecular assembly of molecules, the focus is on site-specific interactions and reactivity of an intermolecular nature. Given a system containing two strong, competing intermolecular interactions, can we tip the balance in favor of one interaction in order to control the resulting 3-D structure? Can crystal engineers use tools forged by synthetic organic chemists to construct supramolecular architectures? More specifically, can we use substituent effects to 'switch' intermolecular interactions on and off by altering the electronic environment of individual molecules? A grand challenge for crystal engineering is to develop synthetic protocols that are robust, versatile, and transferable and that, ultimately, can approach the level of reliability that name reactions offer to conventional chemical synthesis. At the same time, it is also important to set realistic goals and targets. The majority of chemical reactions tend to work on a narrow set of substrates, under specific reactions conditions or with tailor-made catalysts, and they have invariably been developed and optimized *via* lengthy procedures. Also, chemical synthesis can be achieved in multi-step reactions where it is possible to isolate, characterize, and purify intermediates and products along the way, but supramolecular synthesis is generally restricted to one-pot reactions. Hence, when measuring the relative success of a supramolecular synthetic protocol, it is reasonable to recall that, in chemical synthesis, which is correctly viewed as a mature science, many chemical reactions still fail to deliver the desired product in high yields. Thus, even though there are close similarities between covalent and non-covalent synthesis, Scheme 1.1, practitioners of the latter variety often face unique challenges.

1.3 History

There is general agreement that the history of co-crystals dates back to 1844 when Friedrich Wöhler combined solutions of quinone and hydroquinone resulting in a green solid, quinhydrone, which he proposed contained both reactants in 1 : 1 stoichiometry.[5] For several decades, there was disagreement

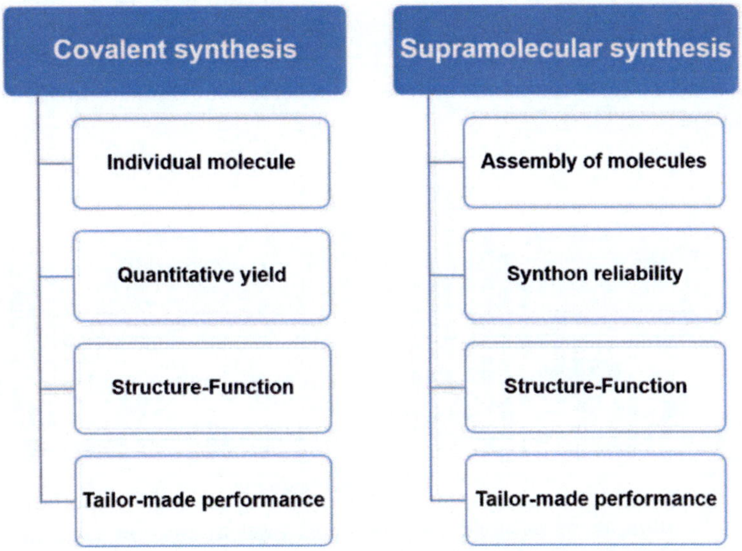

Scheme 1.1 Covalent *vs.* supramolecular synthesis.

about the exact nature of the chemical composition of Wöhler's product and, remarkably, it took over a century before the composition and structure of this crystalline material could be unambiguously verified through single-crystal X-ray diffraction.[6,7] Interestingly, a triclinic polymorph of quinhydrone was reported as late as in 1968.[8] A close examination of the literature from the late 19th and early 20th centuries reveals the presence of literally hundreds of co-crystals (although they would invariably appear under different names) and Stahly has published an excellent comprehensive summary of co-crystals reported prior to the year 2000.[9] Numerous review articles have also provided extensive coverage of the field as it has developed since the beginning of the new millennium.[10] The focus on co-crystals as a readily identifiable research owes a lot to Etter's groundbreaking work in the late 1980s and early 1990s,[11,12] as well as to Desiraju's seminal book on 'crystal engineering'.[13] The way that the field has developed since the early 1990s is readily illustrated by an examination of the relevant literature, Scheme 1.2.

The data clearly indicates that we are still in the midst of an ongoing expansion of the field, and there are no signs that interest in, or contributions to, this field have reached a plateau.

1.3.1 Crystal Engineering

'Crystal engineering' is the knowledge-based design and synthesis of *predictable* and *controllable* supramolecular architectures with desired properties. Schmidt used this term in 1971 as part of his study on photodimerization reactions in crystalline cinnamic acids.[14] The meaning of the term has broadened since then to include solid-state supramolecular chemistry, and a more comprehensive definition has been provided by Desiraju, "*the understanding of intermolecular interactions in the context of crystal packing and the utilization of such understanding in the design of new solids with desired physical and chemical properties*".[13] Thus, control over the

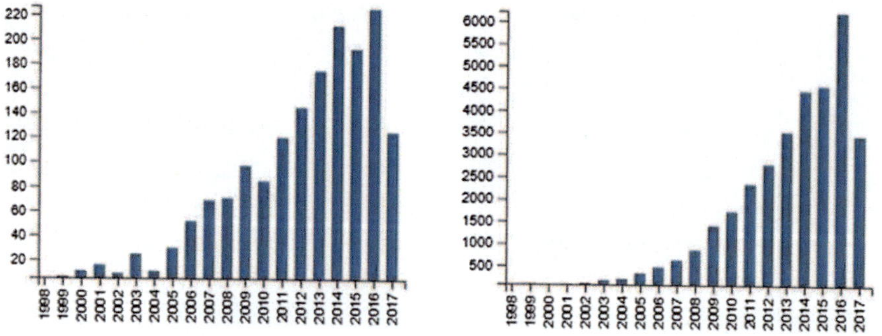

Scheme 1.2 Results from a SciFinder topics search on 'co-crystals' (July 2017). Total number of publications (left) and total number of citations (right). Publications on 'co-crystals' as related to antibodies/DNA/RNA/proteins *etc.* have been filtered out.

arrangement of molecules in the solid state, can enable us to control many of the bulk-properties of crystalline materials.[15]

In this chapter, we will provide a brief overview of the history behind the nomenclature of co-crystals, as well as highlight some practical and versatile supramolecular design strategies based on site-specific strong intermolecular interactions for the directed assembly of co-crystals of considerable complexity. We will also provide an overview of available computational tools in the design of co-crystals. Last, we will highlight some applications of co-crystallization where the bulk properties of solid molecular materials have been optimized.

1.4 Nomenclature

1.4.1 Can We Make It Without a Definition?

Oddly enough, the main topic of this book has not yet been defined in a completely unambiguous manner, and substances that can be referred to as co-crystals have also been classified as 'molecular complexes',[16] 'multi-component molecular crystals',[17,18] and 'ionic co-crystals'.[19] In 2012, forty-six authors[20] published a perspective article titled, 'Polymorphs, Salts, and Co-crystals: What's in a Name?' where they offered the following definition of a co-crystal; "*co-crystals are solids that are crystalline single phase materials composed of two or more different molecular and/or ionic compounds generally in a stoichiometric ratio*". It is worth noting that this suggested definition is broader than what appeared in a draft guidance given by the FDA ("*Solids that are crystalline materials composed of two or more molecules in the same crystal lattice*"),[21] which includes solvates, but excludes inclusion compounds and solid solutions. Suggestions have been made that the term 'co-crystal' is not even necessary; however, given the fact that interest in the synthesis, characterization, and applications of co-crystals has grown extremely rapidly with no signs of abating, the term is undoubtedly here to stay. At this point it may be appropriate to evoke an old Swedish proverb, 'kärt barn har många namn', which loosely translated means 'a beloved child has many names'.

Definitions of scientific terms can be helpful and informative, but attempts at crafting an exact description of the meaning of a substance can produce unnecessary scientific restrictions. There can also be considerable differences between the intent behind a definition; is it meant to be descriptive or prescriptive? The latter may be more useful in regulatory affairs, but a descriptive definition is more likely to provide an educational or instructive outline of the scientific framework surrounding a particular concept or event.

We recognize that there will always be exceptions to any definition and, in some cases, it is simply not possible to classify a substance in one way or another due to experimental limitations or ambiguities related to theoretical models. However, it is helpful for this book to at least have a working definition of the term 'co-crystal' that covers the vast majority of cases and systems that will be examined and presented in the following chapters. Again, we recognize that it is impossible to offer a definition that allows us to accurately classify every single crystalline solid because we do not always have the required experimental data in hand to allow us to make reliable distinctions between two types of solids. Furthermore, some crystalline solids contain such a variety of components (a crystal structure of β-C-arylglucoside contains ethanol, water, a zwitterionic and a neutral version of the target molecule),[22] that they will likely defeat even the most earnest attempt at providing a legally or scientifically unambiguous definition.

Hence, for the purpose of this book, we will primarily focus on crystalline materials that contain two or more neutral organic building blocks that are present in definite stoichiometric amounts. If one of the components in the lattice is a water molecule or a solvent molecule, the solid will be referred to as a hydrate, or a solvate, respectively. If one or more reactants that appear in the crystal are liquids at reaction temperature, the outcome is classified as a co-crystal. Finally, if a liquid acts as reactant and solvent simultaneously, the outcome is also a co-crystal (Scheme 1.3).

Understandably, it is possible to combine labels and classifications across the chart shown in Scheme 1.3 in a variety of ways, leading to sub-groups such as hydrated/solvated salts/co-crystals or ionic co-crystals. However, we are deliberately adopting a descriptive 'definition' that outlines the scientific space, which is appropriate for the vast majority of cases that are presented and discussed throughout this book.

1.4.2 Salt or Co-crystal?[23]

Since there is considerable debate about definitions and nomenclature surrounding co-crystals, it is not surprising that salts represent a particularly interesting and relevant group of multi-component solids. If co-crystals are viewed as multi-component solids containing only neutral species, and salts contain anions and cations, then we may face difficulties when experimental

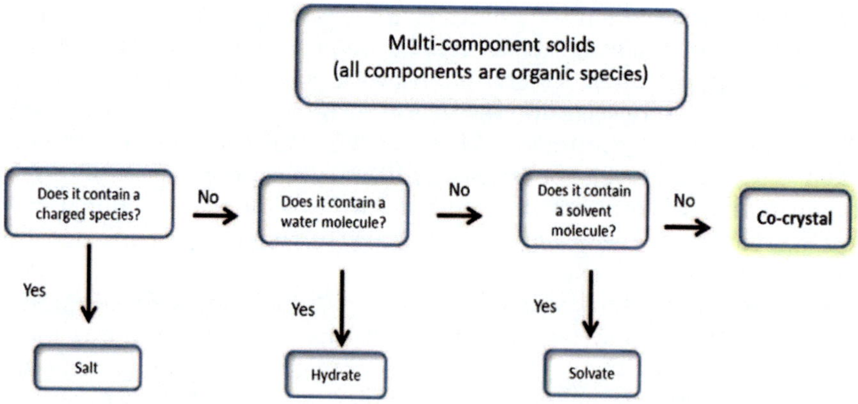

Scheme 1.3 Simplified flowchart for classifying organic multi-component solids.

methods struggle to unambiguously establish if a proton has been fully transferred from an acid to a base.[24] In such cases, it is not possible to determine which category of multi-component solids the compound belongs to. Consequently, definitions and nomenclature notwithstanding, it is probably more important (from all but a patent perspective) to find out if co-crystals and salts display any notable differences in terms of their ease of synthesis, properties, and/or behavior. In this context, it has been demonstrated that carboxylate salts are considerably more likely to (a) exist as solvates/hydrates; or (b) display unexpected chemical compositions (by including a neutral acid molecule in the lattice), than are co-crystals comprising only neutral carboxylic acids (Table 1.1).[25]

The proposed explanation for the discrepancy in structural behavior between co-crystals of carboxylic acids and their deprotonated counterparts is that the carboxylate anion is a far more powerful and demanding hydrogen-bond acceptor, which is not typically content with accepting a single $COO^{-}\cdots{}^{+}H–X$ hydrogen bond. Consequently, if the goal is to synthesize a multi-component solid with specific chemical composition or stoichiometry, neutral multi-component solids, *i.e.* co-crystals, as opposed to salts, offer considerable advantages in terms of predictability.

1.5 Design of Co-crystals

Given the abundance of papers published on co-crystals during the last twenty-five years, one may deduce that the synthesis of such multi-component crystalline solids is almost trivial. However, a heteromeric system is unlikely to form unless interactions between different molecules are favored over interactions between like molecules (which essentially corresponds to a negative deviation from Raoult's law). If the co-crystal is prepared from solution, it is often necessary that the two components have similar solubilities in the solvent of choice, otherwise the least soluble reactant is likely to precipitate by itself, leaving the more soluble reactant behind (this is of course the practical equivalent of a standard 're-crystallization' where impurities are removed from the desired solute). It may also be beneficial to use a co-former (co-crystal former) that is known to be polymorphic (as long as it meets other established criteria) because such molecules have already demonstrated an ability to exist in alternative packing arrangements and in different solid-state architectures. This structural flexibility (or structural

Table 1.1 Lattice contents in approximately 100 co-crystals and 130 salts extracted from the CSD.[25] Reprinted with permission from C. B. Aakeröy, M. E. Fasulo and J. Desper, *Mol. Pharm.*, 2007, **4**, 317–322, Copyright 2007 American Chemical Society.

Outcome	Co-crystals, %	Salts, %
Solvate/hydrate	5	19
Unexpected stoichiometry	1	14
Unexpected lattice content (total)	6	33

infidelity) means that the molecule is not thermodynamically locked into a single packing mode, thus making it more likely to coexist within a multi-component solid.

1.5.1 Etter's Rules and Graph Set Notation

Thanks to Etter's systematic work on hydrogen-bonded co-crystals in the late 1980s,[11,12] we have some empirical hydrogen-bonding guidelines for how such intermolecular interactions may appear in a system with multiple possible outcomes. Etter proposed that:

1. All good proton donors and acceptors are used in hydrogen bonding.
2. Six-membered-ring intramolecular hydrogen bonds form in preference to intermolecular hydrogen bonds.
3. The best proton donors and acceptors remaining after intramolecular hydrogen bond formation form intermolecular hydrogen bonds.

One key aspect of this set is that we need some way of determining what is the 'best' hydrogen-bond donor (or acceptor) and Aakeröy and co-workers have shown that a relative ranking of donor/acceptor strength can be achieved by associating each binding site with a value derived from a calculated molecular electrostatic potential surface.[26]

Etter *et al.* also introduced graph set notation for describing hydrogen-bonded motifs/synthons, as well as for differentiating between different motifs on the basis of number of hydrogen-bond donors and acceptors present in the crystal structure.[11,27] It is depicted as:

$$G^{\mathrm{a}}_{\mathrm{d}}(n)$$

where **G** denotes the observed pattern, ***n*** is the degree or the total number of atoms in the motif, **a** is the number of acceptors, and **d** is the number of donors. ***G*** can denote one of the four possible patterns: (i) **S** (self) for intramolecular hydrogen-bonds; (ii) **C** (chain) for infinite chains; (iii) **R** (ring) for intermolecular rings; and (iv) **D** (discrete) for finite structures. The terms **a** and **d** are omitted when **a** = **d** = 1, and ***n*** is also omitted in the case of **D** when there is only one hydrogen bond. Some examples of different graph set notations are given below (Figure 1.1).

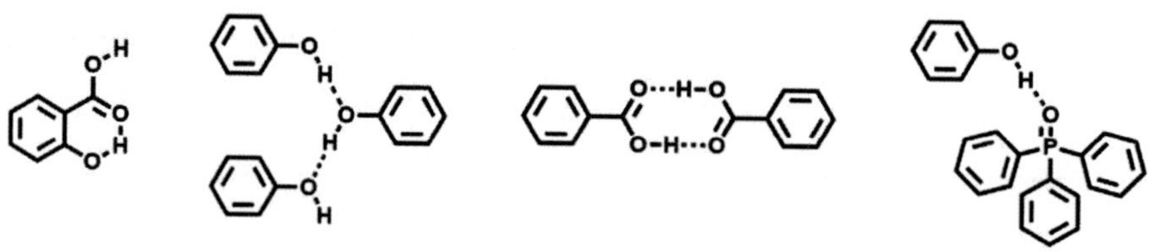

S(6) C(2) $R^2_2(8)$ D

Figure 1.1 Examples of graph set notations.

1.5.2 Supramolecular Synthons

One of the most important contributions to the development of practical co-crystal synthesis is undoubtedly the introduction of the term 'supramolecular synthon', which was defined by Desiraju as "*structural units within supermolecules which can be formed and/or assembled by known or conceivable synthetic operations involving intermolecular interactions.*"[28] The 'synthon' plays a key role in non-covalent synthesis and it is in some ways equivalent to the way in which reaction mechanisms form the cornerstone of covalent synthesis. A synthon, which in order to be practically useful, should be robust enough to withstand competition from other possible intermolecular interactions, and is not the same as a 'rare' or 'unusual' or 'interesting' intermolecular contact observed in a few crystal structures. If, however, specific and recognizable intermolecular interactions appear time after time, they can in principle be transferred intact from one supramolecular system to another; they become functioning 'synthons'. This in turn can offer the means whereby supramolecular synthesis can be performed with a high degree of reliability and reproducibility. A useful and practical synthetic protocol for the assembly of co-crystals requires that we can connect discrete molecular reactants into specific architectures with predetermined chemical composition and metrics. The foundation of any such process generally relies on site-specific intermolecular forces and synthons that favor heteromeric over homomeric interactions capable of producing selective and hierarchical self-assembly. Common synthons include the carboxylic acid–carboxylic acid[29] and oxime–oxime homodimers,[30] and the carboxylic acid–pyridine heterodimer[31] (Figure 1.2).

Supramolecular synthons are an integral part of supramolecular synthesis in crystal engineering, and it is important to identify the limits and conditions under which these synthons can be used to reliably assemble extended and complex multi-component solids.

1.5.3 Binary Co-crystals

Hydrogen- and halogen-bonds are strong intermolecular interactions that have been recognized as important intermolecular linkers.[32,33] Other weaker intermolecular forces or interactions such as C–H···N, C–H···O, C–H···F,

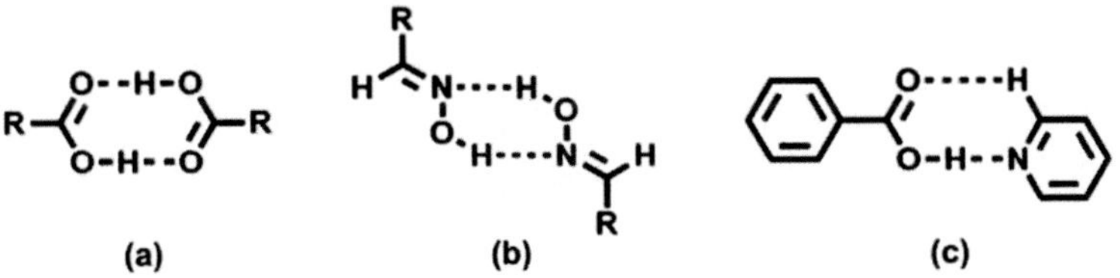

Figure 1.2 Examples of supramolecular synthons: (a) carboxylic acid–carboxylic acid dimer, (b) oxime–oxime dimer, and (c) carboxylic acid–pyridine heterodimer.

C–H···π and π···π are now considered as important structure directing vectors in crystal engineering,[34] but the scope of this book only allows us to discuss a fraction of the many studies conducted over the years. Hence, we will focus our efforts on examining how hydrogen- and halogen-bonded chemical entities prefer to bind to each other.

1.5.3.1 *Hydrogen-bonded Co-crystals*

Hydrogen bonds have been thoroughly studied in supramolecular synthesis, due to their strength and directionality when compared to other intermolecular interactions.[32] Hydrogen-bond complementarity (geometry, and number of hydrogen-bond donors and acceptors) coupled with a knowledge of robust supramolecular synthons can offer reliable tools for constructing supramolecular architectures,[35] and many self-complementary homomeric interactions are well documented, of which the carboxylic acid,[36] amide[37] and oxime[38] dimers are notable examples. The role of these robust supramolecular synthons in co-crystallization is not only to bring different molecules together, but also to impart a degree of directionality and dimensionality to the 3-D structure of solids. This is also true for biological systems where many examples of directed assembly *via* non-covalent interactions can be found.

It is also imperative for the assembly of molecular co-crystals that the heteromeric intermolecular interactions are dominant over the homomeric interactions. This, in principle, can lead to 0-D, 1-D, 2-D, and 3-D supramolecular architectures depending on the building blocks and on the intermolecular interactions present in the crystal structure. Examples of 0-D hydrogen-bonded binary co-crystals include heteromeric carboxylic acid:carboxylic acid dimers (1 : 1),[39] carboxylic acid:amide dimers (1 : 1),[40] carboxylic acid:pyridine dimers (1 : 1),[41] and carboxylic acid:2-aminopyrimidine trimers (2 : 1) (Figure 1.3).[42]

Common examples of hydrogen-bonded co-crystals containing infinite 1-D motifs are bipyridine:dihydroxybenzene,[43] 2-aminopyridine:dicarboxylic acids,[44] diols:diamines,[45] and 2-aminopyrimidine:dicarboxylic acids.[46] 2-D supramolecular architectures built on hydrogen-bond based heteromeric synthons include: piperazine:carboxylic acid,[47] trithiocyanuric acid:bipyridine,[48] triazine:uracil,[49] and isonicotinamide:dicarboxylic acid.[50]

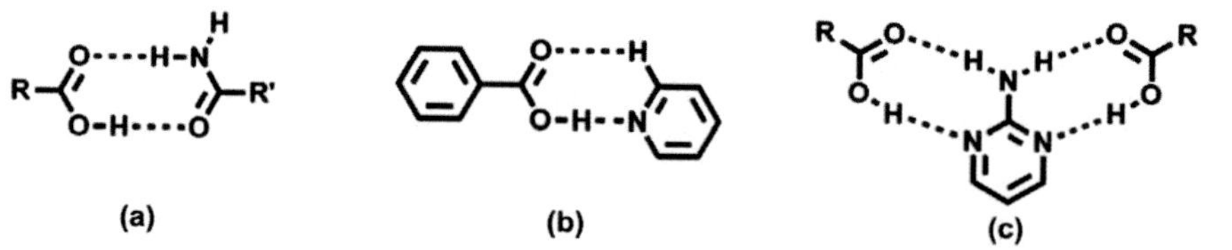

Figure 1.3 Hydrogen-bonded co-crystals: (a) acid–amide dimer; (b) acid–pyridine dimer; and (c) acid–aminopyrimidine trimer.

A more detailed examination of hydrogen-bonded supramolecular synthons in the assembly of molecular co-crystals is presented in Chapter 2 of this book.

1.5.3.2 Halogen-bonded Co-crystals

Halogen-bonds (XB) are known to have comparable directionality and strength to many hydrogen bonds.[51] Consequently, during the past decade or so, they have been used in the assembly of predictable and directional supramolecular architectures.[33,52] There are three types of halogen bonds (Figure 1.4);[53] (a) a conventional halogen bond between electronegative atoms such as nitrogen, oxygen or sulfur, and the electron-deficient tip of a halogen atom, (b) type-I halogen–halogen contacts due to van der Waals interactions between two halogen atoms (characteristic C–X···X angle = 120°, X = Br, I), and (c) type-II halogen–halogen contacts between the electron-rich and electron-deficient regions on a halogen atom (characteristic C–X···X angle = 90°, X = Cl, Br, I). This type II contact is essentially a 'conventional' halogen bond.

Traditional halogen-bond donors include iodo/bromo-perfluoroalkyls, iodo/bromo-perfluoroaromatics, and iodo/bromoaromatic compounds, whereas the most commonly used halogen-bond acceptors include (but are not limited to) pyridines, nitriles, carbonyls and thiols.[33,54] Typically, 'activation' of halogen-bond donors is achieved by fluorinating the aromatic/aliphatic donors, which in turn leads to an increased propensity for halogen-bonding of the donors, and this is also reflected in shorter X···acceptor bond distances. This observation is consistent with the electron-withdrawing effect of the fluorine atoms, which increases the positive electrostatic potential on the tip of the donor halogen atoms. This 'activation' strategy has been frequently utilized in the design of halogen-bonded co-crystals. For example, Aakeröy and co-workers devised a protocol for stabilizing volatile liquid iodoperfluoroalkanes by co-crystallizing them with a series of *N*-oxide

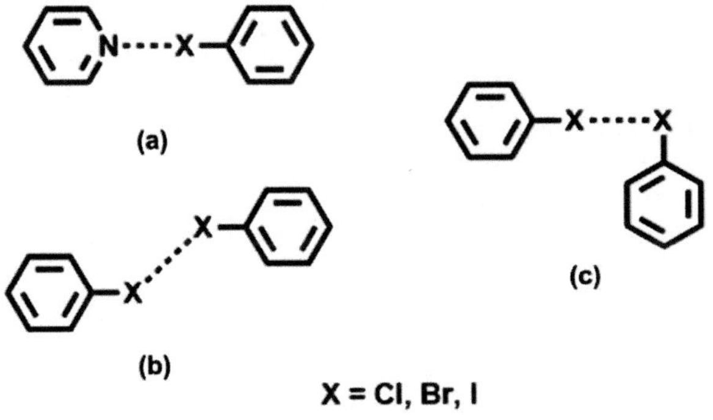

Figure 1.4 (a) Conventional halogen bond; (b) type-I halogen–halogen contacts; (c) type-II halogen–halogen contacts.

and *N*-heteroaromatic bases.[55] Expectedly, the driving force for co-crystallization in this series of co-crystals was the robust N–O···I–C/$N_{aromatic}$···I–C heterosynthon (Figure 1.5).

Another method of 'activating' the traditional halogen-bond donors is by placing the halogen atoms next to an alkyne moiety (Figure 1.6).[56] Diiodoacetylene, C_2I_2, is one of the smallest halogen-bond donors in this class of compounds and has been shown to be a strong ditopic halogen-bond donor.[57] Aakeröy and co-workers have systematically explored the competition between I/Br-ethynyl donors, perfluorinated I/Br donors, and I/Br-phenyl based donors by mapping out the supramolecular landscape of these molecules.[58] They observed that the 'activated' haloethynyl donors were competitive with the traditional perfluorinated halogen-bond donors with co-crystallization following the best donor–best acceptor concept.

A more detailed discussion on the different halogen-bond donors and motifs against a background of supramolecular hierarchy and in crystal engineering is presented in Chapter 3.

1.5.4 Beyond Binary Co-crystals

Although the literature on the synthesis and characterization of co-crystals has grown dramatically in recent years, the vast majority of those studies describe solids and architectures that contain no more than two different discrete building blocks. Despite the fact that many hypothesis-driven protocols for the assembly of such binary co-crystals have been presented, it is still proving to be very difficult to bring together three or more neutral molecules into a crystalline lattice in a predictable manner with a consistent degree of success. Binary co-crystals can be synthesized by considering the intermolecular interactions between the two components involved.

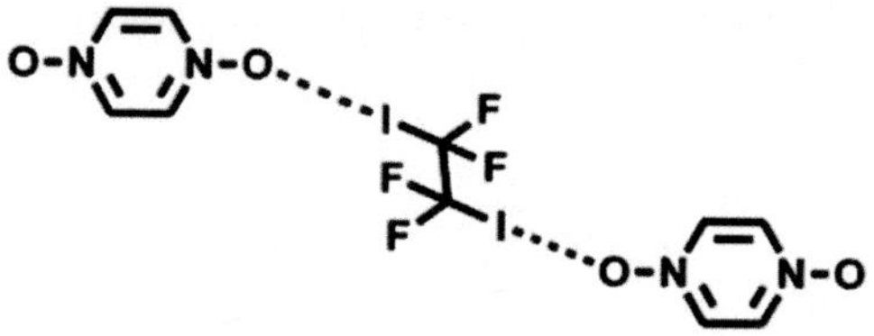

Figure 1.5 1-D chains observed in the binary co-crystal of 1,2-diiodotetrafluoroethane and tetramethylpyrazine *N,N'*-dioxide.[55]

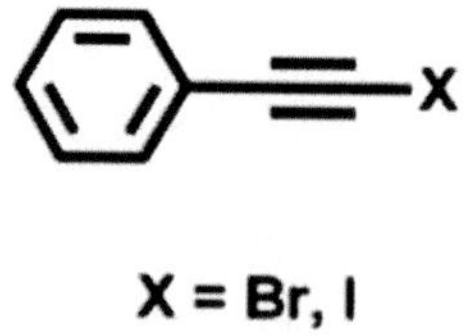

Figure 1.6 'Activated' ethynyl-based halogen-bond donors.

For example, we can assume that co-crystal AB will form if the heteromeric interactions (A···B) are stronger than the homomeric interactions (A···A and B···B). However, supramolecular synthetic complexity increases with an increase in the number of components involved in a co-crystallization experiment. In the synthesis of a ternary co-crystal ABC, the balance and hierarchy between the intermolecular interactions in all three components A, B and C needs to be considered carefully. This is because all binary co-crystal combinations, *i.e.* AB, BC or AC, are also possible along with the desired ternary (ABC) co-crystal. Thus, to achieve predictability in the synthesis of higher-order molecular co-crystals it is imperative to consider and understand the chemical and structural differences in the selected molecules.

Strategies that have been employed successfully tend to build on a hierarchical view of intermolecular interactions by taking advantage of the established guidelines that the best hydrogen-bond donor will bind to the best hydrogen-bond acceptor and the second-best donor to the second-best acceptor. A ranking of donors and acceptors can be carried out by evaluating the electrostatic potential on the relevant binding sites. A simple electrostatic view of assembly can be justified by the fact that the electrostatic component of non-covalent interactions is long-range and directional. An early example of this strategy utilized a combination of isonicotinamide as a central assembly platform and two carboxylic acids of different strengths (and correspondingly different electrostatic potentials on their respective acidic protons).[59] The stronger acid consistently formed a COOH···N_{py} hydrogen bond at one end of the central molecule, and the weaker of the two acids engaged in an amide···acid heterosynthon (Figure 1.7).

This particular approach yields very favorable outcomes, as reflected in a high 'supramolecular yield', which in turn translates into a high frequency of occurrence of a particular intermolecular binding pattern in the presence of potentially disruptive intermolecular interactions. The hierarchical approach was employed in a subsequent study where the assembly hub carried two geometrically unbiased sites that differed only in the electrostatically based ranking.[60] Again, two different co-formers engaged repeatedly with the target to form ternary co-crystals in a manner consistent with the underlying hypothesis (Figure 1.8).

This approach is not restricted to the use of carboxylic acids only, as several different chemical functionalities have been employed in the

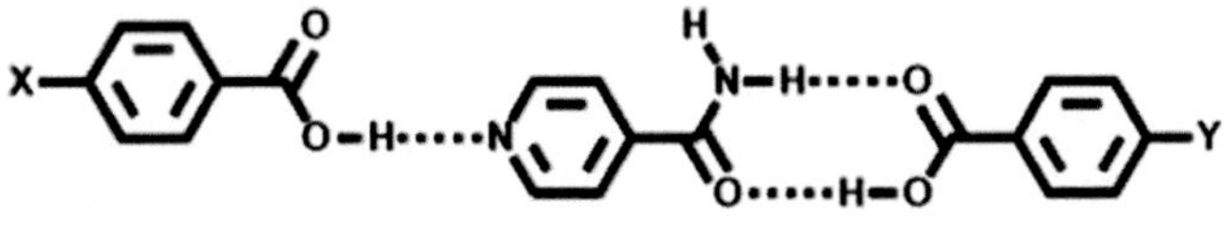

Figure 1.7 An example of a ternary co-crystal assembled *via* a hierarchical strategy using hydrogen bonds of different strengths.[59] The best donor (4-nitrobenzoic acid, left) binds to pyridine, whereas the second-best donor (4-(*N,N'*-dimethyl)-benzoic acid) binds to the amide moiety.

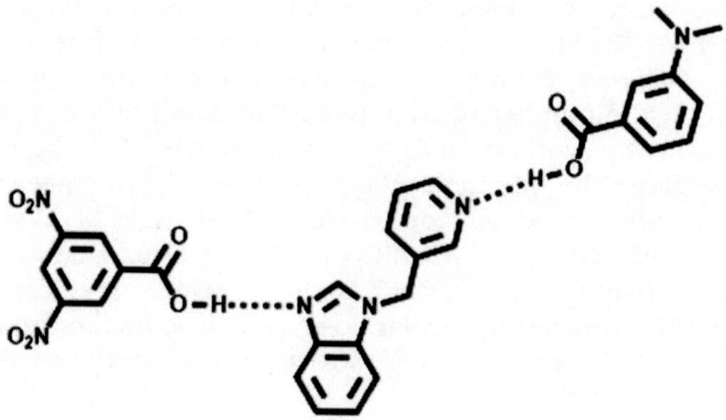

Figure 1.8 Best donor/best acceptor (left) and second-best donor/second-best acceptor (right) combine to produce a ternary co-crystal.[60]

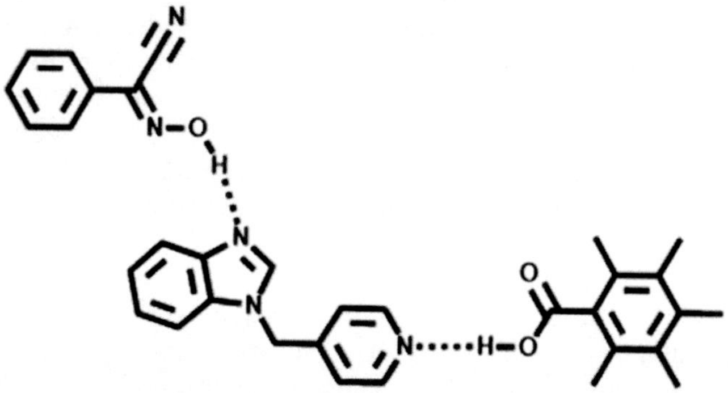

Figure 1.9 The best donor, the oxime, preferentially engages the best acceptor, leaving the second-best acceptor, the pyridyl nitrogen atom, to the carboxylic acid to yield another ternary co-crystal.[61]

synthesis of co-crystals where the primary interactions can be rationalized (and predicted) based on their relative ranking inferred from the molecular electrostatic potentials (Figure 1.9).[61]

It is also possible to employ a strategy that uses a combination of robust hydrogen bonds and geometrically 'biased' molecular building blocks in the controlled synthesis of multicomponent assemblies with desired structural features and composition (Figure 1.10).[62]

By designing molecules that carry mutually complementary hydrogen-bonding moieties (geometrically biased), it is possible to 'override' electrostatically favored interactions, which in turn minimizes unwanted 'synthon crossover'.[63] This mix-and-match strategy can give rise to very useful synthetic protocols that begin to approach the reliability and versatility of covalent synthesis, in terms of successfully preparing a target with predetermined connectivity and metrics.[64]

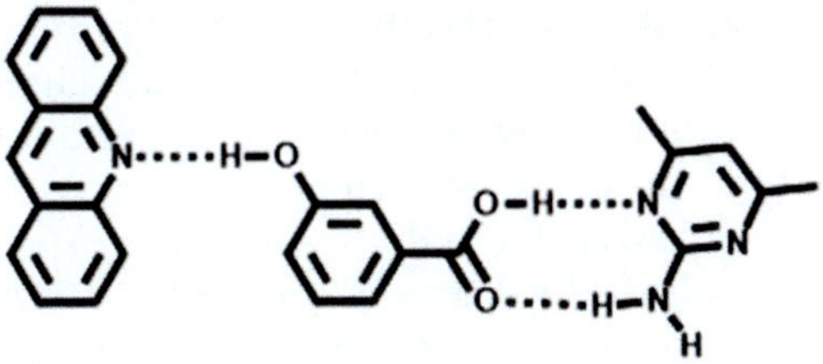

Figure 1.10 The geometric complementarity of the carboxylic acid···aminopyrimidine synthon combined with a robust O–H···N(acridine) hydrogen bond in a ternary co-crystal.[62]

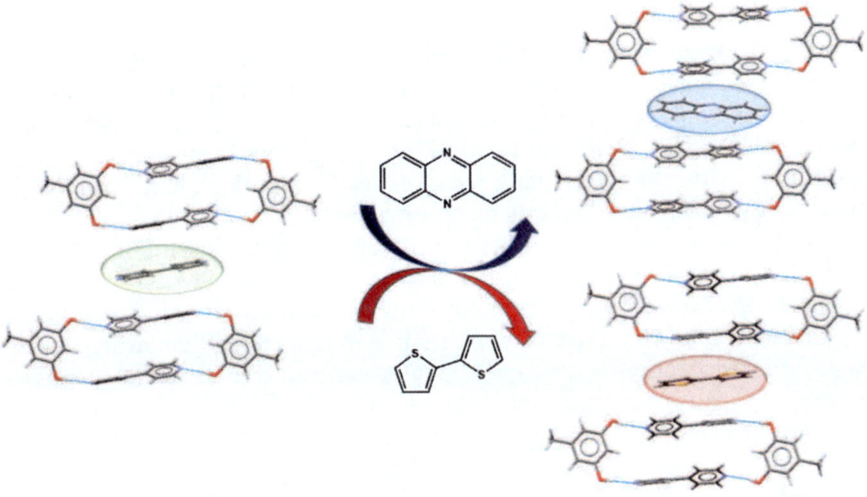

Figure 1.11 Shape and size mimicry combinatorial route to ternary solids.[66]

More recently, a shape-based combinatorial approach has been utilized by Desiraju and co-workers in the assembly of higher-order molecular co-crystals, where one component in a binary co-crystal was partially replaced by a third component such that the molecules participating in stronger hydrogen bonds were left intact, while similarly shaped molecules substituted the weakly interacting molecules in the crystal lattice.[65] A successful example of this strategy is the substitution of weakly bound 4,4′-bipyridine molecules in the bipyridine co-crystals of 2- or 5-methylresorcinol by similarly shaped molecules (Figure 1.11).[66]

Desiraju and co-workers have also employed a combination of their combinatorial technique with a templating strategy to synthesize ternary co-crystals of a polymorphic flavonoid quercetin (known for displaying both synthon and conformational polymorphism).[67] Based on functional-group compatibility of the phenol-pyridine heterosynthon, a series of nitrogen based dibasic co-formers were selected as suitable hydrogen-bond acceptors for the phenolic groups on quercetin. Multiple conformations and synthons are possible in the expected co-crystals of quercetin. Consequently, the virtual library for quercetin co-crystals is vast and the choice of the eventual conformation and synthon being expressed in the actual crystal structure is dictated by the structure of the co-former, the presence of auxiliary molecules and other experimental variables. Guided by their initial library of 24 binary co-crystals/solvates/hydrates of quercetin with different hydrogen-bond acceptors, the authors designed a templating strategy for the isolation of ternary co-crystals of quercetin based on a precise selection of molecular conformations and supramolecular synthons during molecular recognition events. They started with the anhydrous form of quercetin:4,4′-bipyridine co-crystal and using the shape and size mimicry of 2,2′-bis-thiophene relative to 4,4′-bipyridine, they successfully incorporated thiophene molecules into the binary co-crystal, thus leading to a templated ternary co-crystal. A similar templating strategy was also utilized to incorporate larger molecules such as tetrathiofulvalene, pyrene and anthracene into a binary co-crystal of quercetin and 1,2-di(4-pyridyl)ethylene. Also, ternary co-crystals of sulfonamide-based sulfa drugs were synthesized by Nangia and co-workers using a combination of the drug with pyridine carboxamides/cyclic amides and pyridine carboxamides/lactams as co-formers.[68]

The above-mentioned examples of ternary co-crystals highlight the level of difficulty involved in the design of such systems. However, an even more advanced level of complexity is encountered in the design of higher-order co-crystals with four or more components in a single crystalline lattice. In such cases, it is imperative to have even greater control over the balance between hierarchy-based intermolecular interactions.

In recent examples, Desiraju and co-workers have highlighted the use of an analogous combinatorial and templating strategy with polyphenols and *N*-based heterocyclic bases towards the targeted assembly of ten ternary co-crystals and six quaternary co-crystals.[69] They also devised a sequential strategy for the isolation of four- and five-component molecular co-crystals based on the structural inequivalence of different motifs and a careful consideration of the supramolecular hierarchy of strong and weak hydrogen bonds.[70] The premise of this strategy is that if a particular co-former in a co-crystal is present in two different crystal environments, then these differences can be utilized to increase the order of the co-crystal (Figure 1.12). In the first co-crystal AB, the co-formers A and B are present in the same chemical environment. Consequently, neither of the co-formers is receptive to any further change in supramolecular motifs and they are considered as

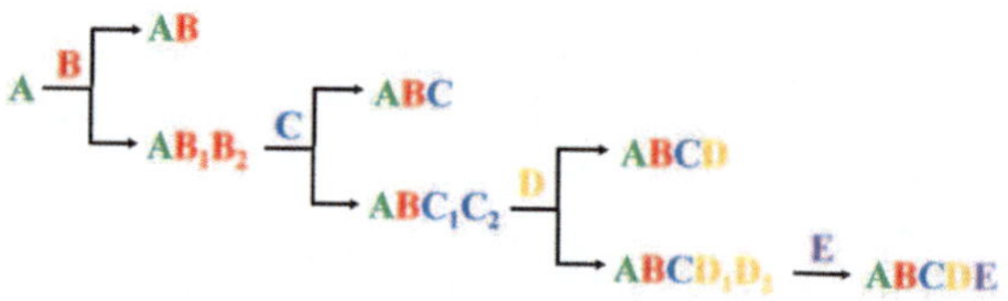

Figure 1.12 Sequential design of a quintinary co-crystal.[70]

synthetic dead ends. In the other binary co-crystal AB_1B_2, co-former B exists in two different crystal environments around its functional groups. Hence the binary co-crystal AB_1B_2 is suitable for supramolecular differentiation.[71]

The next step involves the introduction of co-former C, thus resulting in two ternary co-crystals, *i.e.* ABC (similar crystal environment) and ABC_1C_2 (different crystal environment). Thereafter, the ternary co-crystal ABC_1C_2 is introduced to the fourth component (D), which leads to the isolation of two sets of quaternaries (ABCD and $ABCD_1D_2$). Lastly, quaternary co-crystal $ABCD_1D_2$ is further used for supramolecular differentiation and introduction of co-former E leads to partial displacement of one of the earlier components, thus giving rise to the first reported quintinary co-crystal (ABCDE).[70]

Further advances towards the synthesis of higher-order co-crystals have been made by using combinations of different types of intermolecular interactions (sometimes augmented by geometrically biased binding sites) and hydrogen and halogen bonds can be used as orthogonal synthetic vectors as long as the primary molecular recognition events are designed around a careful combination of geometric and electrostatic complementarity. A two-point contact is more easily incorporated within a hydrogen bond based supramolecular synthon, whereas the vast majority of halogen-bond interactions involve single-point halogen atom···lone pair synthons. This indicates that it may be possible to build more complex supramolecular assemblies with a larger number of different molecules by combining interactions that can be made to operate independently of each other both at the level of molecular recognition, and at the level of overall structural control.

Desiraju and co-workers have demonstrated how hybrid ternary co-crystals based on both hydrogen- and halogen-bonds can be synthesized through a modular synthon-based approach.[72] In one such example, they replaced the amide–amide homosynthon in halobenzamides with the acid–amide heterosynthon between the halobenzamides and aliphatic dicarboxylic acids (Figure 1.13).[73] Thereafter, they incorporated 1,4-dinitrobenzene into the co-crystal lattice *via* I/Br···O_2N interactions.

Recently, Rissanen and co-workers successfully utilized the orthogonality of hydrogen and halogen bonds in the synthesis of more than a dozen ternary co-crystals,[74] where thiourea was employed as a bi-functional 'hub' simultaneously forming N–H···O hydrogen bonds with crown ethers and S···I halogen bonds with fluorinated halogen-bond donors (Figure 1.14).

Although non-covalent interactions are reversible and significantly weaker than most of their covalent counterparts, it is clearly possible to assemble higher-order co-crystals in a reproducible and reliable manner despite the practical limitations intrinsic to one-pot syntheses. An improved awareness and understanding of the balance between intermolecular interactions will kindle new strategies for the synthesis of co-crystals of even greater structural complexity. The ability to incorporate molecular building blocks that span considerable length-scales and chemical diversity will likely create

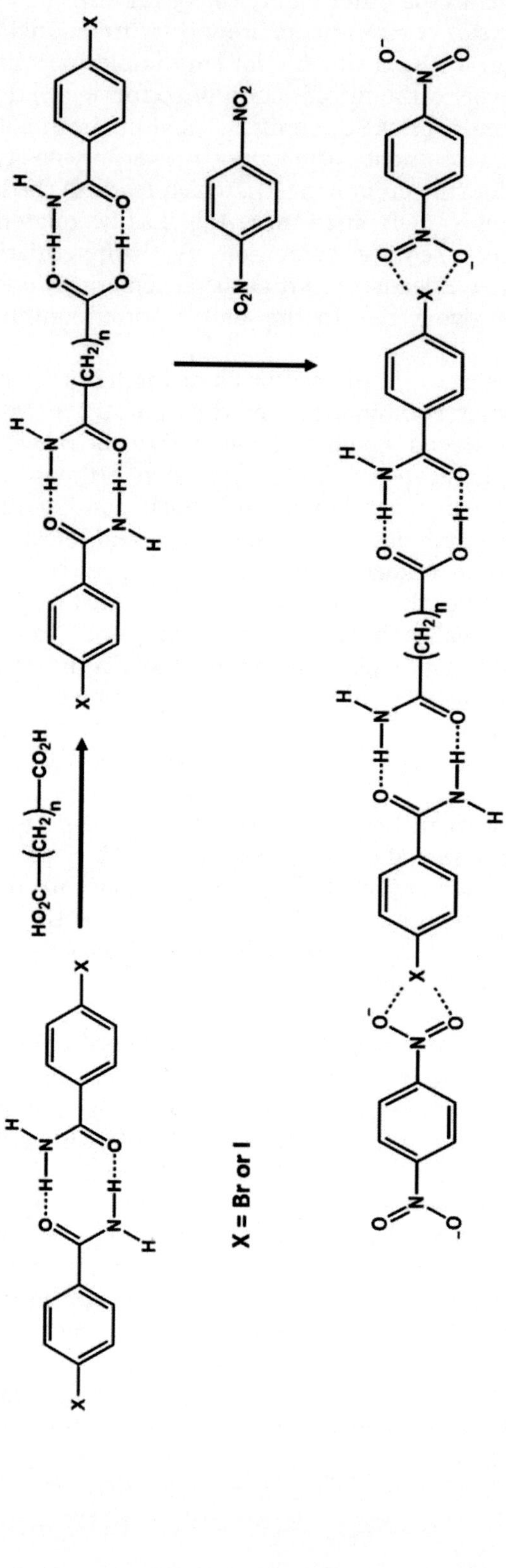

Figure 1.13 Design of hybrid ternary co-crystals based on a combination of hydrogen and halogen bonds.[73]

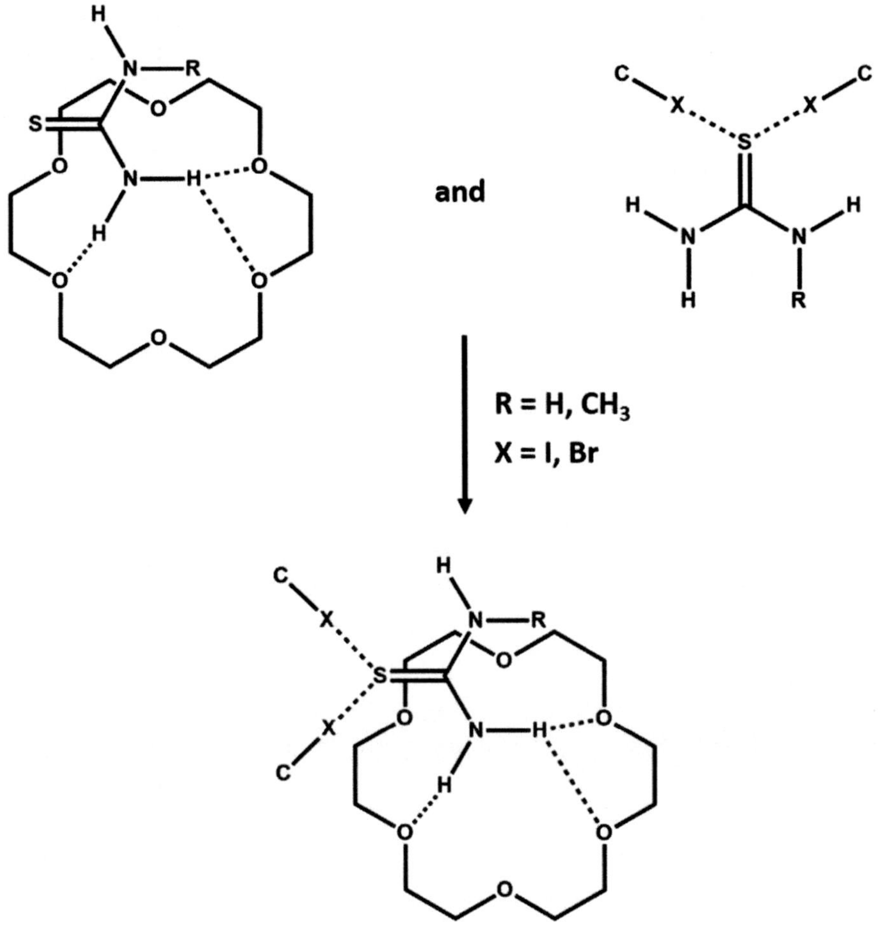

Figure 1.14 Assembly of hybrid ternary co-crystals.[74]

unique opportunities for programmed self-assembly of multi-functional molecular solids.

1.6 Cheminformatics and Co-crystals

The importance of examining large amounts of crystallographic data for identification of trends and patterns of behavior among molecules carrying similar chemical functionalities has long been recognized. The CSD[75] offers the essential starting point for any knowledge-based approach to co-crystal design and synthesis as it contains over 900 000 experimentally determined crystal structures of small organic and metal-organic compounds. Critically, the information in the CSD is curated and checked for errors, which makes it much more than just a depository for experimental data. Systematic analyses of the structural data in the CSD offer valuable information about packing preferences, space-group bias, synthon reliability, and molecular conformational preferences, and consequently such studies can lead the way towards new protocols for co-crystal synthesis that are both robust and versatile.[76] Tools developed by the Cambridge Crystallographic Data Centre also enable the user to readily search and compare structures that can offer new and unique insight into the molecular structure, functional-group composition, conformational flexibility, crystal packing, and morphology of crystalline materials.[77] This knowledge will, in turn, bring us closer to meeting the grand challenge of being able to directly connect molecular structure *via* crystal structure to physical properties of the bulk material.

1.6.1 Intermolecular Contacts: IsoStar and Full Interaction Maps

The IsoStar database is a comprehensive collection of information about non-bonded intermolecular interactions for a wide range of chemical groups.[78] Although, it was initially designed to support molecular invention in the area of rational drug design, it has rapidly expanded to be of interest to both protein crystallographers and crystal engineers.[79] It can highlight the structural landscape that surrounds a wide range of chemical functionalities through the use of an organized library of 3-D scatter plots. The scatter plots are formed by selecting a specific chemical group (known as the central group) and searching for all contacts with any approaching group (to a contact range of the sum of van der Waals radii).[80] The central group is fixed as a reference point (with defined coordinate axes and origin) and the contact groups are superimposed on the plots. Invaluable information about the frequencies and directionalities of these intermolecular contacts can be obtained from these scatterplots. Contoured density surfaces can also be generated from these scatterplots, which in turn are useful for obtaining critical information on geometric preferences of different functional groups. Furthermore, other information such as interaction energies, theoretical *in vacuo* minimum-energy geometries, and statistical data on the frequency of occurrence of different intermolecular interactions can also be obtained.[81]

A complete understanding of the relative stability of different crystalline forms of compounds (polymorphism) is critical not only in drug design, but also in crystal engineering.[82] For example, discovery of stable polymorphs at a later stage can lead to withdrawal of a drug product from the market. A combination of Full Interaction Maps (FIMs) with IsoStar can not only significantly enhance our understanding of different crystalline forms of compounds, but also limit the risk of unexpected appearance of new solid forms.[83,84] Superposing close contacts of preferred intermolecular interaction geometries of specific organic functional groups enables the FIMs functionality to take all relevant interactions and map the space around a single molecule, or collection of molecules (*e.g.* a crystal surface or simulated particle), to show the preferred positions of interactions with specific organic functional groups of different types, such as hydrogen-bond donors and acceptors, and hydrophobic groups (Figure 1.15).[84] This further helps in understanding, not only the stability, conformations and interactions of different polymorphs of a compound, but also the potential hydrogen-bonding sites in molecules.

1.6.2 Hydrogen-bond Motif Searches

Transferability and reproducibility of supramolecular synthons across multiple solid-state networks is important for the predictable design and synthesis of supramolecular architectures. A tool for understanding the competition between different homo- and heterosynthons as well as for discerning the robustness of any synthon is the Motif Search tool in Mercury, which has been optimized for 3-D searching of over 140 predefined commonly observed functional groups.[85] The frequency of occurrence as a percentage of the number of structures in which the motif/functional group is present can be obtained from these searches. For example, aspirin (acetylsalicylic acid) has two potential hydrogen-bonding groups in the form of the carboxylic acid and ester functional groups.[76] Only the carboxylic acid–carboxylic acid $R_2^2(8)$ homosynthon is observed in both polymorphs of aspirin.[86] For the design of heteromeric solids the heterosynthons need to out-compete the homosynthon (frequency of occurrence = 30%). Based on

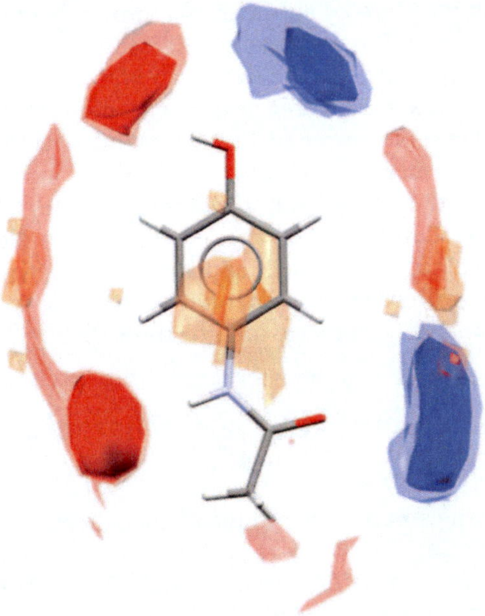

Figure 1.15 Full Interaction Map (FIM) for acetaminophen displaying the potential sites for intermolecular interactions.[83]

functional group compatibility, potential competing functional groups include amides, oximes and aliphatic hydroxyl. A simple comparison of the frequency of occurrences obtained from the motif searches shows that the likelihood of the heterosynthons is much greater and of these the oxime moiety is the best partner (frequency of occurrence = 60%).[76] Thus, in the design of heteromeric molecular solids, it is possible to rely on these motif searches and choose any of the heterosynthons that out-compete the homosynthon.

1.6.3 Multi-component Hydrogen-bond Propensity

The hydrogen-bond propensity (HBP) method has been modified over the years and its applications now include both prediction of polymorphs and the design of heteromeric solids.[87] This method determines the quantitative interaction likelihoods *via* an automated process that is robust and takes into account multiple variables.[76] Probabilities (from 0 to 1) of each hydrogen-bonding interaction for the analyzed specific chemical environment are obtained and the prediction is tailored for the specific functional group present. In the design of heteromeric solids, such as co-crystals *via* non-covalent interactions, the target molecule, as well as its potential co-formers, are selected and each molecule is treated as a separate environment for HBP analysis.[76] The interaction likelihood of all possible hydrogen bonds in each system are determined. The likelihood of the homo-interactions is compared to that of the hetero-interactions, which is then used to determine the multi-component score by subtracting the propensity value of the most-likely homo-interaction from the propensity value of the equivalent hetero-interaction. A positive value for the hydrogen-bond propensity indicates that the hetero-interactions are favored over the homo-interactions.[88] Recently, Aakeröy and co-workers examined the structural chemistry of six thiazole amides using experimental crystallographic data, and a combination of calculated HBPs and hydrogen-bond energies.[89] They observed that both methods correctly identified the main hydrogen-bonded synthon, *i.e.* a pairwise N–H···N dimer, which appeared in all the available crystal structures, thus highlighting the potential of HBPs in the knowledge-based assembly of extended heteromeric architectures (Scheme 1.4).

1.7 Synthesis, Screening and Characterization of Co-crystals

A wide range of experimental techniques[90] have been employed in the synthesis and production of co-crystals, such as electrospray technology,[91] mechanochemistry,[92] extrusion-based methods,[93] solvent-free loading into pores,[94] anvil compression,[95] isothermal suspension conversion,[96] slurry crystallization,[97] and a range of scale-up procedures.[98] Also, in order to optimize protocols for co-crystal synthesis, a range of techniques have been

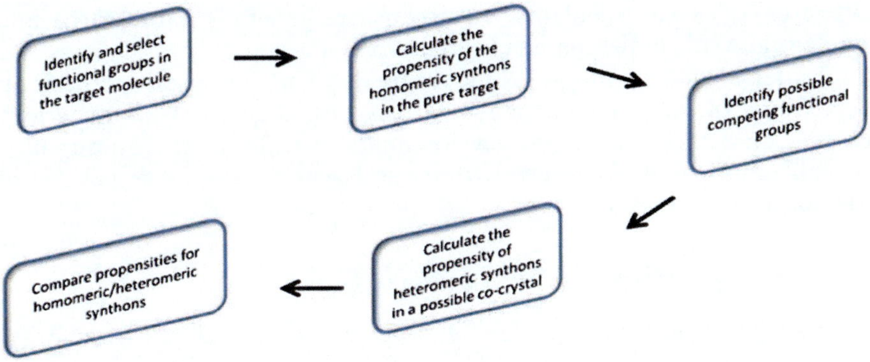

Scheme 1.4 Flowchart for the design of co-crystals based on synthon competition.[76]

employed for improving the process whereby co-formers are identified and selected, such as terahertz spectroscopy,[99] near-infrared spectroscopy,[100] complementary solubility measurements,[101] microfluidic approaches,[102] or, more generally, an improved understanding of mechanisms of co-crystal dissolution.[103]

Effective high-throughput screening for co-crystals of a target compound requires access to experimental techniques that ideally can offer reliable and unambiguous results within a reasonable time frame at a relatively low cost. Powder X-ray diffraction is often used to characterize solids (and to determine if a solid is structurally homogenous or not), but data collections are relatively lengthy, and isostructural compounds and preferred orientation can cause sensitivity issues. Vibrational spectroscopy,[104] especially Raman spectroscopy, provides viable alternatives for co-crystal screening and its focus on bond vibrational energies also allows for the detection of polymorphs. These techniques require small amounts of sample and are non-destructive and relatively fast, thus making them ideal for automated high-throughput systems. Also, Raman spectroscopy and PXRD are complementary techniques at the molecular level, and in combination they can provide an increased understanding of solid-state phenomena.

Several approaches to computational or virtual screening of co-crystals have also been pursued as an alternative to experimental protocols. Fabian[105] identified a set of molecular markers, based on data on co-crystals extracted from the CSD, that were correlated with pairs of molecules that are known to form co-crystals with each other. Shape and polarity were found to be the strongest descriptor for identifying co-crystal formation, whereas relative hydrogen-bond donor/acceptor counts did not show any statistical relationship. Price[106] and Leusen[107] adopted approaches to computational screening based on the assumption that co-crystal formation is driven by thermodynamics, which involves a comparison of calculated lattice energies of known co-crystals with the sum of the lattice energies of the reactants by themselves. Although the results vary somewhat depending upon which crystal structure prediction (CSP) method is employed, and on how the lattice energies are calculated, co-crystals are generally found to be more stable than their individual counterparts.

Chapter 4 briefly highlights different methods for co-crystal synthesis and details the role of mechanochemistry in co-crystal synthesis and characterization. An overview of recent developments in real-time monitoring of co-crystallization by Raman spectroscopy and synchrotron X-ray powder diffraction is also provided.

1.8 Applications of Co-crystals

Co-crystals have been used to develop new solid forms with enhanced or tunable properties, *e.g.* reduced hygroscopicity,[108] altered solubility (Chapter 5),[109] and improved stability of energetic materials (Chapter 6).[110] Apart from this they have also been used in ferroelectrics (Chapter 7),[111] non-linear optical materials,[112] chemical separation (Chapter 9),[113] and agrochemicals.[114]

1.8.1 Pharmaceutical Co-crystals

Sub-optimal properties of solid forms of high-value molecular solids continue to bedevil the pharmaceutical industry in particular. Even though many different aspects of the solid state, such as crystallinity, morphology, mechanical properties, *etc.*, can negatively impact performance and ease of formulation, the feature that undoubtedly has received most attention is aqueous solubility. This is not surprising since insufficiently soluble substances are becoming more commonplace in the market,[115] and it is estimated that more than 2/3 of potentially useful drug candidates are classified as having inadequate solubility in water.[116]

Co-crystals offer a simple and cheap alternative route to favorably alter the physicochemical properties of active pharmaceutical ingredients (APIs) and there is no doubt that the majority of reports on applications of co-crystal technology have focused on pharmaceutically active compounds that may display sub-optimal performances due to less than ideal physicochemical properties.[117] In 2016, the FDA revised its guidance on co-crystals of APIs (pharmaceutical co-crystals) stating that they be given a regulatory classification similar to that afforded to polymorphs of an API.[118] Consequently, drugs based on co-crystals will be developed along the same regulatory pathways as polymorphs, and a series of different 'co-crystals' of an ionic compound, meaning the incorporation of different molecules into the lattice containing the ionic form of the API will be treated as polymorphs of that salt. If the co-former is also furnished with some pharmaceutically relevant property, then the 'double' API co-crystals will not be treated as a completely new API, but instead as a combination formulation comprising multiple active ingredients.

A more in-depth analysis of pharmaceutical co-crystals in drug development along with process development strategies such as ternary phase diagrams and critical parameters is presented in Chapter 5.

1.8.2 Co-crystals of Energetic Compounds

Energetic materials that can store and rapidly release large amounts of chemical energy are of great use in both military and civilian fields.[119] For an energetic material to be useful, it is important that along with having a high density and a high heat of formation, they are stable enough to be manufactured, stored and handled safely.[120]

In the past five years or so, co-crystallization has emerged as a promising technology for synthesizing high-performance energetic materials that have the desired balance between high detonation velocity and low sensitivity.[121] Robust and safe methods for the preparation and characterization of energetic co-crystals are still being established, and there are fundamental issues associated with the structure–property relationships of these crystalline materials. For example, the densities of energetic co-crystals tend to lie in between that of the co-formers. Hence, a greater understanding of the intermolecular interactions governing the assembly of the energetic co-crystals is required in order to not only increase the packing coefficient, but also the density. Also, it is important to devise protocols/experiments for accurately determining the performance of these materials, as currently the performance of energetic co-crystals is mostly determined *via* theoretical techniques. Although many challenges lie ahead in the field of co-crystallization of energetic materials, undoubtedly energetic co-crystals will play an important role in the future of high-performance explosives and propellants.

Chapter 6 highlights the recent developments in the field of energetic co-crystals and details the intermolecular interactions, physical parameters and detonation properties of these crystalline materials.

1.9 Summary

In essence, every practical protocol for co-crystal synthesis relies on a bottom-up approach for targeting a crystalline solid with desired chemical composition, as well as with structural features of well-defined topology and dimensionality. The reactants for all these syntheses typically comprise discrete organic molecules that have been functionalized in such a way that programmed self-assembly is possible. However, the use of non-covalent interactions in the self-assembly process brings reversibility to supramolecular synthesis and the formation and isolation of pure products becomes more challenging in comparison to organic synthesis. Thus, to achieve predictable and reliable assembly of supramolecular architectures, it is important to identify and establish robust supramolecular synthons. As discussed in this chapter, a careful consideration of factors such as geometry and electrostatics coupled with cheminformatics-based approaches can enable us to successfully assemble higher-order molecular solids. Lastly, we have shown that by controlling the assembly of co-crystals, one can begin to achieve predictability in the solid-state properties of co-crystals, which in turn increases the applicability of these crystalline materials in various fields.

References

1. (a) O. Bolton, L. R. Simke, P. F. Pagoria and A. J. Matzger, *Cryst. Growth Des.*, 2012, **12**, 4311–4314; (b) S. Kim, Z. Li, Y.-C. Tseng, H. Nar, E. Spinelli, R. Varsolona, J. T. Reeves, H. Lee, J. J. Song, J. Smoliga, N. Yee and C. Senanayake, *Org. Proc. Res. Dev.*, 2013, **17**, 540–548.
2. J.-M. Lehn, *Supramolecular Chemistry: Concepts and Perspectives*, VCH, Weinheim, 1995.
3. G. R. Desiraju, *Nature*, 2001, **412**, 397–400.
4. (a) E. J. Corey, *Chem. Soc. Rev.*, 1988, **17**, 111–133; (b) K. C. Nicolaou and E. J. Sorenson, *Classics in Total Synthesis*, Wiley-VCH, Weinheim, 1996; (c) E. J. Corey, *Angew. Chem., Int. Ed.*, 2009, **48**, 2100–2117.
5. F. Wohler, *Ann. Chem. Pharm.*, 1844, **51**, 145–163.
6. J. S. Anderson, *Nature*, 1937, **140**, 850.
7. H. Matsuda, K. Osaki and I. Nitta, *Bull. Chem. Soc. Jpn.*, 1958, **31**, 611–620.
8. T. Sakurai, *Acta Crystallogr.*, 1968, **B24**, 403–412.
9. G. P. Stahly, *Cryst. Growth Des.*, 2009, **9**, 4212–4229.
10. (a) S. Cherukuvada, R. Kaur and T. N. Guru Row, *CrystEngComm*, 2016, **18**, 8528–8555; (b) J. W. Steed, *Trends Pharmacol. Sci.*, 2013, **34**, 185–193; (c) J. D. Wuest, *Nat. Chem.*, 2012, **4**, 74–75; (d) N. Blagden, D. J. Berry, A. Parkin, H. Javed, A. Ibrahim, P. T. Gavan, L. L. De Matosa and C. C. Seaton, *New J. Chem.*, 2008, **32**, 1659–1672.
11. M. C. Etter, *Acc. Chem. Res.*, 1990, **23**, 120–126.
12. (a) M. C. Etter and G. M. Frankenbach, *Chem. Mater.*, 1989, **1**, 10–12; (b) M. C. Etter, *J. Phys. Chem.*, 1991, **95**, 4601–4610.
13. G. R. Desiraju, *Crystal Engineering: The Design of Organic Solids*, Elsevier, Amsterdam, 1989.
14. G. M. J. Schmidt, *Pure Appl. Chem.*, 1971, **27**, 647–678.
15. (a) C. B. Aakeröy, B. M. T. Scott, M. M. Smith, J. F. Urbina and J. Desper, *Inorg. Chem.*, 2009, **48**, 4052–4061; (b) L. Garzon-Tovar, A. Duarte-Ruiz and K. Wurst, *Inorg. Chem. Commun.*, 2013, **32**, 64–67; (c) T. Mes, S. Cantekin, D. W. R. Balkenende, M. M. M. Frissen, M. A. J. Gillissen, B. F. M. De Waal, I. K. Voets, E. W. Meijer and A. R. A. Palmans, *Chem. Eur. J.*, 2013, **19**, 8642–8649.
16. G. R. Desiraju, *CrystEngComm*, 2003, **5**, 466–467.
17. A. D. Bond, *CrystEngComm*, 2007, **9**, 833–834.
18. J. D. Dunitz, *CrystEngComm*, 2003, **5**, 506.
19. D. Braga, F. Grepioni, L. Maini, S. Prosperi, R. Gobetto and M. R. Chierotti, *Chem. Commun.*, 2010, **46**, 7715–7717.
20. S. Aitipamula, R. Banerjee, A. K. Bansal, K. Biradha, M. L. Cheney, A. R. Choudhury, G. R. Desiraju, A. G. Dikundwar, R. Dubey, N. Duggirala, P. P. Ghogale, S. Ghosh, P. K. Goswami, N. R. Goud, R. R. K. R. Jetti, P. Karpinski, P. Kaushik, D. Kumar, V. Kumar, B. Moulton, A. Mukherjee, G. Mukherjee, A. S. Myerson, V. Puri, A. Ramanan, T. Rajamannar, C. M. Reddy, N. Rodriguez-Hornedo, R. D. Rogers, T. N. Guru Row, P. Sanphui, N. Shan, G. Shete, A. Singh, C. C. Sun, J. A. Swift,

R. Thaimattam, T. S. Thakur, R. K. Thaper, S. P. Thomas, S. Tothadi, V. R. Vangala, N. Variankaval, P. Vishweshwar, D. R. Weyna and M. J. Zaworotko, *Cryst. Growth Des.*, 2012, **12**, 2147–2152.
21. https://www.fda.gov/downloads/Drugs/GuidanceComplianceRegulatoryInformation/Guidances/UCM516813.pdf.
22. P. P. Deshpande, J. Singh, A. Pullockaran, T. Kissick, B. A. Ellsworth, J. Z. Gougoutas, J. Dimarco, M. Fakes, M. Reyes, C. Lai, H. Lobinger, T. Denzel, P. Ermann, G. Crispino, M. Randazzo, Z. Gao, R. Randazzo, M. Lindrud, V. Rosso, F. Buono, W. W. Doubleday, S. Leung, P. Richberg, D. Hughes, W. N. Washburn, W. Meng, K. J. Volk and R. H. Mueller, *Org. Process Res. Dev.*, 2012, **16**, 577–585.
23. C. C. Seaton, *CrystEngComm*, 2014, **16**, 5878–5886.
24. S. L. Childs, G. P. Stahly and A. Park, *Mol. Pharm.*, 2007, **4**, 323–338.
25. C. B. Aakeröy, M. E. Fasulo and J. Desper, *Mol. Pharm.*, 2007, **4**, 317–322.
26. (a) C. B. Aakeröy, T. K. Wijethunga and J. Desper, *New J. Chem.*, 2015, **39**, 822–828; (b) C. B. Aakeröy, K. N. Epa, S. Forbes, N. Schultheiss and J. Desper, *Chem. Eur. J.*, 2013, **19**, 14998–15003.
27. J. Bernstein, R. E. Davis, L. Shimoni and N.-L. Chang, *Angew. Chem., Int. Ed. Engl.*, 1995, **34**, 1555–1573.
28. G. R. Desiraju, *Angew. Chem., Int. Ed. Engl.*, 1995, **34**, 2311–2327.
29. W. C. Hamilton and J. A. Ibers, *Hydrogen Bonding in Solids*, W. A. Benjamin, New York, 1968.
30. C. B. Aakeröy, A. S. Sinha, K. N. Epa, P. D. Chopade, M. M. Smith and J. Desper, *Cryst. Growth Des.*, 2013, **13**, 2687–2695.
31. (a) P. Sudhakar, R. Srivijaya, B. R. Sreekanth, P. K. Jayanthi, P. Vishweshwar, M. J. Babu, K. Vyas and J. Iqbal, *J. Mol. Struct.*, 2008, **885**, 45–49; (b) C. B. Aakeröy, A. M. Beatty and B. A. Helfrich, *J. Am. Chem. Soc.*, 2002, **124**, 14425–14432.
32. (a) C. T. Seto and G. M. Whitesides, *J. Am. Chem. Soc.*, 1991, **113**, 712–713; (b) C. B. Aakeröy and K. R. Seddon, *Chem. Soc. Rev.*, 1993, **22**, 397–407; (c) E. R. T. Tiekink, J. Vittal and M. J. Zaworotko, *Organic Crystal Engineering: Frontiers in Crystal Engineering*, Wiley, John & Sons, Inc., Chichester, UK, 1st edn, 2010.
33. (a) G. Cavallo, P. Metrangolo, R. Milani, T. Pilati, A. Priimagi, G. Resnati and G. Terraneo, *Chem. Rev.*, 2016, **116**, 2478–2601; (b) P. Metrangolo and G. Resnati, *Cryst. Growth Des.*, 2012, **12**, 5835–5838.
34. (a) Y. Choi, T. Kim, S. Jang and J. Kang, *New J. Chem.*, 2016, **40**, 794–802; (b) G. R. Desiraju, *Chem. Commun.*, 2005, 2995–3001; (c) K. Biradha, *CrystEngComm*, 2003, **5**, 374–384.
35. (a) J.-M. Lehn, *Science*, 2002, **295**, 2400–2403; (b) J. B. Wittenberg and L. Isaacs, Complementarity and Preorganization, *Supramolecular Chemistry: From Molecules to Nanomaterials*, 2012, vol. 1, 25–43.
36. F. H. Herbstein, in *Comprehensive Supramolecular Chemistry*, ed. D. D. MacNicol, F. Toda and R. Bishop, Pergamon Press, New York, 1996, vol. 6, pp. 61–83.

37. J. C. Mareque Rivas and L. Brammer, *New J. Chem.*, 1998, **22**, 1315–1318.
38. (a) V. Bertalosi, G. Gilli and A. Veronese, *Acta. Crystallogr., Sect. B.*, 1982, **B38**, 502–511; (b) E. A. Bruton, L. Brammer, F. C. Pigge, C. B. Aakeröy and D. S. Leinen, *New J. Chem.*, 2003, **27**, 1084–1094.
39. J. A. R. P. Sarma and G. R. Desiraju, *J. Chem. Soc., Perkin Trans. 2*, 1985, 1905–1912.
40. C. B. Aakeröy, J. Desper and B. A. Helfrich, *CrystEngComm*, 2004, **6**, 19–24.
41. T. Sugiyama, J. Meng and T. Matsuura, *J. Mol. Struct.*, 2002, **611**, 53–64.
42. S. Skovsgaard and A. D. Bond, *CrystEngComm*, 2009, **11**, 444–453.
43. C. Glidewell, G. Ferguson, R. M. Gregson and A. J. Lough, *Acta, Crystallogr., Sect. C.: Cryst. Struct. Commun.*, 1999, **55**, 2133–2136.
44. F. Garcia-Tellado, S. J. Geib, S. Goswami and A. D. Hamilton, *J. Am. Chem. Soc.*, 1991, **113**, 9265–9269.
45. O. Ermer and A. Eling, *J. Chem. Soc., Perkin Trans. 2*, 1994, 925–944.
46. M. C. Etter and D. A. Adsmond, *J. Chem. Soc., Chem. Commun.*, 1990, 589–591.
47. T.-J. M. Luo and G. T. R. Palmore, *Cryst. Growth Des.*, 2002, **2**, 337–350.
48. V. R. Pedireddi, S. Chatterjee, A. Ranganathan and C. N. R. Rao, *J. Am. Chem. Soc.*, 1997, **119**, 10867–10868.
49. F. H. Beijer, R. P. Sijbesma, J. A. J. M. Vekemans, E. W. Meijer, H. Kooijman and A. L. Spek, *J. Org. Chem.*, 1996, **61**, 6371–6380.
50. P. Vishweshwar, A. Nangia and V. M. Lynch, *Cryst. Growth Des.*, 2003, **3**, 783–790.
51. (a) C. B. Aakeröy, N. Schultheiss, J. Desper and C. Moore, *CrystEngComm*, 2007, **9**, 421–426; (b) T. Friščić, E. Meštrović, D. S. Šamec, B. Kaitner and L. Fábián, *Chem. Eur. J.*, 2009, **15**, 12644–12652; (c) G. Lapadula, N. Judaš, T. Friščić and W. Jones, *Chem. Eur. J.*, 2010, **16**, 7400–7403; (d) C. B. Aakeröy, P. D. Chopade, C. Ganser and J. Desper, *Chem. Commun.*, 2011, **47**, 4688–4690.
52. (a) A. Mukherjee, S. Tothadi and G. R. Desiraju, *Acc. Chem. Res.*, 2014, **47**, 2514–2524; (b) C. B. Aakeröy and C. L. Spartz, *Top. Curr. Chem.*, 2015, **358**, 155–182.
53. (a) V. R. Pedireddi, D. S. Reddy, G. S. Goud, D. C. Craig, A. D. Rae and G. R. Desiraju, *J. Chem. Soc. Perkin Trans. II*, 1994, **11**, 2353–2360; (b) P. Metrangolo and G. Resnati, *Chem. Eur. J.*, 2001, **7**, 2511–2519; (c) *Halogen Bonding: Fundamentals and Applications, Structure and Bonding*, ed. P. Metrangolo and G. Resnati, Springer, Berlin, 2007; (d) L. Brammer, G. M. Espallargas and S. Libri, *CrystEngComm*, 2008, **10**, 1712–1727.
54. (a) M. Fourmigué and P. Batail, *Chem. Rev.*, 2004, **104**, 5379–5418; (b) K. Rissanen, *CrystEngComm*, 2008, **10**, 1107–1113; (c) G. Cavallo, P. Metrangolo, T. Pilati, G. Resnati, M. Sansotera and G. Terraneo, *Chem. Soc. Rev.*, 2010, **39**, 3772–3783.

55. C. B. Aakeröy, T. K. Wijethunga, J. Benton and J. Desper, *Chem. Commun.*, 2015, **51**, 2425–2428.
56. R. W. Troff, T. Mäkelä, F. Topić, A. Valkonen, K. Raatikainen and K. Rissanen, *Eur. J. Org. Chem.*, 2013, 1617–1637.
57. C. Perkins, S. Libri, H. Adams and L. Brammer, *CrystEngComm*, 2012, **14**, 3033–3038.
58. C. B. Aakeröy, M. Baldrighi, J. Desper, P. Metrangolo and G. Resnati, *Chem. Eur. J.*, 2013, **19**, 16240–16247.
59. C. B. Aakeröy, A. M. Beatty and B. A. Helfrich, *Angew. Chem., Int. Ed.*, 2001, **40**, 3240–3242.
60. C. B. Aakeröy, J. Desper and J. F. Urbina, *Chem. Commun.*, 2005, 2820–2822.
61. C. B. Aakeröy, J. Desper and M. M. Smith, *Chem. Commun.*, 2007, 3936–3938.
62. D. A. Adsmond, A. S. Sinha, U. B. R. Khandavilli, A. R. Maguire and S. E. Lawrence, *Cryst. Growth Des.*, 2016, **16**, 59–69.
63. (a) C. B. Aakeröy, P. D. Chopade and J. Desper, *Cryst. Growth Des.*, 2011, **11**, 5333–5336; (b) H. R. Khavasi and A. A. Tehrani, *CrystEngComm*, 2013, **15**, 5813–5820.
64. B. R. Bhogala, S. Basavoju and A. Nangia, *Cryst. Growth Des.*, 2005, **5**, 1683–1686.
65. (a) N. A. Mir, R. Dubey, S. Tothadi and G. R. Desiraju, *CrystEngComm*, 2015, **17**, 7866–7869; (b) R. Dubey and G. R. Desiraju, *Angew. Chem., Int. Ed.*, 2014, **53**, 13178–13182.
66. S. Tothadi, A. Mukherjee and G. R. Desiraju, *Chem. Commun.*, 2011, **47**, 12080–12082.
67. R. Dubey and G. R. Desiraju, *IUCrJ*, 2015, **2**, 402–408.
68. (a) G. Bolla and A. Nangia, *Chem. Commun.*, 2015, **51**, 15578–15581; (b) G. Bolla and A. Nangia, *IUCrJ*, 2016, **3**, 152–160.
69. R. Dubey, N. A. Mir and G. R. Desiraju, *IUCrJ*, 2016, **3**, 102–107.
70. N. A. Mir, R. Dubey and G. R. Desiraju, *IUCrJ*, 2016, **3**, 96–101.
71. T. Smolka, R. Boese and R. Sustmann, *Struct. Chem.*, 1999, **10**, 429–431.
72. S. Tothadi and G. R. Desiraju, *Chem. Commun.*, 2013, **49**, 7791–7793.
73. S. Tothadi, P. Sanphui and G. R. Desiraju, *Cryst. Growth Des.*, 2014, **14**, 5293–5302.
74. F. Topić and K. Rissanen, *J. Am. Chem. Soc.*, 2016, **138**, 6610–6616.
75. C. R. Groom, I. J. Bruno, M. P. Lightfoot and S. C. Ward, *Acta Cryst.*, 2016, **B72**, 171–179.
76. P. A. Wood, N. Feeder, M. Furlow, P. T. A. Galek, C. R. Groom and E. Pidcock, *CrystEngComm*, 2014, **16**, 5839–5848.
77. (a) C. R. Groom and F. H. Allen, *Angew. Chem., Int. Ed.*, 2014, **53**, 662–671; (b) F. H. Allen and R. Taylor, *Chem. Soc. Rev.*, 2004, **33**, 463–475.
78. I. J. Bruno, J. C. Cole, J. P. M. Lommerse, R. S. Rowland, R. Taylor and M. L. Verdonk, *J. Comput.- Aided Mol. Des.*, 1997, **11**, 525–537.

79. (a) C. J. Radoux, T. S. G. Olsson, W. R. Pitt, C. R. Groom and T. L. Blundell, *J. Med. Chem.*, 2016, **59**, 4314–4325; (b) F. H. Allen, A. J. Cruz-Cabeza, P. A. Wood and D. A. Bardwell, *Acta Crystallogr., Sect. B: Struct. Sci., Cryst. Eng. Mater.*, 2013, **69**, 514–523.
80. S. Motherwell, Cheminformatics and Crystallography. The Cambridge Structural Database, in *Cheminformatics Developments: History, Reviews and Current Research*, IOS Press, Amsterdam, 2004, vol. 1, pp. 129–174.
81. (a) R. Taylor, F. H. Allen and J. C. Cole, *CrystEngComm*, 2015, **17**, 2651–2666; (b) R. Taylor, *CrystEngComm*, 2014, **16**, 6852–6865.
82. D. A. Snider, W. Addicks and W. Owens, *Adv. Drug Delivery Rev.*, 2004, **56**, 391–395.
83. P. A. Wood, T. S. G. Olsson, J. C. Cole, S. J. Cottrell, N. Feeder, P. T. A. Galek, C. R. Groom and E. Pidcock, *CrystEngComm*, 2013, **15**, 65–72.
84. N. Feeder, E. Pidcock, A. M. Reilly, G. Sadiq, C. L. Doherty, K. R. Back, P. Meenan and R. Docherty, *J. Pharm. Pharmacol.*, 2015, **67**, 857–868.
85. (a) C. F. Macrae, I. J. Brunu, J. A. Chisholm, P. R. Edgington, P. McCabe, E. Pidcock, L. Rodriguez-Monge, R. Taylor, J. van de Streek and P. A. Wood, *J. Appl. Crystallogr.*, 2008, **41**, 466–470; (b) J. A. Chisholm and W. D. S. Motherwell, *J. Appl. Crystallogr.*, 2004, **37**, 331–334.
86. (a) Y. Kim, K. Machida, T. Taga and K. Osaki, *Chem. Pharm. Bull.*, 1985, **33**, 2641–2647; (b) P. Vishweshwar, J. A. McMahon, M. Oliveira, M. L. Peterson and M. J. Zaworotko, *J. Am. Chem. Soc.*, 2005, **127**, 16802–16803.
87. (a) P. T. A. Galek, L. Fábián, W. D. S. Motherwell, F. H. Allen and N. Feeder, *Acta Crystallogr., Sect. B: Struct. Sci.*, 2007, **63**, 768–782; (b) P. T. A. Galek, F. H. Allen, L. Fábián and N. Feeder, *CrystEngComm*, 2009, **11**, 2634–2639; (c) P. T. A. Galek, L. Fábián and F. H. Allen, *CrystEngComm*, 2010, **12**, 2091–2099.
88. (a) M. Majumder, G. Buckton, C. F. Rawlinson-Malone, A. C. Williams, M. J. Spillman, E. Pidcock and K. Shankland, *CrystEngComm*, 2013, **15**, 4041–4044; (b) E. Nauha and J. Bernstein, *Cryst. Growth Des.*, 2014, **14**, 4364–4370.
89. B. Sandhu, A. McLean, A. S. Sinha, J. Desper, A. A. Sarjeant, S. Vyas, S. M. Reutzel-Edens and C. B. Aakeröy, *Cryst. Growth Des.*, 2018, **18**, 466–478.
90. S. A. Ross, D. A. Lamprou and D. Douroumis, *Chem. Commun.*, 2016, **52**, 8772–8786.
91. S. Patil, J. Kulkarni and K. Mahadik, *Ind. Eng. Chem. Res.*, 2016, **55**, 8409–8414.
92. (a) F. Fischer, D. Lubjuhn, S. Greiser, K. Rademann and F. Emmerling, *Cryst. Growth Des.*, 2016, **16**, 5843–5851; (b) T. Friščić and W. Jones, *Cryst. Growth Des.*, 2009, **9**, 1621–1637; (c) N. Shan, F. Toda and W. Jones, *Chem. Commun.*, 2002, 2372–2373.
93. (a) S. Li, T. Yu, Y. Tian, C. P. McCoy, D. S. Jones and G. P. Andrews, *Mol. Pharm.*, 2016, **13**, 3054–3068; (b) D. E. Crawford, L. A. Wright, S. L. James and A. P. Abbott, *Chem. Commun.*, 2016, **52**, 4215–4218.

94. (a) E. Skorupska, A. Jeziorna and M. J. Potrzebowski, *J. Phys. Chem. C*, 2016, **120**, 13169–13180; (b) E. Skorupska, P. Paluch, A. Jeziorna and M. J. Potrzebowski, *J. Phys. Chem. C*, 2015, **119**, 8652–8661.
95. R. Lee, A. J. Firbank, M. R. Probert and J. W. Steed, *Cryst. Growth Des.*, 2016, **16**, 4005–4011.
96. D. M. Croker and Å. C. Rasmuson, *Org. Process Res. Dev.*, 2014, **18**, 941–946.
97. N. Takata, K. Shiraki, R. Takano, Y. Hayashi and K. Terada, *Cryst. Growth Des.*, 2008, **8**, 3032–3037.
98. L. Zhao, V. Raval, N. E. B. Briggs, R. M. Bhardwaj, T. McGlone, I. D. H. Oswald and A. J. Florence, *CrystEngComm*, 2014, **16**, 5769–5780.
99. S. P. Delaney and T. M. Korter, *J. Phys. Chem. A*, 2015, **119**, 3269–3276.
100. M. Allesø, S. Velaga, A. Alhalaweh, C. Cornett, M. A. Rasmussen, F. van den Berg, H. Lopez, de Diego and J. Rantanen, *Anal. Chem.*, 2008, **80**, 7755–7764.
101. J. H. ter Horst, M. A. Deij and P. W. Cains, *Cryst. Growth Des.*, 2009, **9**, 1531–1537.
102. S. Goyal, M. R. Thorson, G. G. Z. Zhang, Y. Gong and P. J. A. Kenis, *Cryst. Growth Des.*, 2012, **12**, 6023–6034.
103. (a) F. Cao, G. L. Amidon, N. Rodriguez-Hornedo and G. E. Amidon, *Mol. Pharm.*, 2016, **13**, 1030–1046; (b) M. P. Lipert and N. Rodríguez-Hornedo, *Mol. Pharm.*, 2015, **12**, 3535–3546; (c) Z. Li and A. J. Matzger, *Mol. Pharm.*, 2016, **13**, 990–995.
104. M. A. Elbagerma, H. G. M. Edwards, T. Munshi, M. D. Hargreaves, P. Matousek and I. J. Scowen, *Cryst. Growth Des.*, 2010, **10**, 2360–2371.
105. L. Fabian, *Cryst. Growth Des.*, 2009, **9**, 1436–1443.
106. N. Issa, S. A. Barnett, S. Mohamed, D. E. Braun, R. C. B. Copley, D. A. Tocher and S. L. Price, *CrystEngComm*, 2012, **14**, 2454–2464.
107. H. C. S. Chanab, J. Kendricka, M. A. Neumannc and F. J. J. Leusen, *CrystEngComm*, 2013, **15**, 3799–3807.
108. S. Aitipamula, V. R. Vangala, P. S. Chow and R. B. H. Tan, *Cryst. Growth Des.*, 2012, **12**, 5858–5863.
109. (a) M. Przybyłek, D. Ziółkowska, K. Mroczyńska and P. Cysewski, *Cryst. Growth Des.*, 2017, **17**, 2186–2193; (b) M. Banik, S. Prasad Gopi, S. Ganguly and G. R. Desiraju, *Cryst. Growth Des.*, 2016, **16**, 5418–5428; (c) Y. Chen, L. Li, J. Yao, Y.-Y. Ma, J.-M. Chen and T.-B. Lu, *Cryst. Growth Des.*, 2016, **16**, 2923–2930; (d) S. L. Childs, P. Kandi and S. R. Lingireddy, *Mol. Pharm.*, 2013, **10**, 3112–3127; (e) S. J. Bethune, N. Schultheiss and J.-O. Henck, *Cryst. Growth Des.*, 2011, **11**, 2817–2823; (f) L. H. Thomas, A. R. Klapwijk, C. Wales and C. C. Wilson, *CrystEngComm*, 2014, **16**, 5924–5932; (g) N. R. Goud, R. A. Khan and A. Nangia, *CrystEngComm*, 2014, **16**, 5859–5869; (h) C. B. Aakeröy, S. Forbes and J. Desper, *CrystEngComm*, 2014, **16**, 5870–5877; (i) S. Aitipamula, A. B. H. Wong, P. S. Chow and R. B. H. Tan, *CrystEngComm*, 2014, **16**, 5793–5801.

110. O. Bolton, L. R. Simke, P. F. Pagoria and A. J. Matzger, *Cryst. Growth Des.*, 2012, **12**, 4311–4314.
111. S. Horiuchi, R. Kumai and Y. Tokura, *Chem. Commun.*, 2007, 2321–2329.
112. (a) T. Miyano, T. Sakai, I. Hisaki, H. Ichida, Y. Kanematsu and N. Tohnai, *Chem. Commun.*, 2016, **52**, 13710–13713; (b) S.-J. Kwon, O.-P. Kwon, M. Jazbinsek, V. Gramlich and P. Günter, *Chem. Commun.*, 2006, 3729–3731.
113. P. Metrangolo, T. Pilati, G. Resnati and A. Stevenazzi, *Chem. Commun.*, 2004, 1492–1493.
114. M. Baldrighi, D. Bartesaghi, G. Cavallo, M. R. Chierotti, R. Gobetto, P. Metrangolo, T. Pilati, G. Resnati and G. Terraneo, *CrystEngComm*, 2014, **16**, 5897–5904.
115. (a) C. A. Lipinski, *J. Pharmacol. Tox. Met.*, 2000, **44**, 235–249; (b) T. Takagi, C. Ramachandran, M. Bermejo, S. Yamashita, L. X. Yu and G. L. Amidon, *Mol. Pharm.*, 2006, **3**, 631–643.
116. C. W. Pouton, *Eur. J. Pharm. Sci.*, 2006, **29**, 278–287.
117. (a) N. K. Duggirala, M. L. Perry, Ö. Almarsson and M. J. Zaworotko, *Chem. Commun.*, 2016, **52**, 640–655; (b) J.-R. Wang, Q. Yu, W. Dai and X. Mei, *Chem. Commun.*, 2016, **52**, 3572–3575; (c) Ö. Almarsson and M. J. Zaworotko, *Chem. Commun.*, 2004, 1889–1896; (d) N. Blagden, S. J. Coles and D. J. Berry, *CrystEngComm*, 2014, **16**, 5753–5761; (e) M. K. Sarangi, *Farmacia*, 2014, **62**, 824–839; (f) P. Vishweshwar, J. A. McMahon, J. A. Bis and M. J. Zaworotko, *J. Pharm. Sci.*, 2006, **95**, 499–516.
118. FDA Draft Guidance: Regulatory Classification of Pharmaceutical Co-crystals. Dated August 2016. (http://www.fda.gov/downloads/Drugs/GuidanceComplianceRegulatoryInformation/Guidances/UCM516813.pdf).
119. (a) D. M. Badgujar, M. B. Talawar, S. N. Asthana and P. P. Mahulikar, *J. Hazard. Mater.*, 2008, **151**, 289–305; (b) D. Fischer, T. M. Klapötke and J. Stierstorfer, *Angew. Chem., Int. Ed.*, 2014, **53**, 8172–8175; (c) J. Zhang and J. M. Shreeve, *Dalton Trans.*, 2016, **45**, 2363–2368; (d) N. Fischer, D. Fischer, T. M. Klapötke, D. G. Piercey and J. Stierstorfer, *J. Mater. Chem.*, 2012, **22**, 20418–20422; (e) J. Zhang and J. M. Shreeve, *J. Am. Chem. Soc.*, 2014, **136**, 4437–4445; (f) P. Yin, J. Zhang, D. A. Parrish and J. M. Shreeve, *Chem. Eur. J.*, 2014, **20**, 16529–16536; (g) J. Zhang, S. Dharavath, L. A. Mitchell, D. A. Parrish and J. M. Shreeve, *J. Am. Chem. Soc.*, 2016, **138**, 7500–7503.
120. (a) A. A. Dippold and T. M. Klapötke, *J. Am. Chem. Soc.*, 2013, **135**, 9931–9938; (b) M. B. Talawar, S. K. Jangid, T. Nath, R. K. Sinha and S. N. Asthana, *J. Hazard. Mater.*, 2015, **300**, 307–332.
121. J. Zhang and J. M. Shreeve, *CrystEngComm*, 2016, **18**, 6124–6133.

CHAPTER 2

The Role of Hydrogen Bonding in Co-crystals

SUMAN BHATTACHARYA, KRISHNA S. PERAKA AND MICHAEL J. ZAWOROTKO*

Bernal Institute, Department of Chemical Science, University of Limerick, Limerick, Republic of Ireland
*Email: xtal@ul.ie

2.1 Introduction

Supramolecular chemistry, defined by Lehn as the '*chemistry beyond the molecule*',[1] involves the study of the mechanism of formation of molecular assemblies driven by intermolecular interactions, and the chemical, physical and biological properties of such assemblies. The synthesis of molecular assemblies is typically achieved through the process of *self-assembly* that is controlled by non-covalent intermolecular interactions. The composition of molecular assemblies is controlled by non-covalent interactions and, because these interactions are typically directional, they also control the local and long-range structure. A molecular crystal can therefore be regarded as a type of molecular assembly that is in essence a repeating 3D pattern of molecules controlled by directional non-covalent interactions. In this context, Dunitz asserted that '*a crystal is a perfect supermolecule—a super molecule par excellence*'.[2] That the crystal structure of a molecular compound influences the physicochemical properties of a crystalline solid has been a driving force for the design and discovery of new crystalline solids. Indeed, the design and synthesis of crystalline solids using knowledge of their molecular recognition properties, *crystal engineering*, has become a paradigm for preparing new crystalline materials with desired physicochemical properties. The term *crystal engineering* was introduced by Pepinsky in 1955[3] and implemented by Schmidt in the context of his work in organic solid-state photochemical reactions.[4] In 1989, Desiraju defined crystal engineering as *'the understanding of intermolecular interactions in the context of crystal packing and the utilisation of such understanding in the design of new solids with desired physical and chemical properties'.*[5] *Crystal engineering* has since evolved to encompass the full range of supramolecular interactions, including coordination bonds, and its focus has shifted from the design of composition and structure to control the properties and applications of the resulting crystalline materials. Research into crystal structure prediction (CSP) is complementary to crystal engineering and has evolved in parallel. CSP remains a challenging task despite advances in computing power, methodology and our understanding of intermolecular interactions. Considering the many developments in the field that have occurred over the last two or three decades, it might be appropriate to redefine crystal engineering as *the field of chemistry that studies the design, properties and applications of crystals.*

A number of intermolecular interactions have been recognised as directional in nature and therefore offer potential to control molecular structures as exemplified by hydrogen bonds, halogen bonds and stacking interactions of aromatics. Their interaction energies are now quite well established (Scheme 2.1). Hydrogen bonds, which have been long recognised as enabling to crystal engineering,[6–10] are perhaps the most widely studied intermolecular interaction with respect to supramolecular synthesis. This is because hydrogen bonds offer the following features: appropriate bond strength (not too strong, not too weak); directionality (the structural parameters of hydrogen bonds are well established); ready formation in

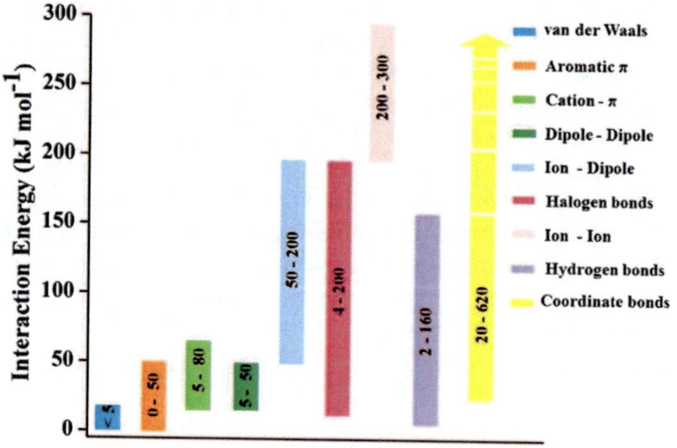

Scheme 2.1 Types of non-covalent interactions that have been exploited in supramolecular chemistry and their range of interaction energies.

appropriate solvents; the prevalence of protic hydrogen atoms at the periphery of organic compounds thanks to important functional groups such as acids, phenols and amines; the large number of functional groups that can serve as hydrogen bond acceptors, *e.g.* aromatic amines, carbonyl moieties. Herein we focus upon hydrogen bonding as a tool for the synthesis and design of co-crystals. Co-crystals are a class of multicomponent crystals comprised of two or more 'co-formers', molecular or ionic compounds that, when pure, exist as solids or low volatility liquids under ambient conditions.[11]

2.2 Multicomponent Molecular Crystals

Multicomponent molecular crystals are crystals that are comprised of more than one molecular species in the crystal lattice. They encompass several crystal form types (Scheme 2.2) including co-crystals,[12,13] hydrates,[14–17] solvates,[18,19] inclusion complexes,[20–22] organic salts[23–25] and solid solutions (mixed crystals).[26–28] Hydrates, solvates and organic salts are of interest because they have long played an important role in pharmaceutical science. With respect to hydrates, they were classified by Morris and Rodriguez-Hornedo based on the type of intermolecular interactions exhibited by water molecules in the crystal lattice.[29] Channel hydrates[30,31] contain self-interacting water molecules within channels in the crystal lattice. Isolated site hydrates[32] have no direct interactions between water molecules, rather they interact with other molecules and as such tend to strongly influence the

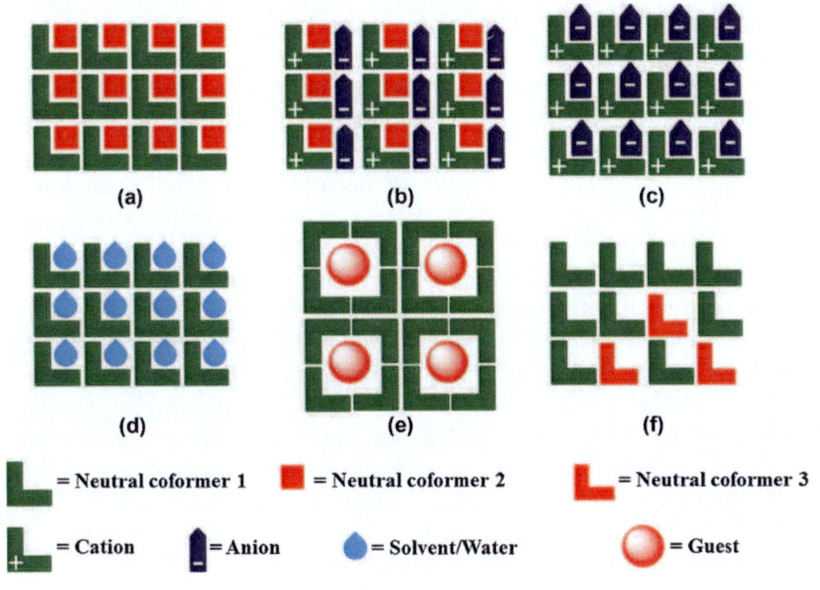

Scheme 2.2 A schematic representation of types of multicomponent crystal forms: (a) molecular co-crystal (MCC), (b) ionic co-crystal (ICC), (c) salt, (d) solvates/hydrates, (e) inclusion complexes and (f) solid solutions.

crystal packing. This classification is equally applicable for solvates. Channel hydrates and solvates can be stoichiometric or non-stoichiometric since the crystal packing is in effect determined by a host molecule. Although the structure and composition of isolated site solvates and hydrates can often be conceptually related to co-crystals, solvates and hydrates are not the focus of this chapter and will only be discussed in the context of co-crystal hydrates and co-crystal solvates.

2.2.1 Co-crystals: Definition, Classification and Synthesis

Co-crystals stand out amongst multicomponent crystals because they are inherently amenable to design by crystal engineering and they are unlikely to occur as the result of serendipity. They are also long known in the scientific literature with notable examples being the ionic co-crystal of NaCl and urea[33,34] and the molecular co-crystal quinhydrone.[35] Quinhydrone was reported by Wöhler in 1844 and subsequently shown in 1960 to be a 1 : 1 co-crystal of quinone and hydroquinone.[36,37] This co-crystal is prototypal in many ways because it is sustained by O–H···O hydrogen bonds that afford a head-to-tail chain comprised of alternating quinone and hydroquinone molecules. Despite their long history, co-crystal nomenclature has been an issue and various terms have been coined such as addition compounds,[38,39] molecular organic compounds,[40] organic molecular compounds,[41,42] hydrogen bonded complexes[43] and heteromolecular crystals.[42] A recently proposed definition for co-crystal came from the consensus of 46 experts in the field. This group defined co-crystals as '*solids that are crystalline single phase materials composed of two or more different molecular and/or ionic compounds generally in a stoichiometric ratio which are neither solvates nor simple salts*'.[11] The components that comprise a co-crystal are then termed *co-formers*. This definition excludes salts (because ions cannot have independent existence), most solvates/hydrates of single component compounds (solvent/water of crystallisation are typically volatile liquids) and solid solutions (non-stoichiometric mixed crystals) from its scope. In this context, it is worthwhile to consider the case of racemic mixtures of chiral compounds. During the course of crystallisation of a racemic mixture, there are two probable outcomes: i) formation of a *racemic conglomerate, i.e. 'an equimolar mixture of crystals each one of which only contains one of the two enantiomers'*.[44] ii) formation of a *racemate, i.e. 'an equimolar mixture of a pair of enantiomers'*.[44] Given that the aforementioned definition of co-crystal requires the co-formers to be different compounds, one could consider a racemate to be a co-crystal of its enantiomers but we do not further address racemates herein.

Co-crystals have gained recent attention mainly thanks to their amenability to crystal engineering and their potential utility, especially in the context of drug substances. With respect to drug substances, pharmaceutical co-crystals,[12,45–48] co-crystals of a biologically active compound (*e.g.* a drug molecule or a nutraceutical compound) and a pharmaceutically acceptable co-former, can exhibit superior properties, *e.g.* stability, solubility, dissolution, bioavailability, *etc. vs.* polymorphs, solvates and hydrates of the same biologically active compound. Pharmaceutical co-crystals are now recognised by the US Food and Drug Administration[49] and several have been approved for use as drug substances in drug products.[50–54] Co-crystallization can also enable solvent-free solid state synthesis[55,56] and address structural challenges like assigning the absolute stereochemistry of a molecule.[57,58] The breadth of scope of co-crystals means that it has become appropriate to sub-classify co-crystals. In 2009, Stahly discussed the history of co-crystals reported in the literature prior to the year 2000.[59] He mentioned compounds reported in the early literature that could be regarded as co-crystals but were comprised of inorganic components such as metal salts of alkali and alkaline earth metals, *e.g.* NaCl, LiCl, $NaNO_2$, NH_4Br, NH_4Cl, *etc.* Paul Pfeiffer, a renowned German chemist of his time, in 1922 published a book titled '*Organishe Molekulverbindungen*'.[60] In the book, Pfeiffer presented a detailed discussion about '*co-crystals*' composed of purely organic materials and also on '*co-crystals*', which were comprised of inorganic components. The Zaworotko group has recently stressed why it is appropriate to classify co-crystals based on the nature of co-formers as being either *molecular co-crystals* (MCCs) or *ionic co-crystals* (ICCs).[61]

2.2.2 Molecular Co-crystals (MCCs)

Molecular co-crystals (MCCs) are crystals comprised of two or more *solid* (under ambient conditions) and *neutral* (molecules or zwitterions with no net charge) co-formers in a stoichiometric ratio. Although hydrogen bonds are the most widely used interaction for the design of MCCs, they can also be sustained by halogen bonds[62–64] and $\pi\cdots\pi$ interactions.[65,66] As noted earlier, perhaps the first report of an MCC in the scientific literature was made by Wöhler in 1844.[35] The prevalence of carboxylic acid moieties or functionality that is complementary to carboxylic acid moieties in biologically active molecules makes carboxylic acids of particular relevance. Further, the large number of pharmaceutically approved carboxylic acids means that perhaps the largest portion of pharmaceutical co-crystal research has focused upon MCCs sustained by carboxylic acid co-formers. Nevertheless, to our knowledge the first examples of MCCs involving biologically active compounds were based upon barbiturates and aminopyridines, as disclosed in a patent published in 1934.[67] The first marketed drug product based upon an MCC, BetaClor®, is also devoid of carboxylic moieties as the active drug substance is a co-crystal of betaine and chloral hydrate.[50]

2.2.3 Ionic Co-crystals (ICCs)

Ionic co-crystals (ICCs) differ from MCCs in two important aspects:

(i) ICCs are sustained by charge-assisted hydrogen bonds or, if metal ions are involved, coordination bonds. Such interactions are less likely to be impacted by solvent and more likely to afford large variations in physicochemical properties.

(ii) ICCs based upon two co-formers must have at least three components, a cation, an anion and an additional molecule or salt. This means that there at least two variables. In general, given that MCCs offer a single variable, ICCs will likely exhibit greater diversity in terms of composition and properties than MCCs.

The term *'ionic co-crystal'* was coined by Braga *et al.* in 2010 in their report on co-crystals formed by barbituric acid and alkali metal halides.[68] However, to our knowledge the earliest report of a crystal form that is an ICC was from Romè de L'Isle in 1783.[33] He described a change in the habit of NaCl crystals when crystallised from aqueous urea (urine). In 1950, single crystals of this compound were studied by Palm and MacGillavry who reported the composition as a 1:1:1 ICC of NaCl, urea and water.[34] ICCs comprised of carboxylate salts and carboxylic acids were reported by Gerhardt, who in 1853 described the solid phase obtained from an alcohol solution of stoichiometric amounts of potassium hydrogen benzoate and benzoic acid.[69] In 1949, Robert filed a patent application for *'complex salts'* of streptothricin describing solids composed of streptothricin and inorganic salts of alkaline earth metals.[70] This could be the first report of an ICC based on a pharmaceutically relevant molecule. However, the first ICC to be marketed as a drug product is Depakote®, a 1:1 ICC of valproic acid (Depakene) and sodium valproate. This ICC exhibits enhanced moisture stability compared to sodium valproate.[51,71,72]

2.2.4 Synthesis of Co-crystals

In general, both MCCs and ICCs can be prepared by well-established crystallisation techniques.[73] Crystallisation from a solution that contains co-formers in stoichiometric proportions is the most widely used technique but is harder to control than single-component crystallisation, especially if the co-formers exhibit very different solubility in the selected solvent medium. Other methods that have been successfully employed include the following: growth of a co-crystal from a melt of co-formers;[74,75] sublimation;[76] solvent drop/dry hand grinding[77–79] or ball mill grinding;[80] crystallisation using slurries.[81,82] The factors that promote successful discovery of co-crystals and their crystallisation can be summarised as below:

- Understanding of supramolecular interactions that can occur in the pure co-formers.
- An appropriate library of co-formers that contain complementary functional groups with respect to the target compound.
- Co-formers should ideally have comparable solubility to avoid crystallisation of the least soluble co-former when solvent crystallisation is the method of isolation.

- Proper choice of solvent since solvents that can form strong hydrogen bonds might interfere with the recognition processes that must occur between co-formers.

2.3 Hydrogen Bonds: Discovery and Classification

Hydrogen bonds are the most widely used non-covalent interactions for the design of co-crystals. The early scientific literature on hydrogen bonds included reports from Werner[83] in 1902 and Hantzsch[84] in 1910 and focused upon interactions involving ammonium cations that were referred to as '*nebenvalenz*', *i.e.* minor valence. The concept of '*hydrogen bond*' was developed independently by Huggins, who presented it in a thesis to explain the '*tautomerism*' observed in acetoacetic esters, and by Latimer and Rodebush who emphasised the importance of '*interaction*' in their explanation for the behaviour of certain polar compounds in aqueous media.[85] Latimer and Rodebush proposed that '*a free pair of electrons on one water molecule can exert sufficient force on a hydrogen atom which is bonded to another water molecule, which consequently binds the two molecules together*'. They described the interaction as a '*weak bond*' involving hydrogen nuclei held between two Lewis octets. The term '*hydrogen bond*' was coined by Pauling in the context of the properties of ice and water.[86] Pauling subsequently used hydrogen bonds to explain interactions between peptide nitrogen and oxygen atoms in the context of denaturation of protein chains.[87] A 1951 study of Pauling *et al.* on models of helical polypeptide chains preceded the double helical structure of DNA proposed by Watson and Crick in 1953.

Hydrogen bonding became viewed as a type of Bronsted acid–base reaction in which the associated proton is in an arrested state where it is neither donated by the acid nor accepted by the base.[88] The dissociation energy of a hydrogen bond can range from 1.25–160 kJ mol^{-1} and can be comprised of up to four components: electrostatic, covalent, dispersion–repulsion and polarisation.[89] The definition of hydrogen bonding was adopted by IUPAC in 2011 and is as follows: '*The hydrogen bond is an attractive interaction between a hydrogen atom from a molecule or a molecular fragment X–H in which X is more electronegative than H, and an atom or a group of atoms in the same or a different molecule, in which there is evidence of bond formation*'.[90]

2.3.1 Geometry

Hydrogen bonding is typically presented as **X–H···A** (X = donor, A = acceptor) and can be described in terms of four geometric parameters, $\boldsymbol{d}$, $\boldsymbol{D}$, $\boldsymbol{\theta}$ and φ (Scheme 2.3a). Thanks to the relatively high quality of modern diffraction data, hydrogen atoms can be located accurately enough to obtain reliable $\boldsymbol{d}$ and $\boldsymbol{\theta}$ values. Multifurcated interactions among hydrogen bonds are also quite common (Scheme 2.3b and c), where a donor group is available to two or more independent acceptors or an acceptor finds itself interacting with two or more donors. In all cases, the interaction can be

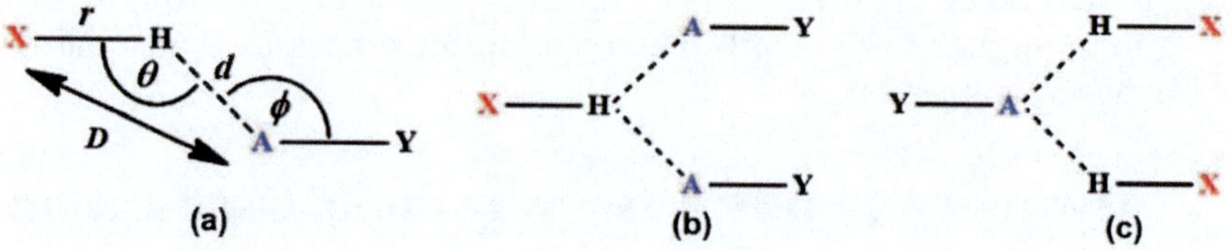

Scheme 2.3 (a) Geometrical parameters associated with a hydrogen bond, (b) schematic representation of bifurcated donor and (c) bifurcated acceptor interactions.

viewed as a hydrogen atom simultaneously attracted to two atoms, **X** and **A**, in such a manner that the hydrogen atom effectively serves as a bridge between them. The interaction has a significant electrostatic contribution that increases with the electronegativity of **X** and **A**.

2.3.2 Classification

Hydrogen bonds can be classified into three broad categories, *very strong*, *strong* and *weak*, based on the interaction energy and the influence they exhibit in determining and controlling the formation of supramolecular structures.[91]

2.3.2.1 'Very Strong' *Hydrogen Bonds*

'*Very strong*' hydrogen bonds exhibit interaction energies of 60–160 kJ mol^{-1} or above. Amongst the strongest hydrogen bonds are those that are charge-assisted, *e.g.* the $[FHF]^-$ anion. They can occur between an acid and its conjugate base ($X–H\cdots X^-$) or a base and its conjugate acid ($X^+–H\cdots X$). In the above cases, the values for ***D*** and ***d*** are short enough to suggest covalent contributions.

2.3.2.2 'Strong' *Hydrogen Bonds*

Strong hydrogen bonds are exemplified by O–H···O and N–H···O hydrogen bonds that sustain α-helix and β-sheet structures in proteins and base pairing in DNA. In the context of crystal engineering, these interactions are very useful and well studied because of the ubiquity of OH and NH moieties in organic chemistry. O–H···O hydrogen bonds are particularly well studied. COOH···COOH dimer and catemeric interactions (mean $D = 2.702$ Å) and OH···OH phenol or alcohol (mean $D = 2.702$ Å) interactions are common. Charge assisted $N^+–H\cdots O^-$ hydrogen bonds occur between ionic moieties, *e.g.* carboxylate···pyridinium ($D = 2.631$ Å), carboxylate···anilinium ($D = 2.677$ Å) and $O–H\cdots Cl^-$ ($D = 3.001$ Å). O–H···N hydrogen bonds are exemplified by acid···pyridine ($D = 2.626$ Å), acid–amine ($D = 2.642$ Å) or phenol···pyridine ($D = 2.761$ Å) interactions. These strong hydrogen bonds are directional and can be exploited to control and tune supramolecular structures. Further, they are readily identified by shifts in IR band frequencies for the X–H moiety and/or large downfield shifts in ^1H NMR spectra. The presence of a strong hydrogen bond can be established from IR spectroscopy. In the case of strong hydrogen bonds, a broad band is typically observed in the region around 1600 cm^{-1} and can be validated by isotopic displacement of hydrogen by deuterium. Structural parameters are also affected such as lengthening of X–H bond distances, shortened H···A distances and X–H···A angles above 170°.

2.3.2.3 'Weak' Hydrogen Bonds

Weak hydrogen bonds with interaction energies of <16 kJ mol^{-1} were not fully recognised until the 1990s.[92–94] In 1962, Sutor proposed the existence of the C–H···O interactions[95] and Taylor and Kennard later provided crystallographic evidence of the existence of such interactions through neutron diffraction studies on 113 crystalline compounds.[96] Kennard and Taylor concluded that interactions like C–H···O, C–H···N, C–H···Cl could be considered hydrogen bonds and that they are attractive in nature. Desiraju published extensively on C–H···O hydrogen bonds in the 1990s.[94,97] These interactions are predominantly electrostatic but are highly influenced by the involvement of various dispersive and charge transfer components associated with the donor and the acceptor moieties. Amongst the strongest interactions of this type are O–H···Ph, C≡C–H···O. Weaker interactions of this type are only slightly stronger than van der Waals interactions, *e.g.* C–H···O, C–H···N and M–H···O (M = metal). C–H···π, S–H··· π, C–H···M are the lowest energy hydrogen bonds, but they can still be directional enough to be structure determining.

2.4 Designing Co-crystals Based on Hydrogen Bonds

2.4.1 Etter's Rules, Hydrogen Bond Patterns and Graph Sets

The ability of hydrogen bonds to sustain networks and molecular assemblies was noted in a series of seminal contributions from Etter.[98–104] The principle outcomes of Etter's work can be summarised as follows:

(a) Identification of different hydrogen bond patterns observed in aggregates of small organic molecules and introduction of the '*graph set*' concept in studying crystal structures.
(b) Formulation of empirical rules which can be instrumental in successfully predicting the connectivity of hydrogen bonds in molecular aggregates in the absence of other strong competing effects.
(c) Identification of organic functionalities that are instrumental in the formation of robust hydrogen bonds.

Etter developed the concept of graph sets for the study of hydrogen bonded networks as follows: i) recognise potential hydrogen bonds; ii) compare the hydrogen bond patterns observed in relevant crystal structures. The application of graph set theory to study molecular aggregates enabled simplification of hydrogen-bonded structures into *structural motifs*—subsets of molecules that exhibit a particular type of hydrogen bonding pattern. The chemical formula of the aggregate became less important, and the aggregate could now be viewed as a '*topology*'. Hydrogen bonded patterns were identified as belonging to one of four basic structural motifs: ring (***R***); chain (***C***); non-cyclic dimer (***D***); self (***S***, intramolecular hydrogen bonded motifs). These motifs were described by notation of the form $\boldsymbol{X^n_m(y)}$ where X is ***R***, ***C***, ***D*** or ***S***, ***m/n*** represents the number of donors and acceptors and ***y*** is the total number of atoms involved in forming the motif (Scheme 2.4a). Examples of some frequently observed graph sets are given in Scheme 2.4b. Graph set analysis can be used to identify '*isographic functional groups*', *i.e.* different functional groups that exhibit the same graph set (Scheme 2.4c).

Etter used graph sets to propose empirical rules correlating functional groups in molecules and hydrogen bond patterns observed in crystal structures.[105] The first three of the 16 rules proposed by Etter apply to any hydrogen-bonded structure:

(a) All good proton donors and acceptors are used in hydrogen bond formation.
(b) Intramolecular hydrogen bonds getting stability due to the formation of a six membered ring would occur in preference to intermolecular hydrogen bonds.
(c) The best proton donors and acceptors, which would remain after formation of the intramolecular hydrogen bonds, would further engage in intermolecular hydrogen bonds.

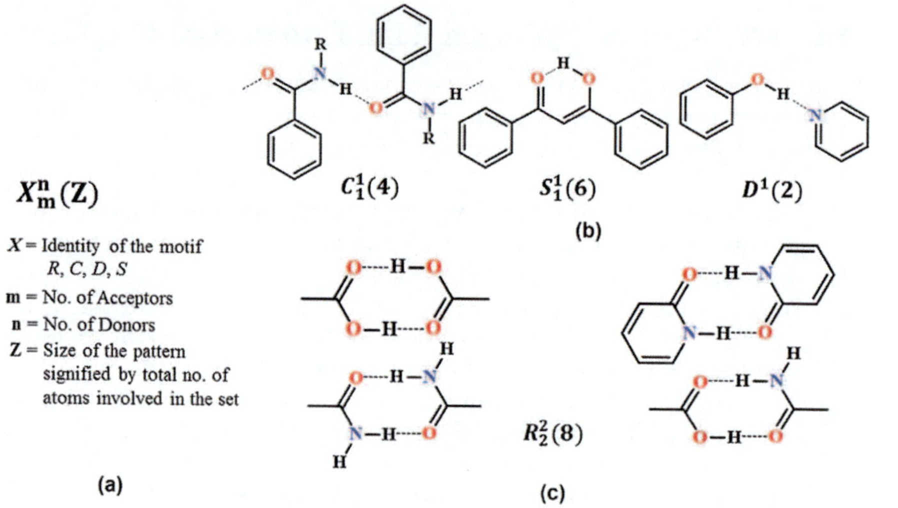

Scheme 2.4 (a) Illustration of the graph set notation, (b) examples of some simple graph sets and (c) isographic graph sets of the type $\boldsymbol{R^2_2(8)}$.

Further rules are more specific to certain systems and functional groups, *e.g.* nitroanilines, diaryl ureas, carboxylic acid co-crystals with 2-aminopyrimidine, co-crystals of nucleotide bases, *etc.* In essence, the hydrogen bond rules formulated by Etter identified the most energetically favourable patterns of intermolecular hydrogen bonding and laid the groundwork for crystal engineering of molecular crystal structures. Diversions from Etter's rules can occur due to steric effects, multiple donor–acceptor sites, ionic interactions, *etc.*

2.4.2 Supramolecular Synthons and Tectons

The concept of '*synthon*' in synthetic chemistry was introduced by Corey in 1965 who defined '*synthons*' as structural units within molecules that can be formed and/or assembled by known or conceivable synthetic operations.[106] In supramolecular chemistry, the synthetic '*targets*' are molecular assemblies like ribbons, squares, channels, *etc.* The '*supramolecular targets*' formed through the interplay of the non-covalent interactions are described topologically rather than in terms of their chemical nature. The molecules in this scenario are in essence '*building units*' or, as termed by Wuest, '*tectons*'.[107–110] The assembly of '*tectons*' is illustrated in Scheme 2.5a.

The concept of '*supramolecular synthon*' has played an important role in crystal engineering of co-crystals and was introduced by Desiraju in 1995.[111]

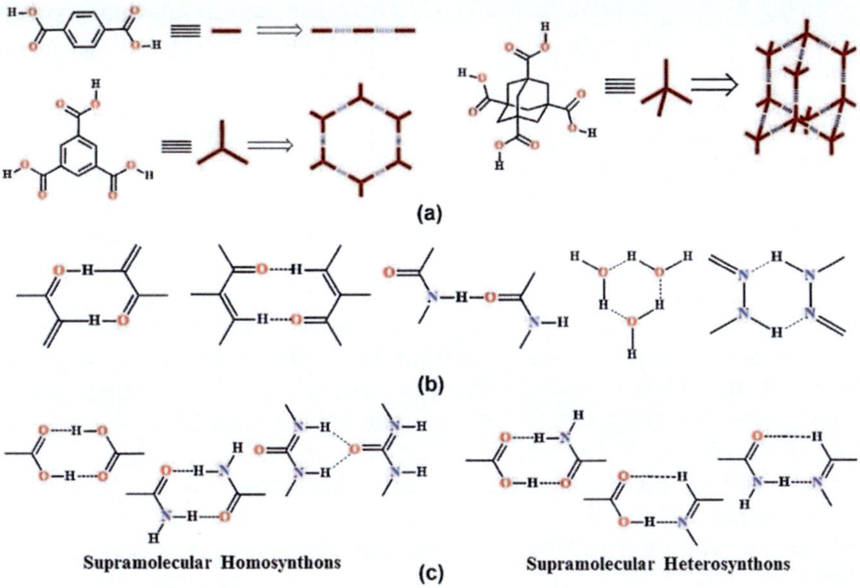

Scheme 2.5 (a) Examples of molecules used as building units or tectons to generate molecular assemblies. (b) Examples of supramolecular synthons commonly observed in crystal structures. (c) Examples of supramolecular homosynthons and supramolecular heterosynthons.

He defined supramolecular synthons as '*structural units within supermolecules which can be formed and/or assembled by known or conceivable synthetic operations involving intermolecular interactions*'. Every non-covalent interaction possesses some degree of directionality and '*supramolecular synthons*' can be viewed as spatial arrangements of such directional molecular interactions, which result in the formation of target molecular assemblies (Scheme 2.5b). Zaworotko proposed the classification of supramolecular synthons into two distinct subgroups: *supramolecular homosynthons* and *supramolecular heterosynthons* (Scheme 2.5c).[112] '*Supramolecular homosynthons*' are formed by self-complementary molecules, *e.g.* carboxylic acid dimers, amide···amide dimers, whereas '*supramolecular heterosynthons*' are formed from two or more different functionalities with complementary moieties, *e.g.* pyridine–carboxylic acid interactions. As detailed herein, supramolecular heterosynthons are particularly relevant to co-crystal design.[113–117]

2.4.3 Supramolecular Synthon Hierarchy

The concepts of graph sets and supramolecular synthons are not alone enough to make the design of crystal structures a routine exercise. It is also necessary to consider the relative strengths of all possible non-covalent interactions and understand how they are affected by competitive and non-competitive environments. Studies that address supramolecular synthon hierarchy can be performed by using existing structural information, and combining the studies with systematic experimental work. The Cambridge Structural Database (CSD), which is managed by the Cambridge Crystallographic Data Centre (CCDC), is a key tool in this context as it is an archive for crystal structures of organic compounds, including coordination complexes and organometallic compounds. The CSD can be used to obtain statistics regarding supramolecular synthons and to search for new supramolecular synthons. With respect to the utility of the CSD, in 1983 Allen *et al.* asserted that '*the systematic analysis of large numbers of related structures is a powerful research technique, capable of yielding results that could not be obtained by any other method*'.[118] Supramolecular synthon hierarchy is such a situation since it involves analysis of the occurrence of a particular supramolecular synthon in the presence and absence of competing supramolecular synthons. Such studies can address the reliability of particular supramolecular synthons for the design of a composition or structure. Fábián stated that '*the formation of synthons is dictated by the strength of hydrogen bond between the given molecule and the co-crystal former rather than the number of hydrogen bond acceptors or donors present*'.[119] In a general sense, if supramolecular heterosynthons are preferred over supramolecular homosynthons then a co-crystal can result if the functional groups are in different compounds. However, there are also co-crystals for which the structure is sustained by supramolecular homosynthons (Figure 2.1).[120,121]

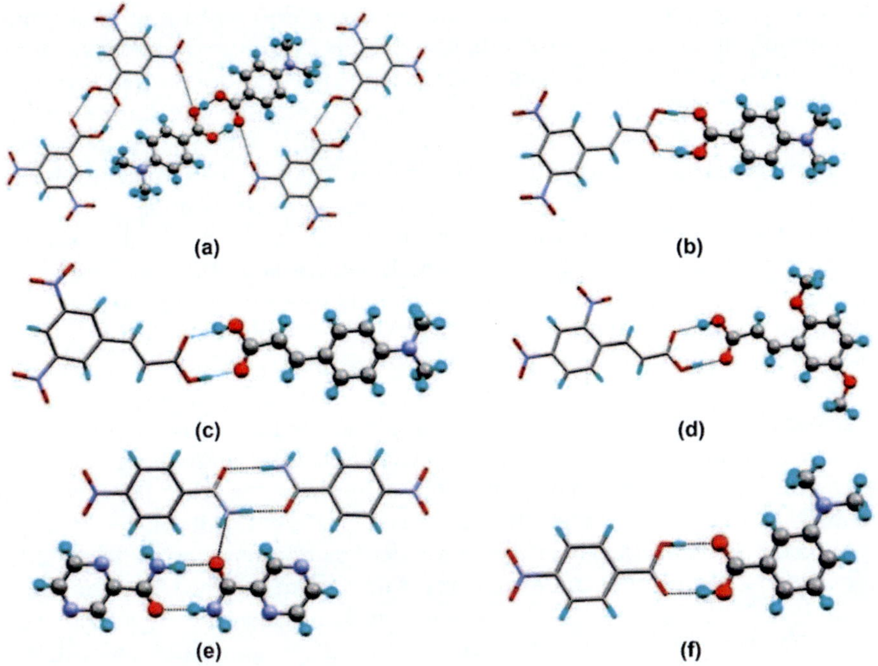

Figure 2.1 Examples of co-crystals sustained by supramolecular homosynthons. (a) GAUTAM, (b) PICDUA, (c) PICFAI, (d) PICFEM, (e) ASAYIC and (f) ASAYEY. In the case of GAUTAM, the structure is sustained by carboxylic acid homodimers. In the case of ASAYIC, the structure is sustained by amide···amide dimers in preference to possible heterosynthons between amide and pyrazine moieties.

A systematic study of supramolecular synthon hierarchy would typically involve the following steps:

(i) Identification of the functional groups that can form supramolecular synthons and the different supramolecular synthons possible from interaction between these functional groups.
(ii) Search the CSD for hits where a particular supramolecular synthon is present (raw search).
(iii) Search the CSD for hits where the particular supramolecular synthon is present in the absence of any competing groups (refined search).
(iv) Design model compound experiments to address the validity of statistics from the CSD study.

Supramolecular synthon hierarchies have been studied by various authors[122–125] and the information obtained from these studies can be utilised to design new crystalline solids where more than one type of supramolecular synthon is feasible.[126,127] The probability of the formation of certain intermolecular hydrogen bonds between different functional groups containing at least one strong hydrogen bond donor group was studied by Infantes *et al.*[128–130] It was evident from these studies that formation of carboxylic acid *supramolecular homosynthons* is more influenced by the presence of other chemical functionalities than amide···amide supramolecular homosynthons. The relative strength of supramolecular homosynthons was concluded to be amides > acids > alcohols.

Zaworotko and coworkers reported several supramolecular synthon hierarchy studies, the first of which addressed the occurrence of hydroxyl···pyridine supramolecular synthons in the presence of co-formers that also contain a cyano moiety.[131] They identified three different supramolecular synthons (Figure 2.2a) that could be expected in the case of a crystal structure containing C≡N, O–H, and the N_{arom} moiety. In the CSD analysis (Figure 2.2b) 136 structures were obtained that contained only OH and N_{arom} moieties; in 135 structures the supramolecular heterosynthon **I** was found to be present while supramolecular homosynthon **III** was found in only 37 structures. These statistics suggest that supramolecular heterosynthon **I** is favoured over supramolecular homosynthon **III**. Another search was performed for structures that contain only C≡N and O–H moieties. Of 77 structures, 57 were found to be sustained by supramolecular heterosynthon **II** whereas in 17 structures supramolecular homosynthon **III** was observed. These observations imply that supramolecular homosynthon **III** is not preferred *vs.* **I** and **II**. In order to determine the

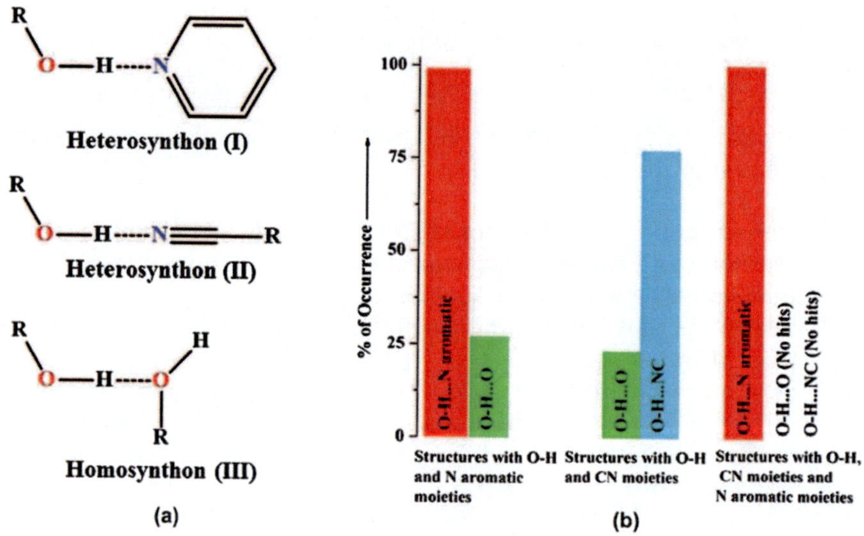

Figure 2.2 (a) Possible supramolecular heterosynthons in structures that include the hydroxyl, cyano and pyridine moieties. (b) CSD analysis of the statistical predominance of the O–H and N_{arom} synthons in terms of percentage of occurrence.

relative hierarchy of the three synthons, it was necessary to study compounds in which all three moieties are present in the absence of any other competing functional groups. Given that only three such crystal structures were archived in the CSD at that point, further experiments were required. 17 relevant co-crystals were prepared, and their structures were found to be sustained by supramolecular heterosynthon **I**, even in the presence of the CN moiety. Based on CSD statistics and the experimental results it was concluded that the relative strength of the three synthons is **I** > **II** > **III**. This study established O–H···N_{arom} (pyridine) as a reliable supramolecular heterosynthon for co-crystal synthesis.

Shattock *et al.* addressed the occurrence of the carboxylic acid··· pyridine heterosynthon in co-crystals in which a co-former also contained a hydroxyl moiety.[132] The preference for supramolecular heterosynthons (66%) over supramolecular homosynthons (34%) was established from a raw CSD search (5690 hits). It was also established from a raw CSD search and subsequent refined CSD search that, among the structures based on carboxylic acid homosynthons, the dimer motif is present in 92% of structures while the catemer motif is present in only 9% of the structures. Various supramolecular heterosynthons involving the carboxylic acid group in structures containing –COOH, –OH and N_{arom} functionalities are presented in Figure 2.3a. It was evident from both raw and refined CSD searches that COOH···N_{arom} heterosynthon (raw 98%, refined 74%) is the most frequently occurring heterosynthon involving carboxylic acids followed by the charge assisted COOH···Cl^- hydrogen bond (raw 100%, refined 64%). The acid···amide heterosynthon (raw 84%, refined 57%) was found to be the third most frequently observed synthon. A similar search of hydroxyl functional groups revealed the presence of OH···OH homosynthon in 26% of the structures. The crystal structures sustained by various heterosynthons were found to involve complementary functionalities like N_{arom} (53%), chloride (73%), amides (53%) and carbonyls (43%). Hence, both carboxylic acids and alcohols tend to yield structures sustained by supramolecular heterosynthons rather than supramolecular homosynthons. This trend was also observed in structures with both carboxylic acid and hydroxyl moieties. The authors focussed particularly on structures with –COOH, –OH and N_{arom} groups and no other acceptor/donor groups. The frequency of occurrence of various supramolecular heterosynthons observed from refined searches is given in Figure 2.3b. Model compound experiments were validated by synthesising a series of 15 co-crystals. It was evident from the co-crystals that carboxylic acid···pyridine and hydroxyl···pyridine supramolecular heterosynthons are equally prominent *vs.* the respective supramolecular homosynthons.

In another study, Kavuru *et al.* investigated carboxylate moieties with a focus on supramolecular synthons formed by carboxylate and hydroxyl groups.[133] The possible supramolecular heterosynthons that can be formed by the carboxylate and hydroxyl groups were first identified

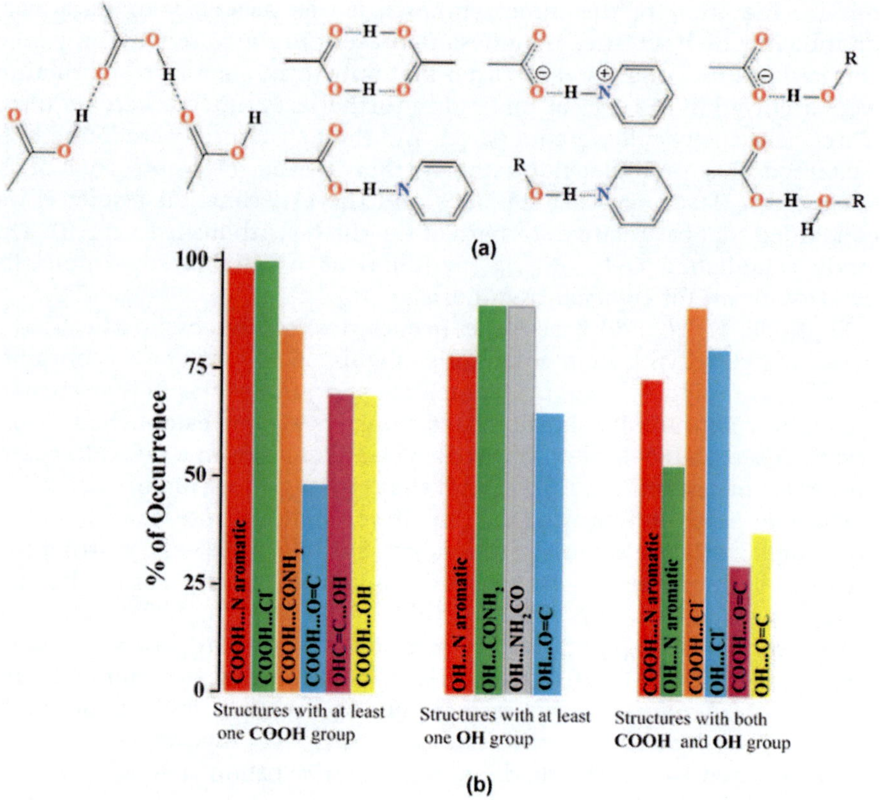

Figure 2.3 (a) Supramolecular synthons formed by carboxylic acid groups with other functional groups in structures containing –COOH, –OH and N_{arom} functionalities. (b) Percentage occurrence of hydrogen bonded supramolecular heterosynthons observed in refined CSD searches.

(Figure 2.4a) and a CSD search yielded 4968 hits containing the carboxylate moiety, of which 23% were found to be zwitterionic. The phenolic group was found in 103 structures where the hydroxyl moieties are not positioned *ortho*- with respect to each other (thereby eliminating hits with intermolecular hydrogen bonds). The $COO^-\cdots O–H$ (phenolic) supramolecular heterosynthon was observed in 58 of the 103 structures. The frequency of occurrence of various supramolecular synthons is presented in Figure 2.4b. Fifteen zwitterionic co-crystals based on amino acids like L-ascorbic acid with various nutraceuticals and polyphenols like quercetin and resveratrol were prepared. The structures of the co-crystals revealed a preference for carboxylate charge assisted hydrogen bonds. It was a general observation that molecules with weakly acidic O–H groups can function as potential co-formers for carboxylates because of their preference for the $COO^-\cdots O–H$ supramolecular heterosynthon. In three of the synthesised co-crystals, gallic acid·nicotinic acid (GALNAC·1.5H_2O), gallic

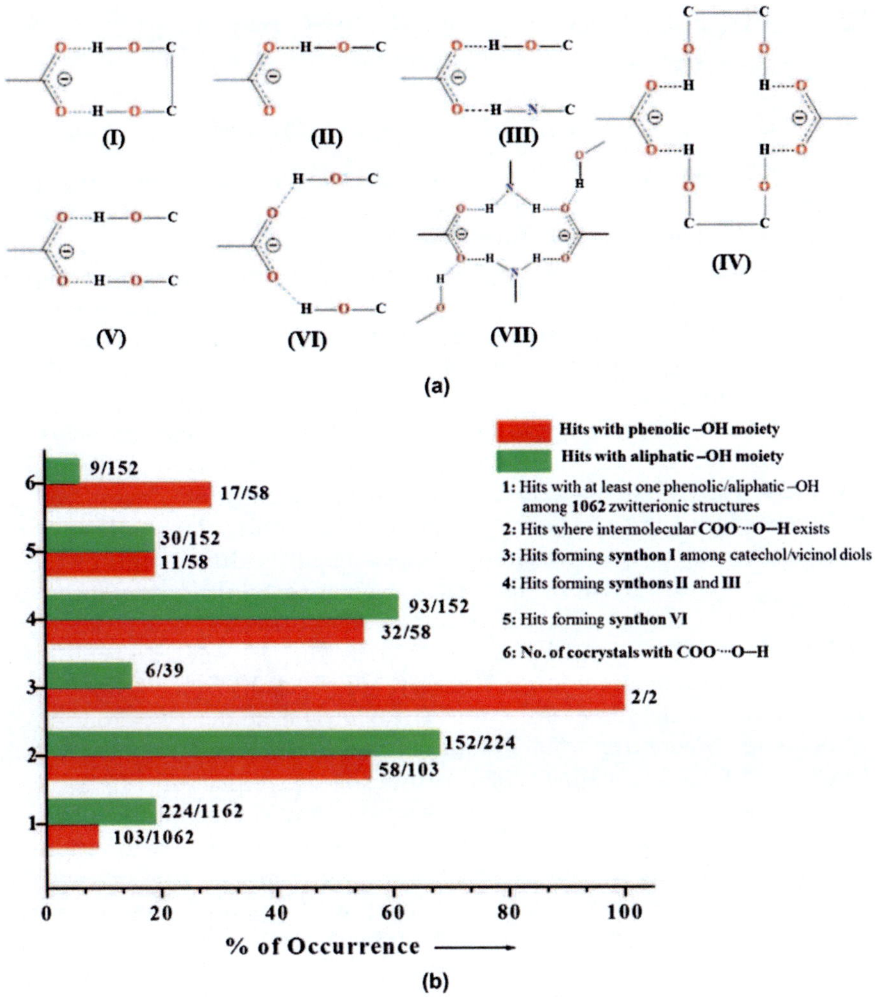

Figure 2.4 (a) Supramolecular synthons expected in structures that include a carboxylate and hydroxyl group. (b) Percentage occurrence of different hydrogen bonded supramolecular heterosynthons observed in the refined searches involving the carboxylate and hydroxyl group.

acid·isonicotinic acid (GALINA·H_2O), and protocatechuic acid·isonicotinic acid PCAINA·H_2O, the charge assisted $COO^{-}\cdots O\text{–}H$ heterosynthon was found to occur over the acid–pyridine supramolecular synthon. The authors thus demonstrated the use of carboxylate zwitterions for synthesising co-crystals based on polyphenols and compounds with weak O–H groups. The CSD analysis and the structural studies confirmed the possibility of the occurrence of these supramolecular synthons, sometimes in preference to otherwise robust hydrogen bonded supramolecular heterosynthons.

2.5 Other Aspects of Crystal Engineering of Co-crystals

2.5.1 Design of 2D and 3D Hydrogen Bonded Networks

Design of 2D and 3D coordination networks is well established thanks to the predictability of metal geometries and relative strength of coordinate bonds.[134–138] However, an analogy can be drawn between coordinate bonds and hydrogen bonds since they both involve a donor and an acceptor. Therefore, just as the coordination sphere around the metal can define a node in a coordination network, the shape of a molecule can serve as a node in a hydrogen-bonded network.[139] 2D networks can be formed by certain organic molecules such as cyanuric acids,[140] melamine,[141] guanidinium salts.[142] Trimesic acid (TMA) is probably the most renowned of such molecules and it self-assembles to form 2D honeycomb (hcb) networks (BTCOAC) thanks to the orientation of the carboxylic acids and the predictability of the carboxylic acid dimer supramolecular synthon in the absence of competing functional groups. The cavity dimensions of the resulting honeycomb network are 14×14 Å (Figure 2.5a).[143] Zaworotko and coworkers demonstrated that 2 : 3 co-crystals of TMA and 4,4′-bipyridine (RAPHAR) and 4,4′-bipyridylethane (SAYMUB) respectively, can exhibit hcb networks of much larger dimensions (Figure 2.5a).[144,145] Similar observations were made for cyclohexane tricarboxylic acid (CTA) and its co-crystals with bpy as reported by Nangia and coworkers.[146,147] The crystal structure of CTA·2H_2O (BANGON) formed by crystallisation of CTA from ethanol is a (6,3) net sustained by carboxylic acid dimer interactions and hydrogen bonds between water and the carboxylic acid moieties (Figure 2.5b). When CTA is co-crystallised with bpy, a structure with an expanded (6,3) net is formed sustained by acid–pyridine and water–acid supramolecular heterosynthons (BANGUT). A similar (6,3) net was also observed in the co-crystal CTA·bpet and is sustained by the acid–pyridine supramolecular heterosynthon (EJUZEO).

A rare example of a primitive cubic, pcu, hydrogen bonded net was exemplified by Men *et al.*, who co-crystallised methanetetrabenzoic acid (H_4MTB) with 4,4′-azopyridine (azpy) (Figure 2.5c).[148] The resulting structure has no acid dimers (**I**) or acid–pyridine supramolecular heterosynthons (**II**). Rather, it is sustained by supramolecular synthon **III** (Figure 2.5c). The four carboxylic acid groups of each H_4MTB molecule were observed to be engaged in two hydrogen bonds. The O–H groups of the carboxylic acid moieties hydrogen bond with two azpy moieties whereas the carbonyl oxygen forms a hydrogen bond with the O–H moiety of two adjacent H_4MTB molecules. H_4MTB molecules are thereby connected with four other H_4MTB molecules through direct hydrogen bonds and two other H_4MTB moieties through the azpy linker. The dimension of the pcu net, 14.7×14.7×26.3 Å, enables 3-fold interpenetration. Co-crystallization of H_4MTB with bpe resulted in a co-crystal isostructural to H_4MTB·azpy. Another co-crystal based pcu net

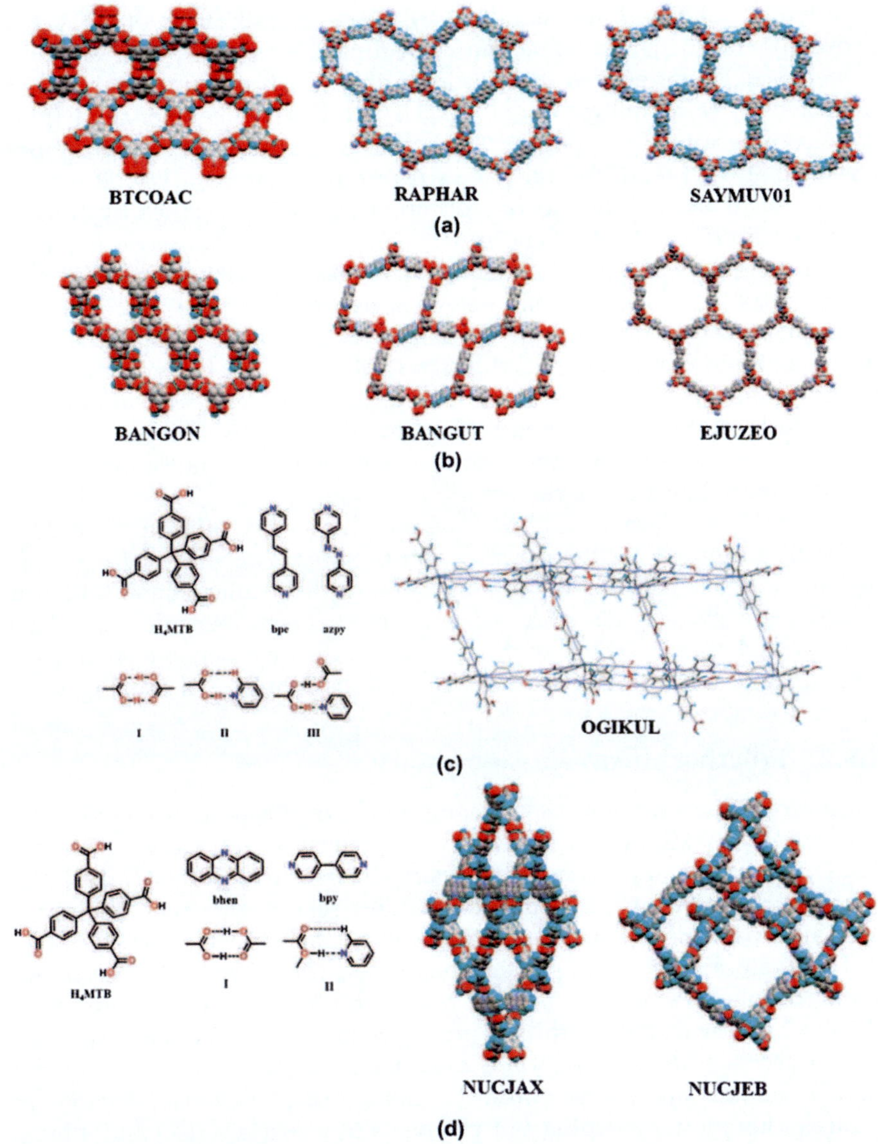

Figure 2.5 (a) Honeycomb (hcb) networks in crystal structures of trimesic acid (BTCOAC) and expanded hcb networks in TMA·bpy (RAPHAR) and TMA·bpet (SAYMUB). (b) (6,3) nets in crystal structures of CTA·$2H_2O$ (BANGON) and CTA·bpy·H_2O (BANGUT) and CTA·bpet (EJUZEO). (c) Synthons and the pcu net observed in co-crystals of H_4TMB and azbp co-formers (OGIKUL). (d) dia nets in co-crystals of H_4TMB·phen (NUCJAX) and H_4TMB·bpy (NUCJEB).

reported by Saha *et al.* was obtained by the co-crystallization of 2,4,6-tris(4-hydroxyphenoxy)-1,3,5-triazine and bipyridine.[149]

Men *et al.* reported two diamondoid, dia, nets based on purely organic molecules by co-crystallization of H_4MTB with linkers like phenazine (phen) and 4,4′-bipyridine (bpy) (Figure 2.5d).[150] The structure of 1 : 1 H_4MTB·phen reveals that each H_4MTB molecule hydrogen bonds to two H_4MTB moieties *via* acid–acid dimers and to two different H_4MTB molecules by O–H···N hydrogen bonds with phen co-formers. The resulting dia network exhibits adamantanoid cages of dimensions of 25.5×60.54×30.81 Å and 7-fold interpenetration. Another dia net was observed in the 1 : 2 co-crystal H_4MTB·bpy reported in the same paper. All four acid groups of H_4MTB hydrogen bond to bpy moieties through O–H···N interactions forming a net of dimension 53.99×53.99×56.99 Å. The resulting void was found to be occupied by 18-fold interpenetration. Another instance of organic dia nets was reported by Nangia and coworkers who studied ammonium salts of cyclohexane-1,3,5-tricarboxylic acid.[151]

The examples discussed above suggest that co-crystallization could be generally employed for the construction of supramolecular hydrogen bonded networks if co-formers with appropriate molecular shape and functionality are chosen, *e.g.* trimesic acid, methanetetrabenzoic acid. These co-formers define the shape of the network while the other co-former, which acts as the linker, can then be used to modulate the size of the network.

2.5.2 Polymorphism in Co-crystals

Polymorphs occur when the same compound exhibits more than one crystal structure.[152,153] McCrone defined polymorphism as *'the ability of a given element or compound to crystallise as more than one distinct crystal species'.*[154] Polymorphs open up an avenue for studies related to structure–property relationships. This is because polymorphs exhibit different crystal packing and therefore different physicochemical properties, *e.g.* mechanical strength, solubility and dissolution rate, stability towards moisture, photochemical reactivity. Polymorphism among co-crystals can be categorised into four classes – *synthon polymorphs*, *conformational polymorphs*, *packing polymorphs* and *tautomeric polymorphs*.[155] Among these, *tautomeric polymorphism* is rare although a dimorphic 1 : 1 co-crystal of piroxicam (PX) and 4-hydroxybenzoic acid (4HBA) has been reported.[156] It should be noted that a 1 : 1 co-crystal of two co-formers is not a polymorph of a 2 : 1 co-crystal of the same co-formers.

Synthon polymorphism occurs when different solid forms of a co-crystal are sustained by different supramolecular synthons.[157,158] Polymorphic 2 : 1 co-crystals of 4-hydroxybenzoic acid (4HBA) and tetramethylpyrazine (TMP) were reported by Sreekanth *et al.*[159] Form I (ODOBIT) of 4HBA·TMP exhibits a structure sustained by acid–acid dimers and hydroxyl–pyrazine supramolecular heterosynthons whereas form II (ODOBIT01) is sustained by acid–pyridine and hydroxyl–carbonyl heterosynthons (Figure 2.6a). Form I

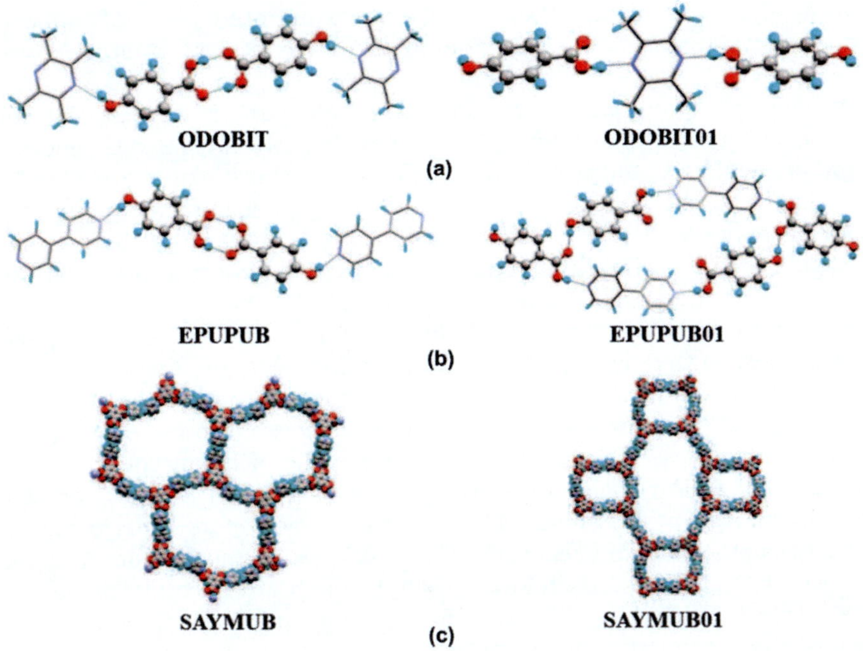

Figure 2.6 Synthon polymorphism in dimorphic molecular co-crystals (a) 4HBA·TMP (ODOBIT and ODOBIT01), (b) 4HBA·bpy (EPUPUB and EPUPUB01) and (c) conformational polymorphism observed in 2 : 3 co-crystals of TMA and the bpea : hcb network in SAYMUB and (10,3)-connected net in SAYMUB01.

deviates from the expected hydrogen bond synthon hierarchy, *i.e.* a homosynthon is formed over a possible heterosynthon. A similar situation is observed in the case of the dimorphic co-crystals of 4-hydroxybenzoic acid and 4,4′-bipyridine (bpy) reported by Mukherjee *et al.* (Figure 2.6b).[160] The structure of form I (EPUPUB) is sustained by acid–acid dimers and hydroxyl–pyridine heterosynthons whereas in form II (EPUPUB01) exhibits hydroxyl–carbonyl and acid–pyridine heterosynthons.

Conformational polymorphism results from different molecular conformations[161,162] and can be observed in molecules with torsional freedom that exhibit several low energy conformations. If the energy required for rotation about single bonds is comparable to lattice energy differences between possible polymorphs, then conformational polymorphism is expected. The co-crystal of trimesic acid (TMA) and 1,2-bis(4-pyridyl)ethane (bpea) forms polymorphs sustained by acid–pyridine heterosynthons.[145] Form I (SAYMUB) forms a (6,3) honeycomb network with a cavity of 39×27 Å. The honeycomb network undergoes parallel interpenetration. Form II of TMA·bpea (SAYMUB01) is an 18-fold interpenetrated (10,3)-a net (Figure 2.6c). A set of three (10,3)-a nets undergoes 3-fold interpenetration, and each set then undergoes interpenetration with two other sets to afford

9-fold interpenetration. Two 9-fold interpenetrated nets of opposite handedness further interpenetrate to form an 18 fold interpenetrated (10,3)-a net.

Polymorphs of co-crystals are considered to be *packing polymorphs* if they exhibit different crystal packing even though there are no differences in supramolecular synthons or molecular conformations. The dimorphic 2 : 1 co-crystals of salicylic acid and *N,N′*-diacetylpiperazine are examples of packing polymorphism.[163] Orthorhombic form I, NUKXEX, has a layered structure with the salicylic acid molecules oriented almost in plane with the *N,N′*-diacetylpiperazine. In the case of form II, NUKXEX01, the structure is not layered, and the salicylic acid molecules are oriented perpendicularly with respect to the plane of the *N,N′*-diacetylpiperazine moiety.

Polymorphism is a challenge in the context of drug substances and Almarsson and Zaworotko suggested that co-crystallization of drug substances can be useful in reducing the propensity of a compound to form polymorphs and/or solvates/hydrates.[164] Their assertion was supported by experimental evidence obtained from crystallisation studies performed on carbamazepine and piracetam.[165,166] In 2014 Aitipamula reported the existence of 114 co-crystals archived in CSD, which are polymorphic, and in 15% of these cases the polymorphs formed concomitantly.[155] However, one cannot draw general conclusions based upon these limited studies. Systematic approaches like high-throughput screening and the use of *in situ* atomic force microscopy for experimental screening of polymorphic forms are required and can be routinely conducted.

2.5.3 Co-crystal Solvates and Hydrates

Hydrates demand special attention because of their prevalence in pharmaceutical compounds.[167] More generally, the small size of the water molecule and its ability to form strong hydrogen bonds with different functionalities makes lattice water incorporation a common phenomenon in molecular solids.[14] As mentioned earlier, Morris and Rodriguez-Hornedo classified hydrates based on the interaction of water molecules with the crystal lattice:[29] (i) in *channel hydrates* self-interacting water molecules exist in channels of the crystal lattice, (ii) *isolated site hydrates* exhibit no direct interaction between water molecules in the lattice, and (iii) *metal ion associated hydrates* contain water molecules that interact with metal ions. Clarke *et al.* reported a study on 11 *co-crystal hydrates* and attempted to correlate the thermal stability of the hydrate with the observed crystal structures.[168] The co-crystal hydrates were categorised into four different groups based upon their thermal stability: (i) loss of water below 100 °C, (ii) loss of water between 100 °C and 120 °C, (iii) loss of water above 120 °C, and (iv) dehydration concurrent with melting of the crystal. The first category of hydrates all exist as channel hydrates, but otherwise there was no correlation between structure and thermal stability. The authors noted that due to the high number of possible H-bond environments possible for water molecules it could be difficult to find structure–property correlations.

In 2010 Aitipamula and coworkers reported that among 3624 co-crystals reported in CSD (CSD version 2009), 451 form co-crystal hydrates (12.5%).[169] Among these 451 co-crystal hydrates, 396 structures are pure hydrates whereas 55 also contain solvent molecules. According to their statistics, the occurrence of hydrates among co-crystals is higher than that of single component structures (6835/135 997, 5%). In an updated CSD search reported by the Zaworotko group, the frequency of hydrates among single component structures was found to be 6.5% (16 710/257 442).[32] One might be tempted to conclude that hydrate propensity increases when a co-crystal is formed. However, the statistical information obtained from the CSD with respect to hydrate propensity does not reflect all crystallisation outcomes and experimental conditions are not archived so, for example, whether or not there was exposure to humidity during crystallisation cannot be searched for.

In solvates, solvent molecules can likewise form part of the network[170,171] or exist in lattice voids like channels[172,173] or cages.[174] Saha *et al.* reported a dimorphic methanol solvate of trimesic acid and 4, 4′-methylenebis(2,6-dimethylaniline).[175] The dimorphic co-crystal solvates AWAXOM and AWAXOM01 are synthon polymorphs sustained by different supramolecular synthons. AWAXOM is a hydrogen bonded hexagonal network sustained by two different supramolecular synthons involving the co-formers trimesic acid and 4,4′-methylenebis(2,6-dimethylaniline) and MeOH (Figure 2.7a). AWAXOM01 is comprised of hydrogen-bonded sheets sustained by carboxylic acid dimers (Figure 2.7b).

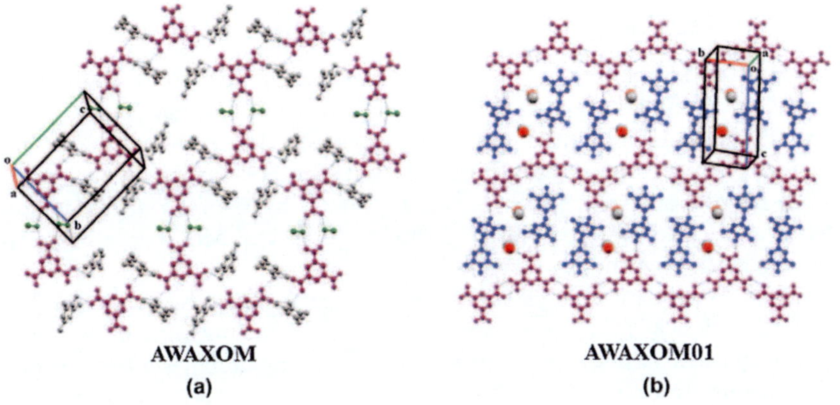

Figure 2.7 (a) Hexagonal hydrogen bonded network in the structure of AWAXOM through solvent mediated synthons. Only the hydrogen-bonded part of MBDA is shown for clarity. (b) The two-dimensional hydrogen bonded structure of AWAXOM01.

2.5.4 Crystalline Inclusion Compounds (CICs)

Crystalline inclusion compounds (CICs) form when a molecule (the *host*) crystallises with intrinsic voids or has the capability of generating voids within the crystal lattice. A second molecule (the *guest*) then occupies the voids. Nassimbeni proposed classification according to the nature of the void as *clathrands* or *cavitands*.[20] The first category consists of host molecules like crown ethers, calixarenes and cyclodextrins where the associated void is inherent to the molecule itself and hence intrinsic. The second class of CICs consists of non-porous small molecules that can self-assemble to form molecular networks that contain void spaces. In either case, the formation of a CIC is governed by crystal packing of the host molecule and interactions between the host and the guest. The second category of CICs mentioned above is quite significant in the crystal engineering context because they are amenable to fine-tuning and design.[176,177]

Peraka *et al.* demonstrated the use of 5-hydroxyisophthalic acid (HPA) for generating a series of CICs with co-formers, namely citral, 2-phenyl-2-butenal, (+)-camphor, (±)-camphor, isoamyl acetate, maltol and vanillin.[178] The authors demonstrated that certain guest molecules enable HPA molecules to generate guest accessible voids leading to the formation of CICs. Seven CICs sustained by hydrogen bonds between HPA and guests were reported. The guest molecules, aldehydes, ketones or esters, occupy the voids within HPA honeycomb nets and were supported by O–H···O interactions between the phenolic group of HPA and the carbonyl moieties of the guest molecule. In the structure of HPA·vanillin, guest molecules lie inside the honeycomb net resulting in AAAA type stacking (Figure 2.8a) whereas for HPA·citral the guest molecule projects out of the plane of the honeycomb net leading to ABAB stacking (Figure 2.8b).

2.5.5 Molecular Co-crystals with More Than Two Co-formers

Designing and synthesising molecular co-crystals with more than two co-formers is a challenge. A full understanding of supramolecular synthon hierarchy becomes essential and co-formers should be chosen based upon hierarchy. An essential condition is that the supramolecular synthons should be orthogonal to each other, *i.e.* occurrence of a particular supramolecular synthon should not be influenced by the occurrence of another supramolecular synthon.

The first report on co-crystals made from more than two co-formers was made by Aakeröy and coworkers in 2001.[115] The authors demonstrated the hierarchical superiority of the acid···pyridine heterosynthon over the acid···amide heterosynthon, thereby resulting in formation of three 1 : 1 : 1 co-crystals BUDZUV, BUFBIP and BUFQAO (Figure 2.9a).

The first report of 4-component (quaternary) co-crystals was made by Nangia and coworkers in 2005.[179] The authors exploited 1,3,5-cyclohexanetricarboxylic acid (H_3CTA) to design several ternary co-crystals and a

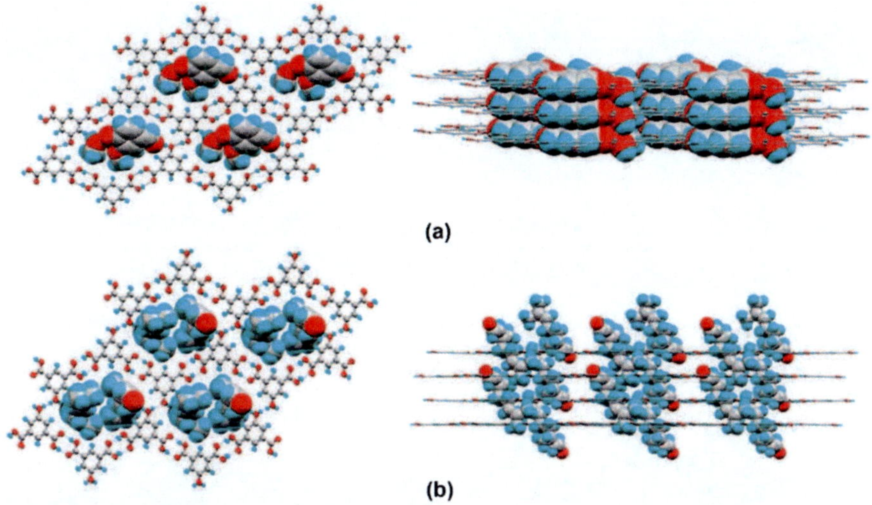

Figure 2.8 Hydrogen bonded honeycomb network in the structure of (a) HPA·Vanillin and AAAA type stacking observed in the structure of HPA·Vanillin CIC and (b) HPA·citral and ABAB type stacking observed in the structure of HPA·citral.

quaternary co-crystal with bent 4,4′-bipyridines. The quaternary co-crystal (KAVDOB) is comprised of H_3CTA, 1,2-bis(4-pyridyl)ethane (bpy-eta), tetramethylenebipyridine (bpy-bu) and 1,4-dichlorobenzene (Figure 2.9b). H_3CTA and bipyridine linkers bpy-eta and bpy-bu assembled to form a 3-component hydrogen bonded network with a cavity occupied by a dichlorobenzene molecule.

Rissanen and Topić reported a series of ternary co-crystals using the orthogonality of hydrogen bond and halogen bond interactions.[180] They chose crown ethers as hydrogen bond acceptors, thioureas as hydrogen bond donors, and different Br/I based perfluorobenzenes as halogen bond donors. Exploiting the orthogonality of hydrogen bond and halogen bond interactions enabled 15 ternary co-crystals sustained by hydrogen bonds and halogen bonds. Two of the reported ternary co-crystals, OQITAL and OQIJIJ, are shown in Figure 2.10a.

Desiraju and coworkers reported the first 5-component co-crystals by applying a strategy based on the supramolecular synthon approach.[181] The authors started with the hypothesis that if a co-former of a lower order crystal can exist in two different structural environments, the differences between the structural environment can be exploited to add more co-formers to the crystal lattice. The key is to start with two co-formers, A and B, for which different binaries are possible – AB and AB1B2 (Figure 2.10b). The binary AB could be regarded as a synthetic dead end as no further co-formers can be added. The binary AB1B2 represents the situation where the functional groups in B exist in two environments B1 and B2. The next co-former C

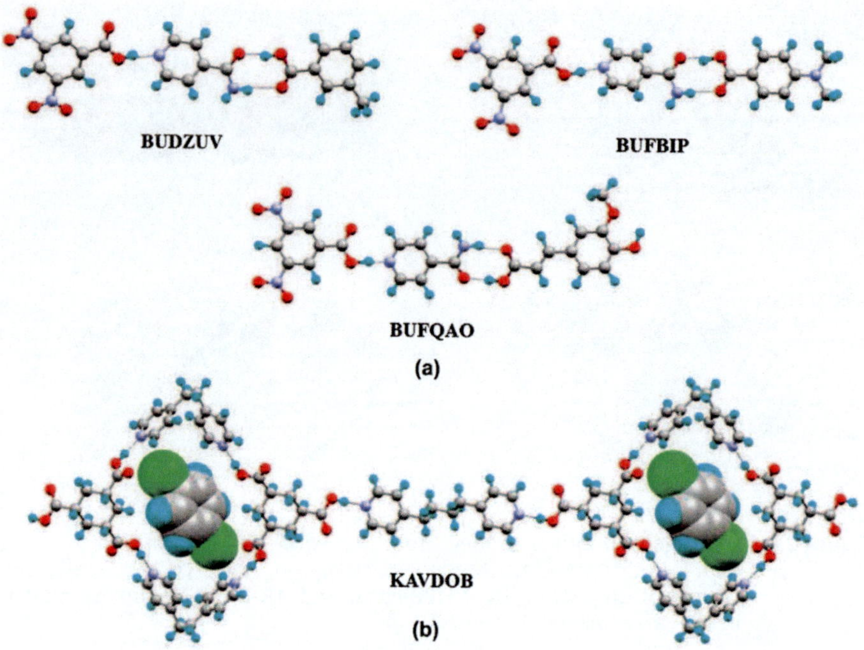

Figure 2.9 (a) 1:1:1 Ternary co-crystals of 3,5-dinitrobenzoic acid, isonicotinamide, and 4-(dimethylamino)benzoic acid (BUDZUV), 3,5-dinitrobenzoic acid, isonicotinamide and 3-methylbenzoic acid (BUFBIP) and 3,5-dinitrobenzoic acid, isonicotinamide and 4-hydroxy-3-methoxycinnamic acid (BUFQAO). (b) The hydrogen-bonded network in KAVDOB, a quaternary co-crystal comprised of a hydrogen bonded host network involving H_3CTA, bpy-eta and bpy-bu. 1,4-dichlorobenzene is the fourth co-former.

should be able to differentiate between B1 and B2 and preferentially combine to any one of them, *i.e.* either with B1 or B2, to afford a ternary system—ABC or ABC1C2. The same strategy can be used to introduce other co-formers. The authors reported a series of 10 quaternary and a quintinary co-crystal (Figure 2.10c).

2.6 Applications of Co-crystals

2.6.1 Pharmaceutical Co-crystals

Perhaps the most important application of co-crystals lies in the field of pharmaceutical science. Co-crystals of a drug substance can be synthesised with suitable co-formers and the resulting solid can be used as an alternative drug substance with enhanced physicochemical properties. Drug substances must be suitable for use in a drug product, *i.e.* their solubility, dissolution rate, stability (moisture and pH), bioavailability and pharmacokinetic behaviour must be acceptable. The vast majority of drug molecules have

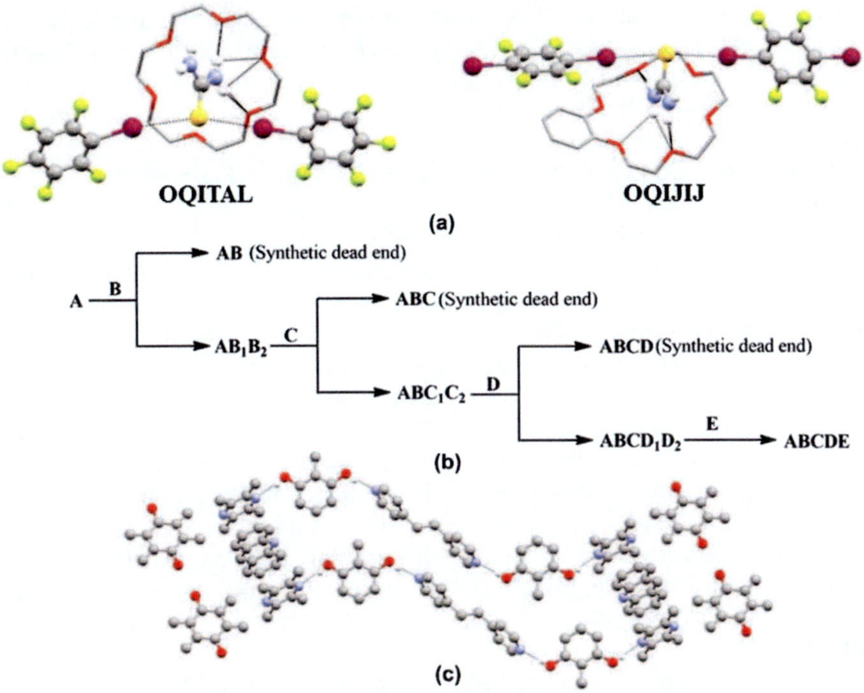

Figure 2.10 (a) The ternary co-crystal OQITAL made from co-formers 18-Crown-6, thiourea and bis(1,2,3,4,5-pentafluoro-6-iodobenzene and OQIJIJ made from co-formers 2,3,5,6,8,9,11,12,14,15-decahydro-1,4,7,10,13,16-benzohexaoxacyclooctadecine, thiurea and 1,2,4,5-tetrafluoro-3,6-diiodobenzene (OQIJIJ). (b) Flow chart for the synthetic strategy of co-crystals with more than two formers adopted by Desiraju and coworkers. (c) Supramolecular synthons in quintinary co-crystal of 2-methylresorcinol, tetramethylpyrazine, 1,2-bis(4-pyridyl)ethylene, acridine and 2,3,5-trimethyl-1,4-benzoquinone.

potential H-bonding functionalities like –COOH or –OH moieties in their structure. Hence, they are amenable to form hydrogen-bonded co-crystals with appropriate co-formers. An early report on pharmaceutical co-crystals was made by Oswald *et al.* in 2002 in which co-crystals of paracetamol with six different co-formers were reported.[182] In 2003, Walsh *et al.* reported co-crystals of ibuprofen, flurbiprofen and aspirin with bipyridine by exploiting the acid···pyridine supramolecular heterosynthon.[112]

Over the last decade, there have been numerous reports of co-crystals of active pharmaceutical ingredients (APIs)[161–163] and nutraceuticals[183–185] with a variety of low-toxicity co-formers. The selection of a library of co-formers is a critical step since co-formers should be inherently safe, complementary with the API and likely to impact the property of interest. Potential co-formers are listed in the *Priority-Based Assessment of Food Additives* (PAFA), which contains the chemical and toxicological information on about 2000 substances. This list includes compounds that are designated as *Generally Recognized as Safe* (GRAS). The *Everything Added to Food* (EAFUS) database in the United States maintains a list of *ca.* 1000 food additives. Herein we present selected case studies on hydrogen bonded pharmaceutical and nutraceutical co-crystals.

Quercetin (QUE) is a well-known dietary flavonoid that suffers from poor solubility and poor bioavailability. Zaworotko and co-workers reported a series of QUE (used as QUE·$2H_2O$) co-crystals with caffeine, isonicotinamide and theobromine (Figure 2.11a).[186] The structures are sustained by supramolecular heterosynthons involving the O–H groups of QUE and aromatic N atoms of the co-formers. The solubilities of the four co-crystals QUECAF, QUECAF·MeOH, QUEINM and QUETBR·$2H_2O$ were found to be 3.627, 2.018, 1.22 and 0.326 mg ml^{-1} respectively in 1 : 1 EtOH–H_2O, much higher than the solubility of QUE·$2H_2O$ (0.267 mg ml^{-1}) in the same medium (Figure 2.11b). In particular, QUECAF and QUECAF·MeOH exhibited solubility enhancements by 8- and 14-fold compared to QUE. Pharmacokinetic studies for QUE·$2H_2O$ and the four new QUE solid forms were conducted by administering the respective solid forms to male Sprague Dawley rats. The pharmacokinetic parameters observed for QUE were quite poor whereas the four co-crystals were found to exhibit much improved pharmacokinetic parameters.

A report on epigallocatechin-3-gallate (EGCg) solid forms, also by Zaworotko and co-workers, addressed the opposite problem to that faced by QUE.[187] EGCg is a polyphenol that is the primary component of green tea and EGCg is considered to offer numerous health benefits. However, it suffers from poor oral bioavailability. Interestingly, EGCg is highly soluble in water, but membrane permeability of EGCg has been found to be low. In order to address this problem, the Zaworotko group synthesised seven new solid forms of EGCg including co-crystals, co-crystal hydrates and mixed solvates with different co-formers (Figure 2.11c). The authors aimed to develop new solid forms of EGCg that would exhibit reduced aqueous solubility *vs.* EGCg. It was hypothesised that a reduction in aqueous solubility could reduce the rate of dissolution and modulate the *in vivo* absorption of EGCg. The structures of the co-crystals, EGCg·INM·$5H_2O$, EGCg·NIC·$9H_2O$, EGCg·INA·$3H_2O$, EGCg·INA, and EGCg·NAC·xH_2O, are sustained by strong hydrogen bonds and their aqueous solubility was found to be less than EGCg (Figure 2.11d). Pharmacokinetic studies conducted upon male Sprague Dawley rats revealed that EGCg·INA·$3H_2O$ and EGCg·INA exhibit better pharmacokinetics than EGCg.

Aakeröy and coworkers demonstrated modulation of the melting point and solubility in co-crystals of the anti-cancer drug hexamethylenebisacetamide (HMBA) with dicarboxylic acid co-formers of increasing chain length (Figure 2.12a).[188] HMBA is known to be an inhibitor of proliferation of lung cancer cells and is used in the treatment of myelodysplastic syndrome. HMBA was co-crystallised with a series of dicarboxylic acids ($n = 4$, 6, 8, 10, 12). The structure of the co-crystals consists of linear alternating chains of

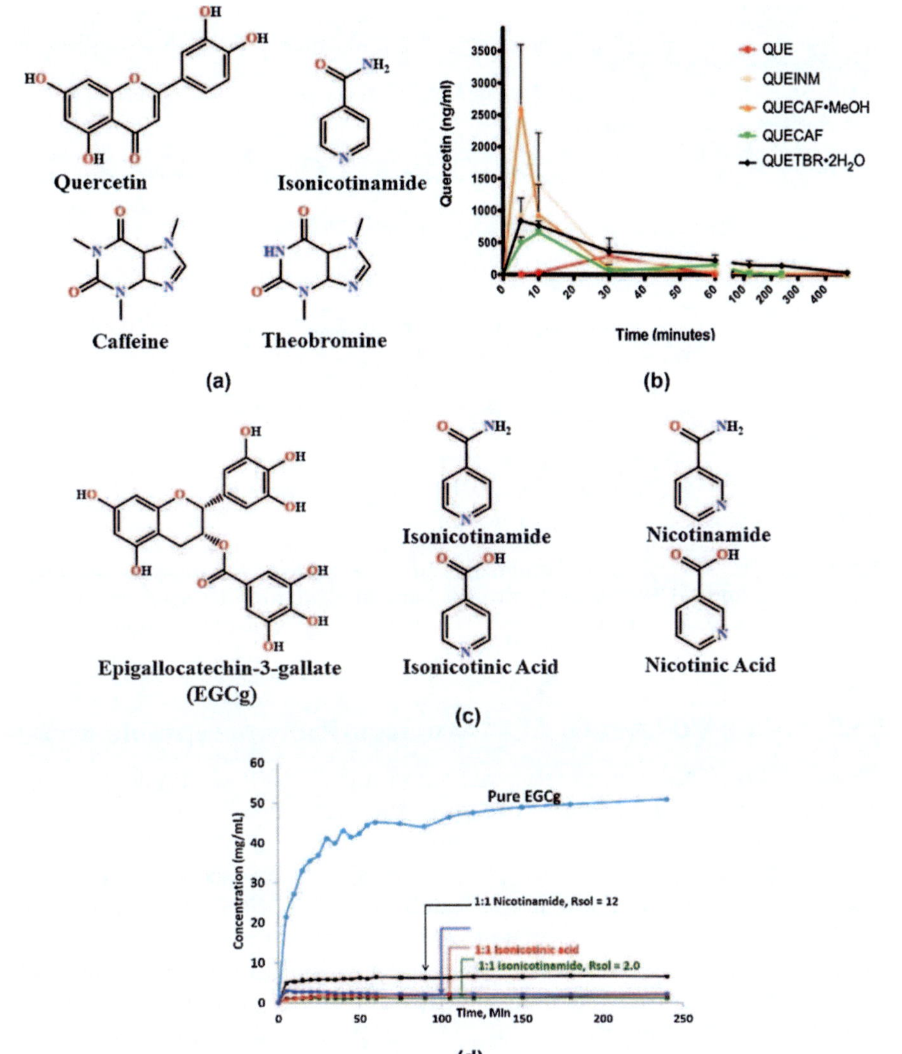

Figure 2.11 (a) Co-crystal co-formers used in the study of QUE·$2H_2O$. (b) Dissolution profiles of QUE·$2H_2O$ and its co-crystals in 1 : 1 ethanol/water mixture. (c) Co-crystal co-formers used in the study of epigallocatechin-3-gallate (EGCg). (d) Aqueous solubilities of EGCg and its co-crystals.

HMBA and co-former molecules connected by the acid···pyridine heterosynthon. The chains assemble to form a layer *via* amide···amide homosynthons (Figure 2.12b). It was observed that the resulting co-crystals exhibit a gradual increase of melting point with increase in the acid chain. The solubility of the co-crystals was found to decrease with increase in the chain of the acid.

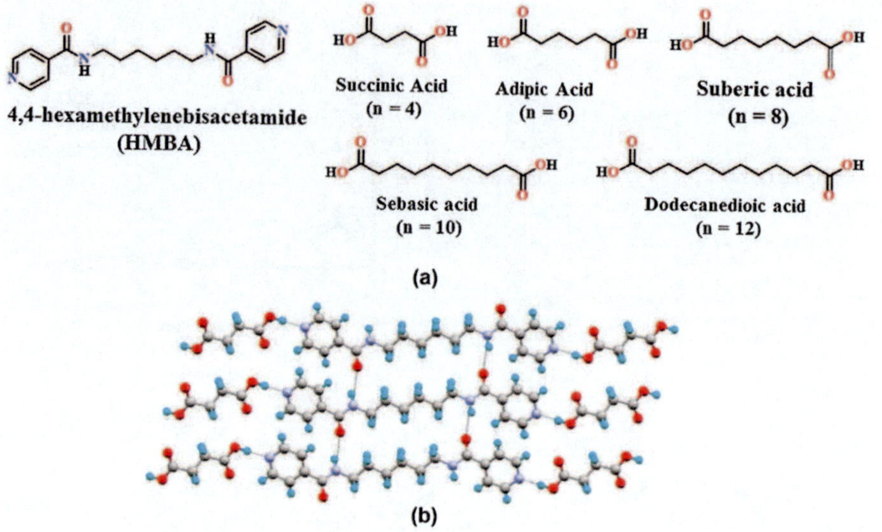

Figure 2.12 (a) Co-crystal co-formers used in the study of hexamethylenebisacetamide (HMBA). (b) Hydrogen bonded structure of HMBA and succinic acid co-crystal aided by O–H···N and N–H···O hydrogen bonds.

2.6.2 Ionic Co-crystals of Pharmaceutically Acceptable Metals

The design of ICCs using metal cations is not yet well established, despite their appearance in the literature over 200 years ago. Part of the reason might be that predicting the coordination modes of alkali/alkaline earth metals, especially Na^+, Li^+, Ca^{2+}, Mg^{2+}, and their complexes is non-routine. However, this does not mean that such ICCs are difficult to isolate and characterise. Further, these metals are pharmaceutically acceptable[189,190] and sodium cations are the most widely used cations in pharmaceutical salts that have been approved as drug products.[23,24] The formation of an ICC is controlled by coordination bonds and/or non-covalent interactions like charge-assisted hydrogen bonds. Hence, it is important to consider co-formers with potential functionality that would not only exhibit metal coordination but also act as acceptor/donor sites for hydrogen bonds. Herein we discuss three cases studies of ICCs based on drug substances.

Childs *et al.* reported a crystal engineering approach to ICCs of the drug substance in Prozac®, fluoxetine hydrochloride (FxHCl), with carboxylic acids co-formers. This was accomplished by exploiting the hydrogen bond accepting capability of the chloride anion of FxHCl.[191] Benzoic acid (*Bz*), succinic acid (*Su*) and fumaric acid (*Fm*) were used as co-formers and the structures of the ICCs were observed to be sustained by charge-assisted O–H···Cl hydrogen bonds (Figure 2.13a). FxHCl·*Bz* and FxHCl·*Fm* were observed to exhibit a lower dissolution rate than that of FxHCl whereas the dissolution rate of FxHCl·*Su* was found to be higher than that of FxHCl

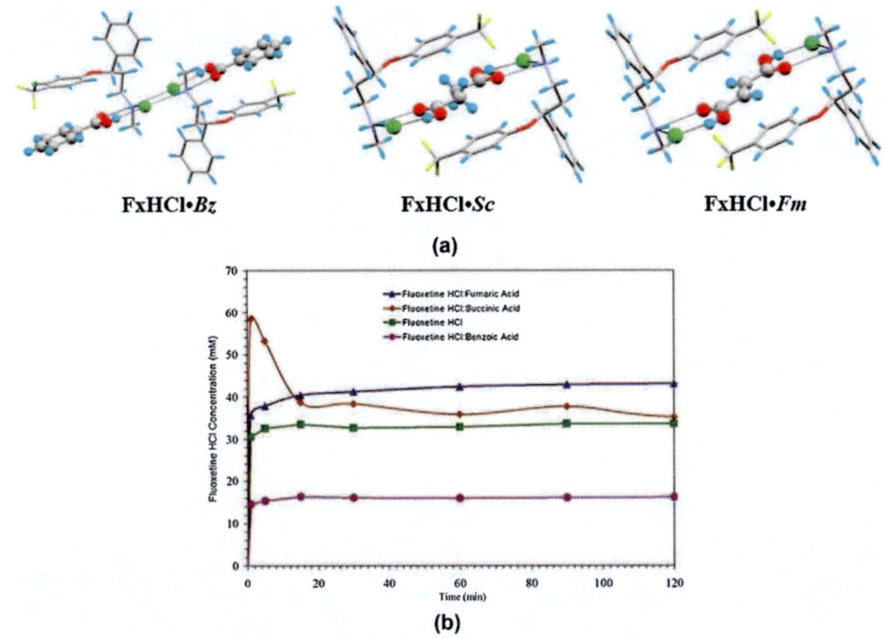

Figure 2.13 (a) Hydrogen bonding in FxHCl·*Bz*, FxHCl·*Su* and FxHCl·*Fm* (*Bz*, *Su* and *Fm* moieties and atoms invovled in hydrogen bonds are shown in ball and stick). (b) Dissolution profiles for FxHCl·*Bz*, FxHCl·*Su*, FxHCl·*Fm* and FxHCl in water.

(Figure 2.13b). The melting point of FxHCl·*Fm* (164.22 °C) was slightly higher than that of FxHCl (158.6 °C), whereas the melting points of FxHCl·*Bz* (133 °C) and FxHCl·*Sc* (136.96 °C) were found to be lower than that of FxHCl.

Duggirala *et al.* reported LiCl based ICCs of glucose that exhibit enhanced stability to humidity compared to LiCl.[192] LiCl is a hygroscopic salt that is deliquescent when exposed to humidity. Two isostructural ICCs, namely LICGLU and LIBGLU, were synthesised by reaction of glucose with lithium chloride and lithium bromide, respectively. The crystal structure of LIBGLU was established by single crystal X-ray diffraction and its isostructurality with LICGLU was confirmed by comparing their respective powder X-ray diffraction patterns. Three crystallographically independent Li^+ cations exhibit coordination numbers of 5, 6 and 5 and association of Li^+ cations and glucose molecules affords a helical chain (Figure 2.14a). These chains are in-turn linked by O–H···O and O–H···Br hydrogen bonds (Figure 2.14b). The role of hydrogen bonding in LICGLU is therefore quite different from that of MCCs and ICCs in which hydrogen bonding is key to the very existence of the co-crystal. LICGLU was found to exhibit superior moisture stability compared to LiCl in a study conducted at 25 °C and 50% RH. This conclusion was supported by a dynamic vapour sorption study, which revealed that the weight gained by LICGLU was much less than LiCl over the same time

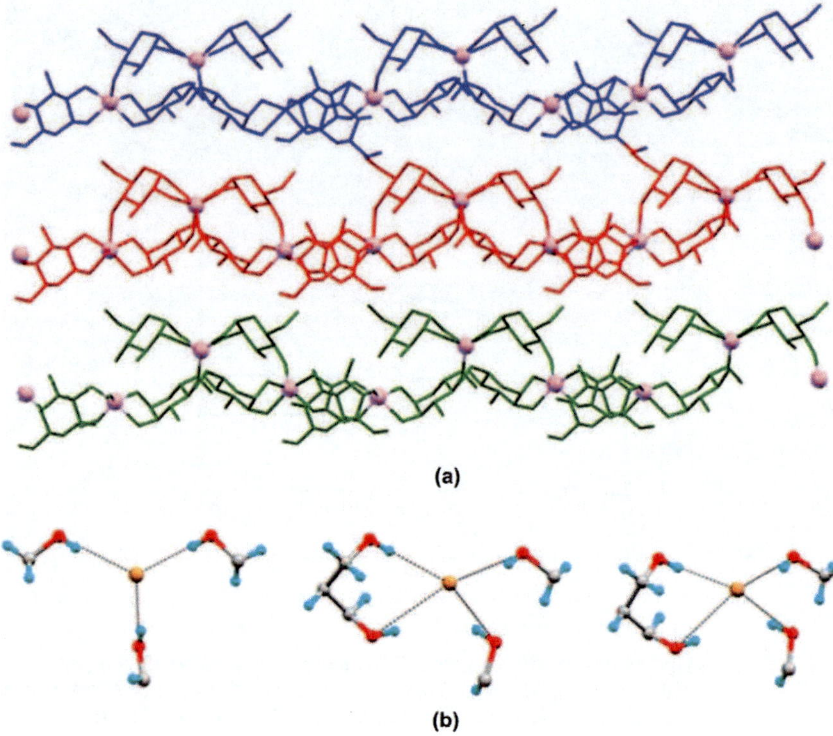

Figure 2.14 (a) Helical chains in LIBGLU comprised of Li^{+} cations and glucose molecules. Bromide anions have been removed for clarity. (b) Hydrogen bonds between the glucose units and bromides in LIBGLU. For clarity, only hydrogen-bonded hydroxyl groups of glucose molecules are shown.

period. A rapid gain in weight was observed when LICGLU was exposed to 40% RH. The pharmacokinetic studies performed on LICGLU revealed that its performance is comparable to that of LiCl.

Braga *et al.* made simultaneous use of the coordination bonds and hydrogen bonds to prepare ICCs of $CaCl_2$ with a number of pharmaceutically relevant amides.[190] The amides used for the study were malonamide, piracetam, nicotinalide and barbituric acid (Figure 2.15a). $CaCl_2$ was chosen for the study because of its low toxicity and suitability for use in drug products. The resulting ICCs were characterised by both powder and single crystal X-ray diffraction. In the structures of the ICCs, the C=O moiety of the co-formers coordinates with the metal cation whereas the N–H moieties were observed to hydrogen bond with chloride anions. Water molecules also coordinate with the calcium cations. The coordination geometry of the Ca^{2+} cations in this series of compounds was somewhat diverse in nature (Figure 2.15b). The ICCs based on piracetam and nicotinamide were found to exhibit improved idiosyncratic drug reactions in physiological condition compared to pure piracetam and nicotinamide.

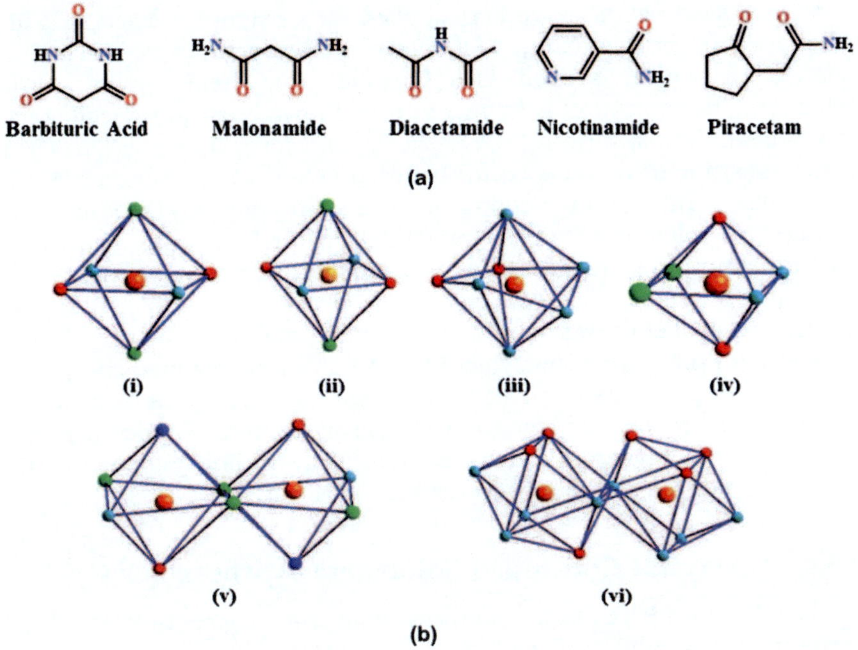

Figure 2.15 (a) Co-formers used by Braga *et al.* for the preparation of ICCs. (b) The coordination polyhedra around Ca^{2+} cations in (i) $(piracetam)_2\cdot CaCl_2\cdot 2H_2O$, (ii) malonamide$\cdot CaCl_2\cdot 2H_2O$, (iii) diacetamide$\cdot CaCl_2\cdot 5H_2O$, (iv) $(nicotinamide)_2\cdot CaCl_2\cdot 2H_2O$ and (v) nicotinamide$\cdot CaCl_2\cdot 2H_2O$ (vi) barbituric acid$\cdot CaCl_2\cdot 5H_2O$.

The thermal stability of both ICCs was also found to be enhanced in comparison to that of the pure components.

2.6.3 Co-crystals of Agrochemicals

Agrochemicals are chemicals that find their use in the agriculture industry and are categorised as pesticides, insecticides, fungicides or herbicides based on their usage. They are marketed as liquid or solid formulations comprising one or more agrochemically active compound. Issues faced with agrochemicals often contrast those faced by pharmaceuticals, *e.g.* achieving higher thermal stability by enhancement of melting points, achieving lower solubility in order to avoid the agrochemical being washed away and stable formulations. Co-crystallization is an effective tool to generate alternate solid forms of agrochemicals with improved (enhanced) melting points, improved (decreased) solubility, better storage and formulation stability. Nissinen and Nauha reported five co-crystals of the agrochemical active thiophanate-ethyl (diethyl-4,4′-(*o*-phenylene)bis(3-thioallophanate)) with pyridine based co-formers in which the structures were sustained by N–H···N hydrogen bonds.[193,194]

In 2005, Aakeröy and coworkers applied for a patent for co-crystals of cyprodinil and pyrimethanil, well-known anilinopyrimidine-based fungicides.[195] Cyprodinil is used as a fungicide for cereals, grapes, various vegetables and crops and is known to exist in two enantiotropically related polymorphic forms, form A (MP 70–72 °C) and form B (MP 74–76 °C). The phase transition from A to B occurs between 15 and 40 °C. This data suggests that under normal storage conditions the solid state of cyprodinil is susceptible to a polymorphic phase transition. Pyrimethanil is an agrochemical used for controlling the growth of grey moulds in fruits that tends to crystallise if formulated as a solid dispersion. It also has a low melting point (96 °C). The challenge was to produce alternate solid forms of cyprodinil and pyrimethanil that exhibit less volatility. The inventors claimed co-crystals of cyprodinil and pyrimethanil with carboxylic acids such as benzoic acid, succinic acid, oxalic acid, maleic acid and pyrazinecarboxylic acid. It was claimed by the inventors that co-crystals of cyprodinil and pyrimethanil exhibit the desired improved properties.

2.6.4 Co-crystal Controlled Solid-state Synthesis (C^3S^3)

Co-crystallization is in essence a process of assembling co-former molecules in the solid state, and if the co-formers are potential reactants then they can be a vehicle for solid-state synthesis. Kohlschütter stated that, in the case of solid-state reactions, reaction occurs between nearest neighbouring molecules and with minimum atomic or molecular mobility.[196] A crystal engineering approach to solid state reactions was reported in 1960 when Schmidt demonstrated stereo specificity of [2 + 2] solid-state photodimerisation in cinnamic acid polymorphs.[197] Schmidt observed that, upon exposure to UV light, α-cinnamic acid produces α-truxillic acid whereas β-cinnamic acid affords β-truxillic acid. He proposed that the feasibility of solid-state photodimerisation would be determined by the crystal structure and postulated the *topochemical principle*, which states that '*the stereochemistry of the dimer is determined by the contact geometry (anti-parallel or parallel) of the nearest neighbouring double bonds provided that the centre to centre distance of these double bonds is of the order of* 4.2 Å'.[4] Herein we present case studies of C^3S^3 being exploited to conduct organic reactions in the solid state.

2.6.4.1 Solid State Nucleophilic Substitution

Etter and coworkers reported a hydrogen bonded co-crystal of 4-chloro-3,5-dinitro of benzoic acid and 4-aminobenzoic acid by the solid state grinding method.[198] They observed that when the co-crystal is heated at 180 °C a nucleophilic substitution leads to the formation of product **A** (Scheme 2.6). Product **A** was also synthesised independently in solution. Interestingly, different polymorphs of product **A** were obtained. The single crystal for the co-crystal was not determined but its structure was projected

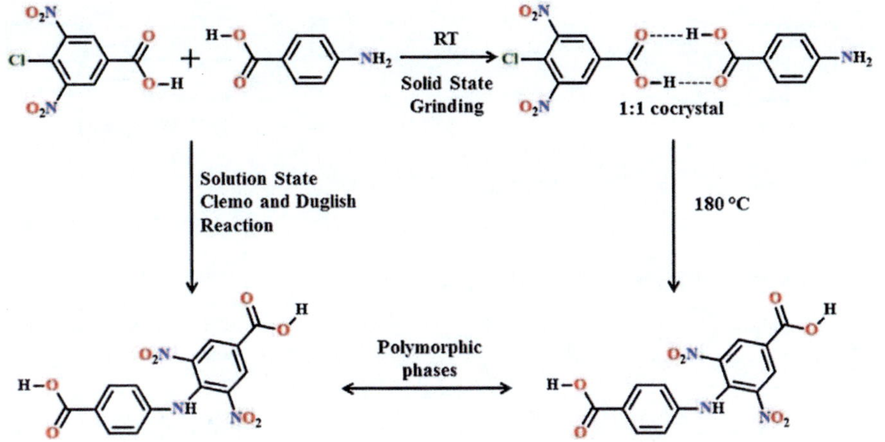

Scheme 2.6 The formation of nucleophilic aromatic substitution product (**A**) *via* a co-crystal of 4-chloro-3,5-dinitrobenzoic acid and 4-aminobenzoic acid.

from the co-crystal of 3,5-dintrobenzoic acid and 4-aminobenzoic acid. The authors could not comment on whether or not the reaction followed the principle of least motion, as postulated by Kohlschütter. They argued that the relatively high reaction temperature would enhance molecular motions, which might in turn result in reorientation of the co-crystal formers.

2.6.4.2 [2 + 2] Solid-state Photodimerisation

A general approach followed in the literature for a [2 + 2] solid-state photodimerisation reaction is to use molecules that act as templates to orient olefin moieties within the distance required by the topochemical principle (Scheme 2.7a). The template molecule is co-crystallised with the olefin former, the co-crystal is exposed to UV and then the photodimer product is formed through a single crystal to single crystal transformation. Several small molecules have been reported by as templates for studying [2 + 2] solid-state photodimerisation reactions (Scheme 2.7b).[199–202] Importantly, the olefin moieties in the crystal should lie within 4.2 Å and preferably be aligned in a parallel manner. However, there are reports of a reaction occuring even though olefins are aligned anti-parallel.[203]

MacGillivray and co-workers demonstrated the templating ability of 1,2,3,4-cyclobutane tetracarboxylic acid (CBTA, a photoproduct of solid state reaction of fumaric acid[56]) to orient *trans*-1,2-bis(4-pyridyl)ethylene (4,4′-bpe) in accordance to the topochemical principle in dimorphic CBTA·4,4′-bpe co-crystals.[204] CBTA forms two polymorphic co-crystals with 4,4′-bpe, which are packing polymorphs, both comprising hydrogen bonded chains (Figure 2.16a and b). Upon UV exposure both polymorphs were found to exhibit photoreactions resulting in the formation of a cyclobutane product (Figure 2.16c). Interestingly, it was observed that the reaction rate is slower

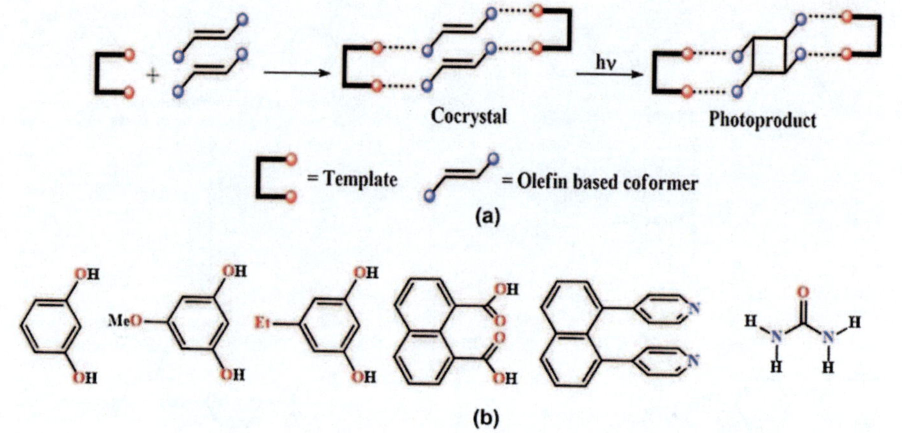

Scheme 2.7 (a) Schematic representation of co-crystal controlled solid state photo-dimerisation process. (b) Some small molecule co-crystal co-formers that have been used as templates for solid state photo dimerisation.

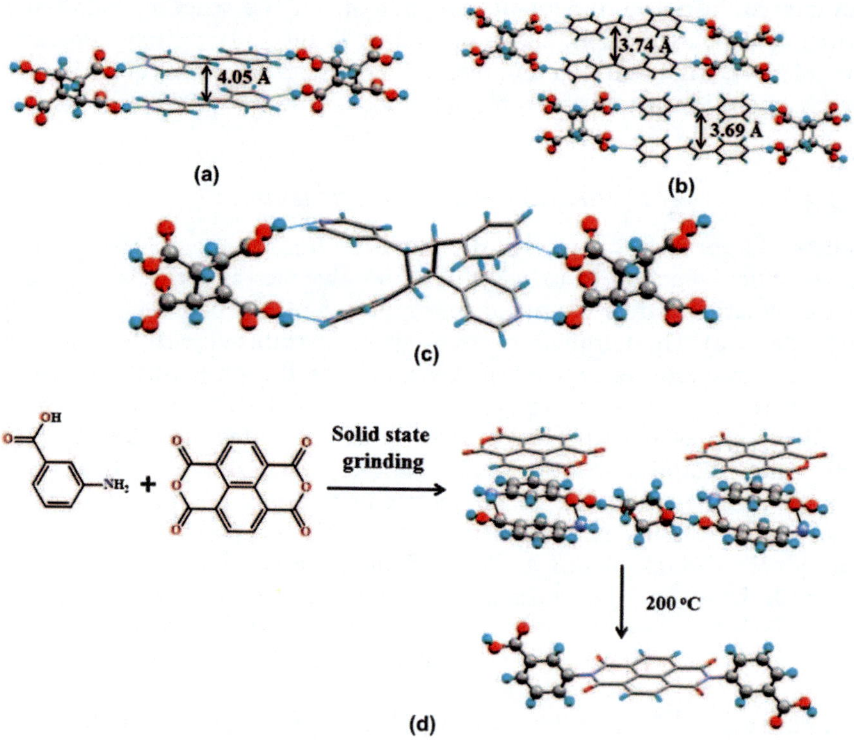

Figure 2.16 The hydrogen bonded chains observed in (a) CBTA·4,4′-bpe form I and (b) CBTA·4,4′-bpe form II. (c) Structure of the resulting photoproduct. (d) C^3S^3 approach for anhydride–amine condensation reactions in NTCDA·ABA co-crystal leading to the synthesis of diimide.

in case of the polymorph in which the stacking distance between the olefin groups is higher (Figure 2.16a).

2.6.4.3 Solid-state Condensation Reactions

Zaworotko and coworkers used C^3S^3 to perform condensation reactions between anhydride and amine-based co-formers resulting in imides.[205] The authors reported the 1 : 2 co-crystal of 1,4,5,8-naphthalenetetracarboxylic dianhydride (NTCDA) and 2-methyl-4-nitroaniline (MNA), which, when heated to 180 °C, affords the respective imide. The structure of the NTCDA·MNA co-crystal was found to be sustained by a charge transfer interaction between the aromatic moieties of NTCDA and MNA along with N–H···O hydrogen bonds involving the amine–nitro synthon between MNA moieties. The carbonyl C atom of NTCDA and the amine N atom are separated by 3.42 Å in the co-crystal, which is well within limits set by the *topochemical principle*. Charge transfer interactions were indicated by the purple colour of the co-crystal. Solvent-drop grinding of the two co-formers gave mixtures of the co-formers which, when heated to 130 °C, afforded the NTCDA·MNA co-crystal. Further heating to 180 °C resulted in the C^3S^3 product. When NTCDA was crystallised with 3-aminobenzoic acid (ABA) a 1 : 1 co-crystal was formed, which crystallises as a 1 : 1 : 1 co-crystal dioxane solvate. When this co-crystal was heated to 200 °C the corresponding imide was formed (Figure 2.16d). The structure of the NTCDA·ABA·dioxane solvate is sustained by a hydrogen bond between the ABA moieties and dioxane forming a 1D chain stacked over layers of NTCDA molecules through charge transfer interactions. The C^3S^3 approach adopted for amine–anhydride based co-formers leading to imides was implemented by the Zaworotko group *via* an anhydride–amine based co-crystal.[206,207]

2.6.5 Co-crystal Based NLO Materials

Materials that exhibit non-linear optical activity (*second-order harmonic generation*, SHG) are used in optical and electro-optical devices. Important requirements for a material to exhibit SHG are: (i) structurally it should be non-centrosymmetric and (ii) it should have a properly aligned dipole such that the polarisation effects are not nullified.[91] Co-crystallization has been implemented to explore and design materials in which achiral molecules crystallise in non-centrosymmetric space groups.[208] Early reports on organic salt based SHG materials were made by Aakeröy *et al.*[209] They advocated the use of organic materials for the synthesis of SHG active materials owing to their stability, solubility and design amenability. In 1992 Aakeröy, Leslie and Seddon discussed the importance of hydrogen bonds in sustaining structures of organic salts based on dihydrogenphosphate by lattice energy calculations. They concluded that 20–25% of the lattice energy of these materials was contributed to by hydrogen bonds. In 1992 Aakeröy *et al.* reported a series of hydrogen bonded tartarate salts of the type [AH][L-tartH]

(A = organic cations).[210] The 3-hydroxypyridine based salt was observed to exhibit 13-fold better SHG activity than α-SiO_2.

Huang *et al.* reported a series of 22 co-crystals based on the phenol···pyridine heterosynthon.[211] Out of the 22 reported crystals, four co-crystals were found to crystallise in non-centrosymmetric space groups and exhibited positive SHG activity. An SHG active system composed of achiral moieties crystallised in a non-centrosymmetric lattice was reported by Koshima and coworkers. The co-crystals were comrpised of 2-amino-5-nitropyridine and achiral benzene sulphonic acids.[212] The crystal structures of the solids were found to be sustained by strong hydrogen bonds between the organic ions supported by π···π interactions. 2-Amino-5-nitropyridine (2A5NP) inherently exhibits a strong dipole due to the presence of the electron withdrawing nitro functionality (hydrogen bond acceptor) and the electron donating amino functionality (hydrogen bond donor). The salts 2A5NP·*p*-$CH_3(C_2H_4)SO_3H$ and 2A5NP·*p*-$HO(C_2H_4)SO_3H$ delivered higher SHG activity compared to standard SHG candidates like *m*-nitroaniline and 4-(dimethylamino)-3-acetamidonitrobenzene.

2.7 Conclusions

The case studies discussed in this chapter suggest that considerable understanding has been gained in co-crystal synthesis and crystal engineering with respect to how to design materials with desired compositions, structures and useful properties. This work in the solid-state parallels the evolution of supramolecular chemistry and has been partly enabled by advances made in the areas of X-ray diffraction and development of computational models that are predictive. Fundamental studies from the 1980s and 1990s enabled the subsequent enormous growth in the number of reports of co-crystals of APIs and agrochemicals in the scientific and patent literature. C^3S^3 is in some ways where it all began thanks to the work of Schmidt but is undergoing a renaissance as it promises solvent-free, waste-free alternatives to solution-based synthesis protocols. However, challenges remain before co-crystals can offer broader utility:

- Systematic studies that afford a better understanding of the hierarchy of hydrogen bonding interactions will enable informatics to facilitate more efficient crystal structure prediction (CSP) of co-crystal structures and, especially, their properties with a higher degree of reliability. This will require faster and more reliable modelling methods validated by experimental studies. Computational methodologies are presently better suited to single molecule systems.
- The CSD is a powerful tool for addressing structural questions that cannot be readily answered using other approaches. However, the information available in the CSD is still not enough to draw conclusions about the relative stability of all supramolecular synthons or motifs that might be exploited for co-crystal synthesis. Further, the absence of

information about the experimental conditions used for crystallisation requires 'old-fashioned' and cumbersome approaches to access information. More systematic studies are required to facilitate filling of the gaps in the information that can be provided by the CSD.

- Crystallisation prcoesses that result in formation of co-crystals and their polymorphs, solvates and hydrates remain poorly understood. This can be attributed to more complex phase diagrams and more variables than exist for single-component crystallisation. Given that physicochemical properties of such phases will be different from each other, this is particularly important in the context of pharmaceutical co-crystal and agrochemical co-crystals. Better understanding of process crystallisation conditions is required in order to manufacture co-crystals at scale.
- The diversity of possible co-crystals for a given co-former is both an advantage and a handicap. The advantage is that fine-tuning of properties can enable discovery by design of the right material for the right application. The handicap is that the large number of experiments needed to investigate hundreds or even thousands of likely co-crystals severely limits the pace of the discovery and development process. If the three earlier challenges can be overcome synergistically then this fourth challenge can be addressed by judicious selection of a small library of co-formers for a particular co-crystal structure with a desired property.

Acknowledgements

MJZ acknowledges Science Foundation Ireland for support (SFI Award 13/RP/B2549). SB thanks the Irish Research Council for an IRC-SET Post-doctoral fellowship (GOIPD/2016/87).

References

1. J.-M. Lehn, *Angew. Chem., Int. Ed. Engl.*, 1988, **27**, 89–112.
2. J. D. Dunitz, *Pure Appl. Chem.*, 1991, **63**, 177.
3. R. Pepinsky, *Phys. Rev.*, 1955, **100**, 971.
4. G. Schmidt, *Pure Appl. Chem.*, 1971, **27**, 647–678.
5. G. R. Desiraju and G. W. Parshall, *Crystal Engineering: The Design of Organic Solids*, Elsevier, 1989.
6. C. B. Aakeröy and K. R. Seddon, *Chem. Soc. Rev.*, 1993, **22**, 397–407.
7. S. Subramanian and M. J. Zaworotko, *Coord. Chem. Rev.*, 1994, **137**, 357–401.
8. J. Bernstein, R. E. Davis, L. Shimoni and N. L. Chang, *Angew. Chem., Int. Ed. Engl.*, 1995, **34**, 1555–1573.
9. C. B. Aakeröy and A. M. Beatty, *Aust. J. Chem.*, 2001, **54**, 409–421.
10. C. B. Aakeröy and A. M. Beatty, *Solid State, Crystal Engineering and Hydrogen Bonds*, 2004.

11. S. Aitipamula, R. Banerjee, A. K. Bansal, K. Biradha, M. L. Cheney, A. R. Choudhury, G. R. Desiraju, A. G. Dikundwar, R. Dubey, N. Duggirala, P. P. Ghogale, S. Ghosh, P. K. Goswami, N. R. Goud, R. K. R. Jetti, P. Karpinski, P. Kaushik, D. Kumar, V. Kumar, B. Moulton, A. Mukherjee, G. Mukherjee, A. S. Myerson, V. Puri, A. Ramanan, T. Rajamannar, C. M. Reddy, N. Rodriguez-Hornedo, R. D. Rogers, T. N. Guru Row, P. Sanphui, N. Shan, G. Shete, A. Singh, C. C. Sun, J. A. Swift, R. Thaimattam, T. S. Thakur, R. K. Thaper, S. P. Thomas, S. Tothadi, V. R. Vangala, N. Variankaval, P. Vishweshwar, D. R. Weyna and M. J. Zaworotko, *Cryst. Growth Des.*, 2012, **12**, 2147–2152.
12. N. Schultheiss and A. Newman, *Cryst. Growth Des.*, 2009, **9**, 2950–2967.
13. C. B. Aakeröy and D. J. Salmon, *CrystEngComm*, 2005, **7**, 439–448.
14. G. R. Desiraju, *J. Chem. Soc., Chem. Commun.*, 1991, 426–428.
15. A. L. Gillon, N. Feeder, R. J. Davey and R. Storey, *Cryst. Growth Des.*, 2003, **3**, 663–673.
16. R. K. Khankari and D. J. Grant, *Thermochim. Acta*, 1995, **248**, 61–79.
17. L. Infantes, L. Fábián and W. S. Motherwell, *CrystEngComm*, 2007, **9**, 65–71.
18. F. H. Herbstein, *Cryst. Growth Des.*, 2004, **4**, 1419–1429.
19. L. Mandelcorn, *Chem. Rev.*, 1959, **59**, 827–839.
20. L. R. Nassimbeni, *Acc. Chem. Res.*, 2003, **36**, 631–637.
21. D. J. Cram and J. M. Cram, *Container Molecules and Their Guests*, Royal Society of Chemistry, 2007.
22. R. Hoss, O. König, V. Kramer-Hoss, U. Berger, P. Rogin and J. Hulliger, *Angew. Chem., Int. Ed. Engl.*, 1996, **35**, 1664–1666.
23. S. M. Berge, L. D. Bighley and D. C. Monkhouse, *J. Pharm. Sci.*, 1977, **66**, 1–19.
24. P. L. Gould, *Int. J. Pharm.*, 1986, **33**, 201–217.
25. D. P. Elder, R. Holm and H. L. de Diego, *Int. J. Pharm.*, 2013, **453**, 88–100.
26. A. I. Kitaigorodsky, in *Mixed Crystals*, Springer, 1984, pp. 181–199.
27. J. Lee, Q. Zhang and F. Saito, *J. Solid State Chem.*, 2001, **160**, 469–473.
28. M. K. Mishra, U. Ramamurty and G. R. Desiraju, *J. Am. Chem. Soc.*, 2015, **137**, 1794–1797.
29. K. Morris and N. Rodriguez-Hornedo, *Encycl. Pharm. Technol.*, 1993, **7**, 393–441.
30. E. Tieger, V. Kiss, G. Pokol, Z. Finta, M. Dušek, J. Rohlíček, E. Skořepová and P. Brázda, *CrystEngComm*, 2016, **18**, 3819–3831.
31. S. P. Stokes, C. C. Seaton, K. S. Eccles, A. R. Maguire and S. E. Lawrence, *Cryst. Growth Des.*, 2014, **14**, 1158–1166.
32. A. Bajpai, H. S. Scott, T. Pham, K.-J. Chen, B. Space, M. Lusi, M. L. Perry and M. J. Zaworotko, *IUCrJ*, 2016, **3**, 430–439.
33. P. J. B. L. Romè de L'Isle, *Crystallography*, 2nd edn, 1783.
34. J. H. Palm and C. MacGillavry, *Acta Crystallogr.*, 1963, **16**, 963–968.
35. F. Wöhler, *Eur. J. Org. Chem.*, 1844, **51**, 145–163.
36. T. Sakurai, *Acta Crystallogr.*, 1965, **19**, 320–330.

37. T. Sakurai, *Acta Crystallogr., Sect. B: Struct. Crystallogr. Cryst. Chem.*, 1968, **24**, 403–412.
38. S. Harkema, J. Bats, A. Weyenberg and D. Feil, *Acta Crystallogr., Sect. B: Struct. Crystallogr. Cryst. Chem.*, 1972, **28**, 1646–1648.
39. S. Harkema and J. Ter Brake, *Acta Crystallogr., Sect. B: Struct. Crystallogr. Cryst. Chem.*, 1979, **35**, 1011–1013.
40. C. Buehler and A. G. Heap, *J. Am. Chem. Soc.*, 1926, **48**, 3168–3172.
41. J. S. Anderson, *Nature*, 1937, **140**, 583–584.
42. S. Pekker, É. Kováts, G. Oszlányi, G. Bényei, G. Klupp, G. Bortel, I. Jalsovszky, E. Jakab, F. Borondics and K. Kamarás, *Nat. Mater.*, 2005, **4**, 764–767.
43. K. Hoogsteen, *Acta Crystallogr.*, 1959, **12**, 822–823.
44. G. P. Moss, *Pure Appl. Chem.*, 1996, **68**, 2193.
45. N. Shan and M. J. Zaworotko, *Drug Discovery Today*, 2008, **13**, 440–446.
46. D. J. Good and N. R. Rodríguez-Hornedo, *Cryst. Growth Des.*, 2009, **9**, 2252–2264.
47. W. Jones, W. S. Motherwell and A. V. Trask, *MRS Bull.*, 2006, **31**, 875–879.
48. A. V. Trask, *Mol. Pharm.*, 2007, **4**, 301–309.
49. https://www.fda.gov/downloads/Drugs/Guidances/UCM281764.pdf, (accessed August 2016).
50. D. O'Nolan, M. L. Perry and M. J. Zaworotko, *Cryst. Growth Des.*, 2016, **16**, 2211–2217.
51. E. M. Meade, *Canadian Patent, CA1136151 A*, 1982.
52. S. Karki, T. Friščić, W. Jones and W. S. Motherwell, *Mol. Pharmaceutics*, 2007, **4**, 347–354.
53. W. T. A. Harrison, H. S. Yathirajan, S. Bindya, H. G. Anilkumar and G. R. Devaraju, *Acta Crystallogr., Sect. C*, 2007, **63**, o129–o131.
54. L. Feng, P. H. Karpinski, P. Sutton, Y. Liu, D. F. Hook, B. Hu, T. J. Blacklock, P. E. Fanwick, M. Prashad and S. Godtfredsen, *Tetrahedron Lett.*, 2012, **53**, 275–276.
55. J. A. Perman, A. J. Cairns, Ł. Wojtas, M. Eddaoudi and M. J. Zaworotko, *CrystEngComm*, 2011, **13**, 3130–3133.
56. T. Friščić and L. R. MacGillivray, *Chem. Commun.*, 2005, 5748–5750.
57. K. S. Eccles, R. E. Deasy, L. Fábián, A. R. Maguire and S. E. Lawrence, *J. Org. Chem.*, 2011, **76**, 1159–1162.
58. M. B. Atkinson, S. S. Mariappan, D.-K. Bučar, J. Baltrusaitis, T. Friščić, N. G. Sinada and L. R. MacGillivray, *Proc. Natl. Acad. Sci.*, 2011, **108**, 10974–10979.
59. G. P. Stahly, *Cryst. Growth Des.*, 2009, **9**, 4212–4229.
60. P. Pfeiffer, *Organische Molekülverbindungen*, F. Enke, 1922.
61. N. K. Duggirala, M. L. Perry, Ö. Almarsson and M. J. Zaworotko, *Chem. Commun.*, 2016, **52**, 640–655.
62. D. Cinčić, T. Friščić and W. Jones, *Chem. - Eur. J.*, 2008, **14**, 747–753.
63. D. Fox, P. Metrangolo, D. Pasini, T. Pilati, G. Resnati and G. Terraneo, *CrystEngComm*, 2008, **10**, 1132–1136.

64. C. B. Aakeröy, T. K. Wijethunga and J. Desper, *CrystEngComm*, 2014, **16**, 28–31.
65. X. Pang, H. Wang, W. Wang and W. J. Jin, *Cryst. Growth Des.*, 2015, **15**, 4938–4945.
66. J. C. Collings, P. S. Smith, D. S. Yufit, A. S. Batsanov, J. A. Howard and T. B. Marder, *CrystEngComm*, 2004, **6**, 25–28.
67. F. V. Heyden, *Fr. Pat., 769,586*, 1934.
68. D. Braga, F. Grepioni, L. Maini, S. Prosperi, R. Gobetto and M. R. Chierotti, *Chem. Commun.*, 2010, **46**, 7715–7717.
69. C. Gerhardt, *Eur. J. Org. Chem.*, 1853, **87**, 149–179.
70. R. L. Peck, *U. S. Pat. No.*, 2,474,758, 1949.
71. E. M. Meade, *U. S. Pat. No.*, 4,988,731, 1991.
72. B. C. Sherman, *U. S. Pat. No.*, 6,077,542, 2000.
73. A. I. Vogel, Textbook of Practical Organic Chemistry, Longmans Scientific and Technical, United Kingdom, 1956, p. 135.
74. S. Li, T. Yu, Y. Tian, C. P. McCoy, D. S. Jones and G. P. Andrews, *Mol. Pharmaceutics*, 2016, **13**, 3054–3068.
75. J. C. Bennion, Z. R. Siddiqi and A. J. Matzger, *Chem. Commun.*, 2017, **53**, 6065–6068.
76. T. Zhang, Q. Yu, X. Li and X. Ma, *J. Cryst. Growth*, 2017, **469**, 114–118.
77. A. V. Trask, J. van de Streek, W. S. Motherwell and W. Jones, *Cryst. Growth Des.*, 2005, **5**, 2233–2241.
78. A. V. Trask, W. S. Motherwell and W. Jones, *Chem. Commun.*, 2004, 890–891.
79. A. N. Sokolov, D. K. Bučar, J. Baltrusaitis, S. X. Gu and L. R. MacGillivray, *Angew. Chem., Int. Ed. Engl.*, 2010, **49**, 4273–4277.
80. V. Pedireddi, W. Jones, A. Chorlton and R. Docherty, *Chem. Commun.*, 1996, 987–988.
81. N. Takata, K. Shiraki, R. Takano, Y. Hayashi and K. Terada, *Cryst. Growth Des.*, 2008, **8**, 3032–3037.
82. S. Zhang, H. Chen and Å. C. Rasmuson, *CrystEngComm*, 2015, **17**, 4125–4135.
83. A. Werner, *Eur. J. Org. Chem.*, 1902, **322**, 261–296.
84. A. Hantzsch, *Eur. J. Inorg. Chem.*, 1910, **43**, 3049–3076.
85. W. M. Latimer and W. H. Rodebush, *J. Am. Chem. Soc.*, 1920, **42**, 1419–1433.
86. L. Pauling, *J. Am. Chem. Soc.*, 1935, **57**, 2680–2684.
87. A. E. Mirsky and L. Pauling, *Proc. Natl. Acad. Sci.*, 1936, **22**, 439–447.
88. T. Steiner, *Angew. Chem. Int. Ed. Engl.*, 2002, **41**, 48–76.
89. G. R. Desiraju, *Acc. Chem. Res.*, 2002, **35**, 565–573.
90. E. Arunan, G. R. Desiraju, R. A. Klein, J. Sadlej, S. Scheiner, I. Alkorta, D. C. Clary, R. H. Crabtree, J. J. Dannenberg and P. Hobza, *Pure Appl. Chem.*, 2011, **83**, 1637–1641.
91. J. L. Atwood and J. W. Steed, *Encyclopedia of Supramolecular Chemistry*, CRC Press, 2004.

92. G. C. Pimentel and A. D. McClellan, *The Hydrogen Bond*, W. H. Freeman and Co., San Francisco, CA, 1960.
93. T. Steiner, *Crystallogr. Rev.*, 1996, **6**, 1–51.
94. G. R. Desiraju, *Acc. Chem. Res.*, 1991, **24**, 290–296.
95. D. J. Sutor, *Nature*, 1962, **195**, 68–69.
96. R. Taylor and O. Kennard, *J. Am. Chem. Soc.*, 1982, **104**, 5063–5070.
97. G. R. Desiraju, *Acc. Chem. Res.*, 1996, **29**, 441–449.
98. M. C. Etter, *J. Am. Chem. Soc.*, 1982, **104**, 1095–1096.
99. M. Etter, Z. Urbańczyk-Lipkowska, S. Baer and P. Barbara, *J. Mol. Struct.*, 1986, **144**, 155–167.
100. M. C. Etter and G. M. Frankenbach, *Chem. Mater.*, 1989, **1**, 10–12.
101. M. Etter, D. Britton and S. M. Reutzel, *Acta Crystallogr., Sect. C: Cryst. Struct. Commun.*, 1991, **47**, 556–561.
102. M. C. Etter, *J. Phys. Chem.*, 1991, **95**, 4601–4610.
103. M. C. Etter, Z. Urbanczyk-Lipkowska, M. Zia-Ebrahimi and T. W. Panunto, *J. Am. Chem. Soc.*, 1990, **112**, 8415–8426.
104. M. C. Etter and D. A. Adsmond, *J. Chem. Soc., Chem. Commun.*, 1990, 589–591.
105. M. C. Etter, *Acc. Chem. Res.*, 1990, **23**, 120–126.
106. E. J. Corey, *Pure Appl. Chem.*, 1967, **14**, 19–38.
107. M. Simard, D. Su and J. D. Wuest, *J. Am. Chem. Soc.*, 1991, **113**, 4696–4698.
108. X. Wang, M. Simard and J. D. Wuest, *J. Am. Chem. Soc.*, 1994, **116**, 12119–12120.
109. J.-H. Fournier, T. Maris, J. D. Wuest, W. Guo and E. Galoppini, *J. Am. Chem. Soc.*, 2003, **125**, 1002–1006.
110. D. Su, X. Wang, M. Simard and J. D. Wuest, *Supramol. Chem.*, 1995, **6**, 171–178.
111. G. R. Desiraju, *Angew. Chem., Int. Ed. Engl.*, 1995, **34**, 2311–2327.
112. R. B. Walsh, M. Bradner, S. Fleischman, L. Morales, B. Moulton, N. Rodriguez-Hornedo and M. Zaworotko, *Chem. Commun.*, 2003, 186–187.
113. N. Shan, A. Bond and W. Jones, *Cryst. Eng.*, 2002, **5**, 9–24.
114. J. J. Kane, R.-F. Liao, J. W. Lauher and F. W. Fowler, *J. Am. Chem. Soc.*, 1995, **117**, 12003–12004.
115. C. B. Aakeröy, A. M. Beatty and B. A. Helfrich, *Angew. Chem., Int. Ed. Engl.*, 2001, **40**, 3240–3242.
116. L. S. Reddy, N. J. Babu and A. Nangia, *Chem. Commun.*, 2006, 1369–1371.
117. C. B. Aakeröy, J. Desper and B. Levin, *CrystEngComm*, 2005, **7**, 102–107.
118. F. H. Allen, O. Kennard and R. Taylor, *Acc. Chem. Res.*, 1983, **16**, 146–153.
119. L. Fábián, *Cryst. Growth Des.*, 2009, **9**, 1436–1443.
120. C. V. K. Sharma, K. Panneerselvam, T. Pilati and G. R. Desiraju, *J. Chem. Soc., Perkin Trans. 2*, 1993, 2209–2216.

121. C. B. Aakeröy, J. Desper and B. A. Helfrich, *CrystEngComm*, 2004, **6**, 19–24.
122. C. B. Aakeröy, P. D. Chopade and J. Desper, *Cryst. Growth Des.*, 2013, **13**, 4145–4150.
123. C. B. Aakeröy, M. Fasulo, N. Schultheiss, J. Desper and C. Moore, *J. Am. Chem. Soc.*, 2007, **129**, 13772–13773.
124. B. Sarma, N. K. Nath, B. R. Bhogala and A. Nangia, *Cryst. Growth Des.*, 2009, **9**, 1546–1557.
125. D.-K. Bučar, R. F. Henry, G. G. Zhang and L. R. MacGillivray, *Cryst. Growth Des.*, 2014, **14**, 5318–5328.
126. S. Tothadi, P. Sanphui and G. R. Desiraju, *Cryst. Growth Des.*, 2014, **14**, 5293–5302.
127. K. Raatikainen and K. Rissanen, *CrystEngComm*, 2009, **11**, 750–752.
128. L. Infantes and W. D. S. Motherwell, *Chem. Commun.*, 2004, 1166–1167.
129. L. Infantes and W. D. S. Motherwell, *Z. Kristallogr. Cryst. Mater.*, 2005, **220**, 333–339.
130. L. Infantes and W. D. S. Motherwell, in *Science of Crystal Structures*, Springer, 2015, pp. 269–280.
131. J. A. Bis, P. Vishweshwar, D. Weyna and M. J. Zaworotko, *Mol. Pharmaceutics*, 2007, **4**, 401–416.
132. T. R. Shattock, K. K. Arora, P. Vishweshwar and M. J. Zaworotko, *Cryst. Growth Des.*, 2008, **8**, 4533–4545.
133. P. Kavuru, D. Aboarayes, K. K. Arora, H. D. Clarke, A. Kennedy, L. Marshall, T. T. Ong, J. Perman, T. Pujari and Ł. Wojtas, *Cryst. Growth Des.*, 2010, **10**, 3568–3584.
134. C. Janiak and J. K. Vieth, *New J. Chem.*, 2010, **34**, 2366–2388.
135. M. O'Keeffe and O. M. Yaghi, *Chem. Rev.*, 2011, **112**, 675–702.
136. L. Carlucci, G. Ciani, D. M. Proserpio, T. G. Mitina and V. A. Blatov, *Chem. Rev.*, 2014, **114**, 7557–7580.
137. S. R. Batten and R. Robson, *Angew. Chem., Int. Ed. Engl.*, 1998, **37**, 1460–1494.
138. M. J. Zaworotko, *Chem. Commun.*, 2001, 1–9.
139. B. Moulton and M. J. Zaworotko, *Chem. Rev.*, 2001, **101**, 1629–1658.
140. A. Ranganathan, V. Pedireddi, G. Sanjayan, K. Ganesh and C. Rao, *J. Mol. Struct.*, 2000, **522**, 87–94.
141. H. Koyano, P. Bissel, K. Yoshihara, K. Ariga and T. Kunitake, *Chem. – Eur. J.*, 1997, **3**, 1077–1082.
142. V. A. Russell and M. D. Ward, *J. Mater. Chem.*, 1997, **7**, 1123–1133.
143. D. J. Duchamp and R. E. Marsh, *Acta Crystallogr., Sect. B: Struct. Crystallogr. Cryst. Chem.*, 1969, **25**, 5–19.
144. C. V. K. Sharma and M. J. Zaworotko, *Chem. Commun.*, 1996, 2655–2656.
145. T. R. Shattock, P. Vishweshwar, Z. Wang and M. J. Zaworotko, *Cryst. Growth Des.*, 2005, **5**, 2046–2049.
146. B. R. Bhogala and A. Nangia, *Cryst. Growth Des.*, 2003, **3**, 547–554.

147. B. R. Bhogala, P. Vishweshwar and A. Nangia, *Cryst. Growth Des.*, 2002, **2**, 325–328.
148. Y.-B. Men, J. Sun, Z.-T. Huang and Q.-Y. Zheng, *CrystEngComm*, 2009, **11**, 2277–2278.
149. B. K. Saha and S. Bhattacharya, *CrystEngComm*, 2010, **12**, 2369–2373.
150. Y.-B. Men, J. Sun, Z.-T. Huang and Q.-Y. Zheng, *CrystEngComm*, 2009, **11**, 978–979.
151. B. R. Bhogala, P. Vishweshwar and A. Nangia, *Cryst. Growth Des.*, 2005, **5**, 1271–1281.
152. D. Braga and J. Bernstein, in *Making Crystals by Design*, Wiley-VCH Verlag GmbH & Co. KGaA, 2007, pp. 293–314, DOI: 10.1002/9783527610112.ch11.
153. J. Bernstein, R. J. Davey and J. O. Henck, *Angew. Chem., Int. Ed. Engl.*, 1999, **38**, 3440–3461.
154. W. C. McCrone, in *Physics and Chemistry of the Organic Solid State*, Interscience Publishers, New York, 1965.
155. S. Aitipamula, P. S. Chow and R. B. Tan, *CrystEngComm*, 2014, **16**, 3451–3465.
156. S. L. Childs and K. I. Hardcastle, *Cryst. Growth Des.*, 2007, **7**, 1291–1304.
157. N. J. Babu, S. Cherukuvada, R. Thakuria and A. Nangia, *Cryst. Growth Des.*, 2010, **10**, 1979–1989.
158. N. R. Goud and A. Nangia, *CrystEngComm*, 2013, **15**, 7456–7461.
159. B. Sreekanth, P. Vishweshwar and K. Vyas, *Chem. Commun.*, 2007, 2375–2377.
160. A. Mukherjee and G. R. Desiraju, *Chem. Commun.*, 2011, **47**, 4090–4092.
161. A. N. Sokolov, J. C. Sumrak and L. R. MacGillivray, *Chem. Commun.*, 2010, **46**, 82–84.
162. S. SeethaLekshmi and T. N. Guru Row, *Cryst. Growth Des.*, 2012, **12**, 4283–4289.
163. S. Skovsgaard and A. D. Bond, *CrystEngComm*, 2009, **11**, 444–453.
164. Ö. Almarsson and M. J. Zaworotko, *Chem. Commun.*, 2004, 1889–1896.
165. M. B. Hickey, M. L. Peterson, L. A. Scoppettuolo, S. L. Morrisette, A. Vetter, H. Guzmán, J. F. Remenar, Z. Zhang, M. D. Tawa and S. Haley, *Eur. J. Pharm. Biopharm.*, 2007, **67**, 112–119.
166. P. Vishweshwar, J. A. McMahon, M. L. Peterson, M. B. Hickey, T. R. Shattock and M. J. Zaworotko, *Chem. Commun.*, 2005, 4601–4603.
167. A. Y. Lee, D. Erdemir and A. S. Myerson, *Annu. Rev. Chem. Biomol. Eng.*, 2011, **2**, 259–280.
168. H. D. Clarke, K. K. Arora, H. Bass, P. Kavuru, T. T. Ong, T. Pujari, L. Wojtas and M. J. Zaworotko, *Cryst. Growth Des.*, 2010, **10**, 2152–2167.
169. S. Aitipamula, P. S. Chow and R. B. Tan, *Cryst. Growth Des.*, 2010, **10**, 2229–2238.
170. J. Galcera, T. Friščić, E. Molins and W. Jones, *CrystEngComm*, 2013, **15**, 1332–1338.

171. S. Bhattacharya, J. Sameena and B. K. Saha, *Cryst. Growth Des.*, 2011, **11**, 905–909.
172. B. K. Saha and A. Nangia, *Cryst. Growth Des.*, 2007, **7**, 393–401.
173. S. Bhattacharya and B. K. Saha, *CrystEngComm*, 2011, **13**, 6941–6944.
174. B. K. Saha, R. K. Jetti, L. S. Reddy, S. Aitipamula and A. Nangia, *Cryst. Growth Des.*, 2005, **5**, 887–899.
175. S. Bhattacharya and B. K. Saha, *Cryst. Growth Des.*, 2011, **11**, 2194–2204.
176. K. Reichenbächer, A. Neels, H. Stoeckli-Evans, P. Balasubramaniyan, K. Müller, Y. Matsuo, E. Nakamura, E. Weber and J. Hulliger, *Cryst. Growth Des.*, 2007, **7**, 1399–1405.
177. A. Jacobs, L. R. Nassimbeni, G. Ramon and B. K. Sebogisi, *CrystEngComm*, 2010, **12**, 3065–3070.
178. K. S. Peraka, M. Lusi, A. Bajpai and M. J. Zaworotko, *Cryst. Growth Des.*, 2017, **17**, 959–962.
179. B. R. Bhogala, S. Basavoju and A. Nangia, *Cryst. Growth Des.*, 2005, **5**, 1683–1686.
180. F. Topić and K. Rissanen, *J. Am. Chem. Soc.*, 2016, **138**, 6610–6616.
181. N. A. Mir, R. Dubey and G. R. Desiraju, *IUCrJ*, 2016, **3**, 96–101.
182. I. D. Oswald, D. R. Allan, P. A. McGregor, W. S. Motherwell, S. Parsons and C. R. Pulham, *Acta Crystallogr., Sect. B: Struct. Sci.*, 2002, **58**, 1057–1066.
183. N. Schultheiss, S. Bethune and J.-O. Henck, *CrystEngComm*, 2010, **12**, 2436–2442.
184. A. S. Sinha, U. R. Khandavilli, E. L. O'Connor, B. J. Deadman, A. R. Maguire and S. E. Lawrence, *CrystEngComm*, 2015, **17**, 4832–4841.
185. M. J. Bevill, P. I. Vlahova and J. P. Smit, *Cryst. Growth Des.*, 2014, **14**, 1438–1448.
186. A. J. Smith, P. Kavuru, L. Wojtas, M. J. Zaworotko and R. D. Shytle, *Mol. Pharmaceutics*, 2011, **8**, 1867–1876.
187. A. J. Smith, P. Kavuru, K. K. Arora, S. Kesani, J. Tan, M. J. Zaworotko and R. D. Shytle, *Mol. Pharmaceutics*, 2013, **10**, 2948–2961.
188. C. B. Aakeröy, S. Forbes and J. Desper, *J. Am. Chem. Soc.*, 2009, **131**, 17048–17049.
189. O. Shemchuk, D. Braga, L. Maini and F. Grepioni, *CrystEngComm*, 2017, **19**, 1366–1369.
190. D. Braga, F. Grepioni, G. I. Lampronti, L. Maini and A. Turrina, *Cryst. Growth Des.*, 2011, **11**, 5621–5627.
191. S. L. Childs, L. J. Chyall, J. T. Dunlap, V. N. Smolenskaya, B. C. Stahly and G. P. Stahly, *J. Am. Chem. Soc.*, 2004, **126**, 13335–13342.
192. N. K. Duggirala, A. J. Smith, Ł. Wojtas, R. D. Shytle and M. J. Zaworotko, *Cryst. Growth Des.*, 2014, **14**, 6135–6142.
193. E. Nauha, E. Kolehmainen and M. Nissinen, *CrystEngComm*, 2011, **13**, 6531–6537.
194. E. Nauha and M. Nissinen, *J. Mol. Struct.*, 2011, **1006**, 566–569.
195. N. George, J. O. Forrest, R. C. Burton and C. B. Aakeröy, *U. S. Pat. Appl. No.,13/696,180*, 2017.

196. V. Kohlschütter, *Z. Anorg. Allg. Chem.*, 1919, **105**, 1–25.
197. M. Cohen and G. Schmidt, *J. Chem. Soc.*, 1964, 1996–2000.
198. M. Etter, G. Frankenbach and J. Bernstein, *Tetrahedron Lett.*, 1989, **30**, 3617–3620.
199. L. R. MacGillivray, J. L. Reid and J. A. Ripmeester, *J. Am. Chem. Soc.*, 2000, **122**, 7817–7818.
200. L. R. MacGillivray, J. L. Reid, J. A. Ripmeester and G. S. Papaefstathiou, *Ind. Eng. Chem. Res.*, 2002, **41**, 4494–4497.
201. T. Friscić, D. Drab and L. MacGillivray, *Org. Lett.*, 2004, **6**, 4647–4650.
202. B. R. Bhogala, B. Captain, A. Parthasarathy and V. Ramamurthy, *J. Am. Chem. Soc.*, 2010, **132**, 13434–13442.
203. L. R. MacGillivray, *J. Org. Chem.*, 2008, **73**, 3311–3317.
204. S. Bhattacharya, J. Stojaković, B. K. Saha and L. R. MacGillivray, *Org. Lett.*, 2013, **15**, 744–747.
205. M. L. Cheney, G. J. McManus, J. A. Perman, Z. Wang and M. J. Zaworotko, *Cryst. Growth Des.*, 2007, **7**, 616–617.
206. J. A. Perman, K. Dubois, F. Nouar, S. Zoccali, Ł. Wojtas, M. Eddaoudi, R. W. Larsen and M. J. Zaworotko, *Cryst. Growth Des.*, 2009, **9**, 5021–5023.
207. M. J. Zaworotko and M. L. Cheney, *U. S. Pat. Appl. No., 11/934,230*, 2007.
208. S. Sharma, A. Kaur, M. S. Hundal and G. Hundal, *Mol. Cryst. Liq. Cryst.*, 2013, **577**, 73–82.
209. C. B. Aakeröy, G. S. Bahra, P. B. Hitchcock, Y. Patell and K. R. Seddon, *J. Chem. Soc., Chem. Commun.*, 1993, 152–156.
210. C. B. Aakeröy, P. B. Hitchcock and K. R. Seddon, *J. Chem. Soc., Chem. Commun.*, 1992, 553–555.
211. K.-S. Huang, D. Britton and S. R. Byrn, *J. Mater. Chem.*, 1997, **7**, 713–720.
212. H. Koshima, M. Hamada, I. Yagi and K. Uosaki, *Cryst. Growth Des.*, 2001, **1**, 467–471.

CHAPTER 3

Design and Structural Chemistry of Halogen-bonded Co-crystals

FILIP TOPIĆ*[a,†] AND KARI RISSANEN*[b,†]

[a] Department of Chemistry, McGill University, 801 Sherbrooke St. W., H3A 0B8 Montreal, Canada; [b] University of Jyvaskyla, Department of Chemistry, Nanoscience Center, P.O. Box 35, FI-40014 Jyvaskyla, Finland
*Email: filip.topic@mcgill.ca; kari.t.rissanen@jyu.fi

3.1 Introduction

In the recent renaissance of halogen bonding,[1,2] one of the major thrusts was made in the field of crystal engineering, with the studies of Metrangolo and Resnati essentially re-discovering the interaction in that context,[3,4] building on the pioneering studies of Hassel and co-workers started in the 1950s.[5–10] Halogen bonding has been defined as a "*net attractive interaction between an electrophilic region associated with a halogen atom in a molecular entity and a nucleophilic region in another, or the same, molecular entity*".[11] A useful concept in the description of halogen bonding is that of a sigma hole (σ-hole),[12] an area of positive electrostatic potential on the halogen atoms which appears along the R–X bond formed by the halogen X (Figure 3.1).

[†]Both authors contributed equally to the chapter.

Figure 3.1 Molecular electrostatic potential of CF_3I, with the most positive areas in red and the most negative areas in blue.
Adapted with permission from *Journal of Molecular Modeling*, Halogen bonding: the σ-hole, **13**, 2006, T. Clark, M. Hennemann, J. S. Murray *et al.*, © Springer-Verlag 2006, with permission of Springer.[12]

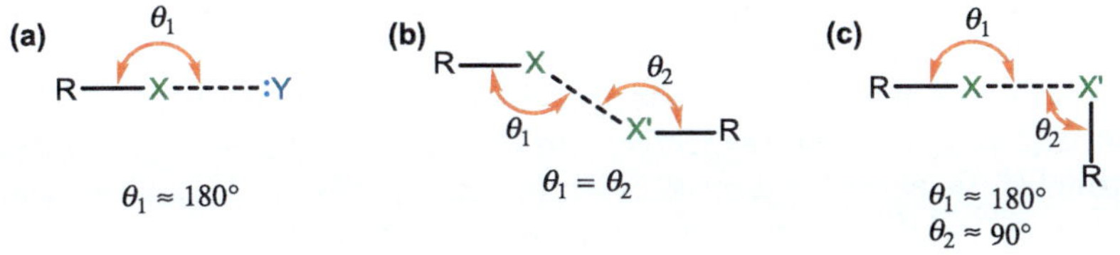

Scheme 3.1 (a) Typical halogen bond with X as the donor and Y as the acceptor. (b) Type I and (c) type II halogen···halogen interaction.

The R–X···Y halogen bond could thus also be described as an interaction between a σ-hole on a halogen atom X (halogen bond donor) and a nucleophilic region (most commonly) of the Lewis base Y (halogen bond acceptor), also emphasizing its predominantly electrostatic nature.[11,13] While greatly reminiscent of hydrogen bonding, halogen bonding was found to be more directional, with R–X···Y angles very close to 180° (Scheme 3.1(a)), which can be easily understood having in mind that the region of the electrostatic potential on a halogen atom (associated with a σ-hole) is typically much narrower than that on a hydrogen atom.[13] The emergence of the σ-hole on a halogen atom X along the R–X bond is accompanied by the appearance of a region of negative electrostatic potential laterally on X. The magnitude of the σ-hole depends on the nature of the halogen atom, namely its polarizability and electronegativity,[13] and generally increases in the order (F <) Cl < Br < I,[12] with the presence of a σ-hole on fluorine a rarity.[14,15] Consequently, the strength of the halogen bonds formed by different halogens in an otherwise identical chemical environment is also expected to increase in the same order. Another variable affecting the magnitude of the σ-hole on a halogen atom is the nature of its chemical environment.[16] Thus *e.g.* the σ-holes on the bromine and iodine atoms respectively increase in CF_3Br and CF_3I compared to CH_3Br and CH_3I, whereas the σ-hole on the chlorine atom, which appears in CF_3Cl, does not even exist in CH_3Cl![12]

With the halogen bonding defined as an attractive interaction between electrophilic and nucleophilic regions of molecular entities, it is possible for the same type of halogen atom to serve both as the electrophilic and the nucleophilic component of a halogen bond, respectively through the σ-hole and the lateral belt of negative electrostatic potential. Two types of such halogen–halogen interactions have been observed, classified as type I (Scheme 3.1(b)) and type II (Scheme 3.1(c)).[17,18] Of these two, only type II conforms to the definition of the halogen bond, as it involves an interaction between the electrophilic σ-hole of one halogen and the nucleophilic lateral belt of the other halogen. Both type I and type II interactions are ubiquitous in the crystal structures of halogenated molecules and have been used in crystal engineering, but rarely so in co-crystallization.

The possibility to tune the magnitude of the σ-hole on a halogen atom X in R–X and, consequently, the strength of the R–X···Y halogen bond by changing the halogen atom X, poses a challenge for the comparison between the systems based on different halogen atoms. One solution to this problem is the normalized distance function $R_{XB} = d(X\cdots Y)/(r_X + r_Y)$, where $d(X\cdots Y)$ is the experimentally obtained distance between X and Y in an R–X···Y halogen bond, and r_X and r_y the respective van der Waals radii of atoms X and Y.[19]

Probably the earliest example of a halogen-bonded co-crystal has been that of nitrogen iodide (**1**) and ammonia (**2**) ($NI_3\cdots NH_3$, **1**·**2**). First described by Colin in 1814,[20] not even the exact composition, let alone its exact structure, was understood at the time. A century and a half later, Bärnighausen and co-workers reported its crystal structure.[21] It showed a polymeric structure, with the NI_3 moieties forming 1-D chains of formula $(NI_3)_n$ through N–I···N halogen bonds. These bonds are symmetric, with the iodine atom situated on an inversion centre, at equal distances ($d(I\cdots N) = 2.30(1)$ Å, $R_{XB} = 0.652$, $<(N\text{–}I\cdots N) = 180°$) from nitrogen atoms (Scheme 3.2). These 1-D chains are further decorated by molecules of ammonia, again engaged in N–I···N halogen bonding ($d(I\cdots N) = 2.53(2)$ Å, $R_{XB} = 0.717$, $<(N\text{–}I\cdots N) = 176.2(6)°$). However, from the modern perspective, the nitrogen iodide **1** can hardly be perceived as a typical halogen bond donor.

3.2 Dihalogen Donors

The simplest species with the potential to serve as halogen bond donors, dihalogen molecules, were also among the first ones studied in the context of halogen bonding,[22,23] with the pioneering structural studies undertaken by Hassel and co-workers in the 1950s. Among the structures reported were those of the co-crystals of elemental chlorine **3** with 1,4-dioxane **4** or elemental bromine **5** with 1,4-dioxane **4** or acetone **6**. In **3**·**4**, the molecule of chlorine acts as a ditopic halogen bond donor, forming one halogen bond through each of

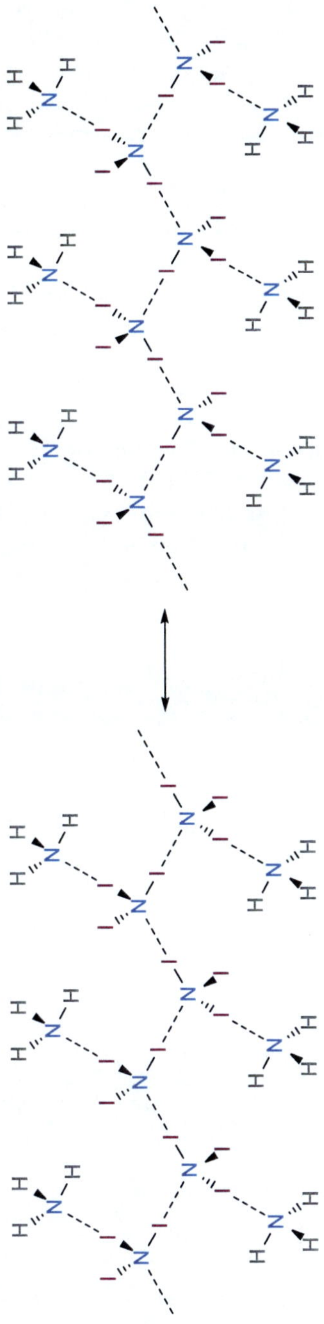

1·2

Scheme 3.2 Halogen-bonded chains in 1·2.

the chlorine atoms (d(Cl···O) = 2.67(3) Å, R_{XB} = 0.817, <(Cl–Cl···O) = 178°).[8] Similarly in **5·4**,[5] the molecule of bromine is engaged in two Br···O halogen bonds (d(Br···O) = 2.71 Å, R_{XB} = 0.804, <(Br–Br···O) = 178.3°). Both **3·4** and **5·4** comprise 1-D chains propagated by respective Cl···O or Br···O halogen bonds (Scheme 3.3). On the other hand, in case of the acetone co-crystal **5·6**,[7,8,24] the oxygen atom of acetone forms two bifurcated Br···O halogen bonds[24] (Scheme 3.3; d(Br···O) = 2.776(6) Å, R_{XB} = 0.824, <(Br–Br···O) = 177.50(10)°, <(Br···O···Br) = 110.6(3)°), giving rise to 1-D zig-zag halogen-bonded chains. Co-crystals comprising discrete complexes have also been described, such as the one between bromine and 4-methoxypyridine **7**, **5·7** (Scheme 3.3), where the molecule of bromine is only engaged in one halogen bond Br···N (d(Br···N) = 2.404(17) Å, R_{XB} = 0.707, <(Br–Br···N) = 177.1(3)°).[25] Similarly, in a co-crystal of bromine with 2(3*H*)-benzothiazolethione **8**, **5·8**, only one atom of bromine is engaged in Br···S halogen bonding (d(Br···S) = 2.2827(13) Å, R_{XB} = 0.625, <(Br–Br···S) = 175.33(4)°).[26] It is interesting to note that the length of the Br–Br bond varies significantly between different co-crystals, and can be correlated with the R_{XB} value for the corresponding halogen bond. Thus, the d(Br–Br) and R_{XB} vary from 2.3090(10) Å and 0.824 in **5·6** to 2.801(2) Å and 0.707 in **5·7** or even 2.8141(7) Å and 0.625 in **5·8**, respectively. Moreover, the co-crystallization of bromine with

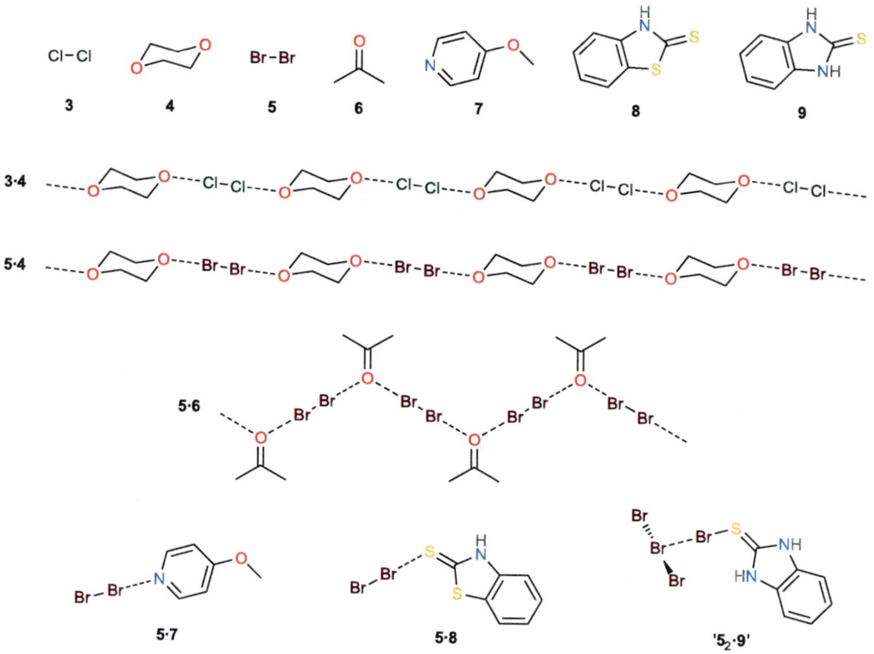

Scheme 3.3 Infinite motifs in co-crystals of elemental chlorine **3** and bromine **5** with 1,4-dioxane **4** and acetone **6**. Discrete motifs in co-crystals of bromine **5** with 4-methoxypyridine **7** and 2(3*H*)-benzothiazolethione **8**, and the salt obtained with 2-benzimidazolethione **9**.

2-benzimidazolethione **9**, instead of the expected co-crystal, results in '$\mathbf{5_2{\cdot}9}$', which is actually a salt comprising a $[\mathbf{9}+\mathrm{Br}]^+$ cation and a Br_3^- anion (Scheme 3.3).[26] This could be seen as a consequence of the very strong covalent character of the Br···S halogen bond in the putative co-crystal of **5** and **9** leading to the heterolytic cleavage of the bromine molecule.

Expectedly, elemental iodine **10** showed similar behaviour to bromine, *e.g.* forming an analogous co-crystal with 1,4-dioxane **4**, **10·4**, sustained by two I···O halogen bonds with both iodine atoms as donors (Scheme 3.4; $d(\mathrm{I}\cdots\mathrm{O}) = 2.808(3)$ Å, $R_{XB} = 0.802$, $<(\mathrm{I{-}I}\cdots\mathrm{O}) = 178.74(8)°$).[27] The iodine connects the molecules of a 1,4-dioxane molecule, acting as a bridge, as also classified by Pennington *et al.*[22] However, such behaviour is by no means the norm, as in many cases only one of the iodine atoms in the molecule acts as a halogen bond donor, forming a simple discrete adduct.[22] One such example is the 2 : 1 co-crystal of iodine and 1,2-bis(4-pyridyl)ethene **11**, $\mathbf{10_2{\cdot}11}$,

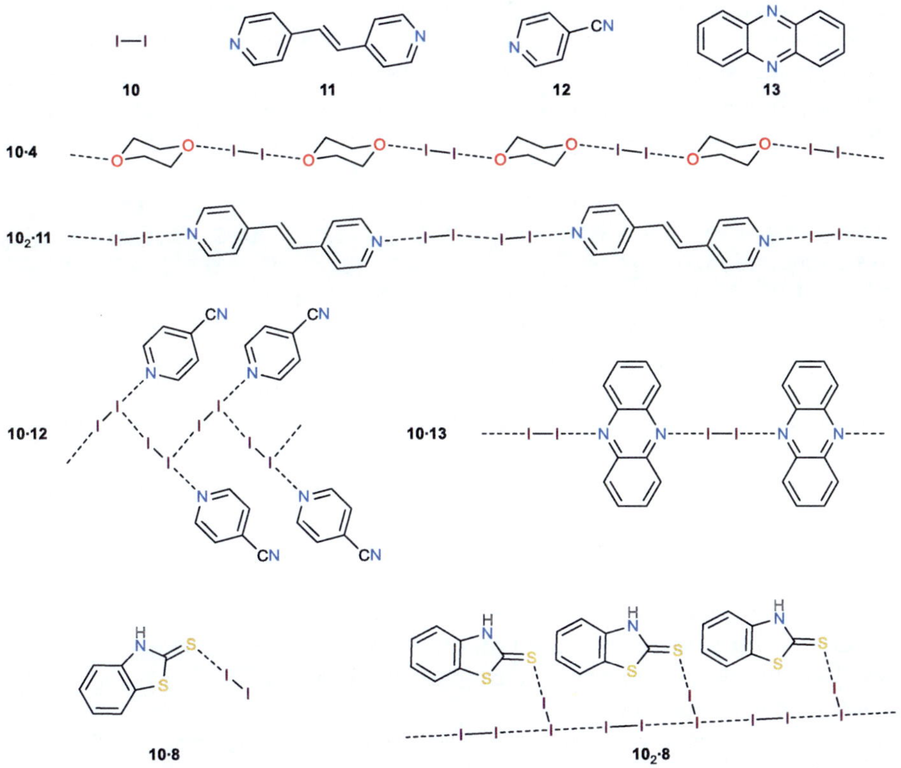

Scheme 3.4 Different interaction modes of elemental iodine **10**: bridging mode in 1-D infinite chains in the co-crystals **10·4** and **10·13**; discrete adducts connected by type I halogen–halogen interactions in $\mathbf{10_2{\cdot}11}$ or type II interactions in **10·12**. Two different co-crystals of **10** and **8**: **10·8** exhibiting a discrete halogen-bonded moiety and $\mathbf{10_2{\cdot}8}$ where the same moiety is incorporated in an infinite motif mediated by type II interactions with an additional iodine molecule.

where **11** forms two halogen bonds, each one with a different iodine molecule, with each of the iodine molecules only participating in one I···N halogen bond (Scheme 3.4; $d(\text{I}\cdots\text{N})=2.422(6)$ Å, $R_{XB}=0.686$, $<(\text{I–I}\cdots\text{N})=178.78(13)°$).[28] Instead of forming additional I···N halogen bonds, these iodine molecules engage in I···I interactions (Scheme 3.4), which could be classified as type I halogen–halogen interactions ($d(\text{I}\cdots\text{I})=3.8853(11)$ Å, $R_{XB}=0.981$, $<(\text{I–I}\cdots\text{I})=166.09(4)°$), but are not considered to be true halogen bonds.[17] In the case of a co-crystal of iodine with 4-cyanopyridine **12**, **10·12** (Scheme 3.4), the molecules of iodine similarly engage in one I···N halogen bond with pyridine ($d(\text{I}\cdots\text{N})=2.554(9)$ Å, $R_{XB}=0.724$, $<(\text{I–I}\cdots\text{N})=175.5(2)°$; $d(\text{I}\cdots\text{N})=2.543(9)$ Å, $R_{XB}=0.720$, $<(\text{I–I}\cdots\text{N})=175.5(2)°$) and one I···I interaction each ($d(\text{I}\cdots\text{I})=3.888(2)$ Å, $R_{XB}=0.982$, $<(\text{I–I}\cdots\text{I})=166.17(5)°$, $<(\text{I}\cdots\text{I–I})=76.74(5)°$; $d(\text{I}\cdots\text{I})=3.922(2)$ Å, $R_{XB}=0.990$, $<(\text{I–I}\cdots\text{I})=162.64(6)°$, $<(\text{I}\cdots\text{I–I})=75.62(5)°$), forming a ribbon-like motif (Scheme 3.4). Here, the geometry of the I···I interactions is consistent with the type II halogen–halogen interactions, which, unlike the type I interactions, conform to the definition of a halogen bond.[11,17,18] Still, the nitrogen-based acceptors can form co-crystals where the iodine acts in bridging mode, as shown in the case of the co-crystal with phenazine **13**, **10·13** (Scheme 3.4; $d(\text{I}\cdots\text{N})=2.982(5)$ Å, $R_{XB}=0.845$, $<(\text{I–I}\cdots\text{N})=180°$; $d(\text{I}\cdots\text{N})=3.098(6)$ Å, $R_{XB}=0.878$, $<(\text{I–I}\cdots\text{N})=180°$; $d(\text{I–I})=2.726(1)$ Å).

Elemental iodine readily forms co-crystals with sulphur-based halogen bond acceptors. For example, a 1:1 co-crystal **10·8** is formed with 2(3*H*)-benzothiazolethione **8** ($d(\text{I}\cdots\text{S})=2.729(6)$ Å, $R_{XB}=0.722$, $<(\text{I–I}\cdots\text{S})=174.19(14)°$, $d(\text{I–I})=3.077(3)$ Å). However, the same components also form a 2:1 co-crystal $\mathbf{10_2{\cdot}8}$ where the **10·8** moieties ($d(\text{I}\cdots\text{S})=2.588(6)$ Å, $R_{XB}=0.685$, $<(\text{I–I}\cdots\text{S})=177.79(13)°$, $d(\text{I–I})=2.969(2)$ Å) are bridged[22] by an additional molecule of iodine by means of two type II interactions ($d(\text{I–I})=2.7504(19)$ Å; first halogen bond: $d(\text{I}\cdots\text{I})=3.423(2)$ Å, $R_{XB}=0.864$, $<(\text{I–I}\cdots\text{I})=174.04(7)°$, $<(\text{I}\cdots\text{I–I})=102.71(6)°$; second halogen bond: $d(\text{I}\cdots\text{I})=3.445(2)$ Å, $R_{XB}=0.870$, $<(\text{I–I}\cdots\text{I})=172.26(7)°$, $<(\text{I}\cdots\text{I–I})=78.46(5)°$).[29] These structures illustrate an important aspect of crystal engineering using elemental iodine as the halogen bond donor, as its propensity to engage in I···I interactions and to incorporate additional iodine molecules can make it difficult to control the co-crystallization outcome.

Halogen bond acceptors based on other chalcogens or pnictogens, such as selenium or phosphorus and arsenic, have also been used with dihalogens. For example, two isostructural co-crystals with iodine **10** and 1,4-dithiane **14** or 1,4-diselenane **15** were described by Chao and McCullough (Scheme 3.5).[30,31] Both co-crystals exhibit a 2:1 stoichiometry, with each of the two iodine molecules forming only one halogen bond in $\mathbf{10_2{\cdot}14}$ ($d(\text{I}\cdots\text{S})=2.867(6)$ Å, $R_{XB}=0.758$, $<(\text{I–I}\cdots\text{S})=177.9(4)°$, $d(\text{I–I})=2.787(2)$ Å)[30] as well as in $\mathbf{10_2{\cdot}15}$ ($d(\text{I}\cdots\text{Se})=2.829(4)$ Å, $R_{XB}=0.729$, $<(\text{I–I}\cdots\text{Se})=180.0(3)°$, $d(\text{I–I})=2.870(3)$ Å).[31] A typical example of a co-crystal with a phosphorus-based halogen bond acceptor is that of iodine **10** and triphenylphosphine **16**, **10·16**.[32] In this 1:1 co-crystal, iodine forms one halogen bond to the phosphorus of **16** ($d(\text{I}\cdots\text{P})=2.480(3)$ Å,

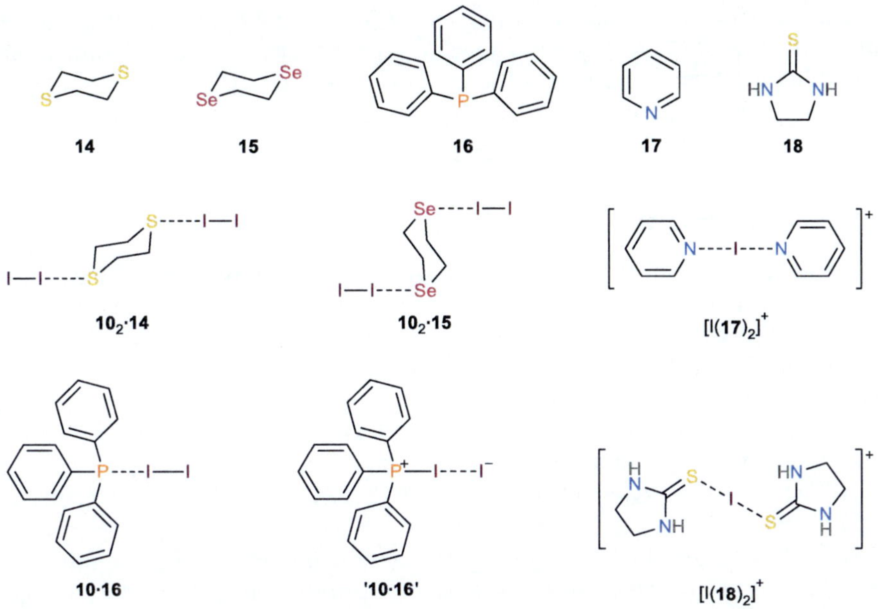

Scheme 3.5 Co-crystals of elemental iodine **10** with 1,4-dithiane **14** and 1,4-diselenane **15**. Two possible representations of the discrete adduct in the structure of the co-crystal/salt of iodine **10** and triphenylphosphine **16**. Structures of the bis(pyridine)iodonium $[I(\mathbf{17})_2]^+$ and bis(imidazolidine-2-thione)iodonium $[I(\mathbf{18})_2]^+$ cations, respectively, resulting from the attempted co-crystallizations between iodine **10** and pyridine **17** or 2-imidazolidinethione **18**.

$R_{XB} = 0.656$, $<(\text{I–I}\cdots\text{P}) = 178.23(6)°$, $d(\text{I–I}) = 3.1607(18)$ Å). Interestingly, the short $d(\text{I}\cdots\text{P})$ and significantly elongated $d(\text{I–I})$ bond distance suggest that the co-crystal might be more accurately described as consisting of a $[Ph_3PI]^+$ cation halogen-bonded to an iodide (I^-) anion (Scheme 3.5). This is supported by the structural study, reported by Cotton and Kibala, of several different products obtained by co-crystallization (or reaction) attempts using **10** and **16**.[33] Namely, the structures of the two 2 : 1 adducts best described by the formula $\mathbf{10}_2$·**16** were interpreted as consisting of $[Ph_3PI]^+$ cations and polyiodide anions. This phenomenon is not limited to the adducts of iodine with phosphines, as similar behaviour has been observed *e.g.* with pyridine- and thione-type halogen bond acceptors, where the heterolytic cleavage of diiodine yields iodonium cations. In the case of unsubstituted pyridine **17**, the reaction with iodine yields bis(pyridine)iodonium triiodide $[I(\mathbf{17})_2]I_3$[9,34] (Scheme 3.5) with an iodine atom coordinated to two pyridines and positioned exactly half way between them ($d(\text{I}\cdots\text{N}) = 2.251(3)$ Å, $R_{XB} = 0.638$, $<(\text{N}\cdots\text{I}\cdots\text{N}) = 180°$).[34] On the other hand, the reaction of iodine with 2-imidazolidinethione **18** similarly results in a mixed iodide (I^-)/triiodide (I_3^-) bis(2-imidazolidinethione)iodonium salt $[I(\mathbf{18})_2](I_3)_{0.5}(I)_{0.5}$ with the iodine atom in a similarly symmetric environment between the sulphur atoms of the two molecules of **18** (d(I···S) = 2.6279(5) Å, R_{XB} = 0.695; d(I···S) = 2.6298(5) Å, R_{XB} = 0.696; <(S···I···S) = 180°).[35] The examples outlined here illustrate another difficulty associated with the use of dihalogens as halogen bond donors, namely their tendency towards heterolytic cleavage and the associated formation of polyhalide species.

3.3 Saturated Halocarbons

On the other hand, very similar, simple saturated halocarbons have been regularly exploited as halogen bond donors for well over 50 years. Namely, they have already been used by Odd Hassel in the pioneering work leading up to his Nobel Prize in Chemistry in 1969. For example, he described the co-crystals of iodoform **19** or bromoform **20** with hexamethylenetetramine **21**. Both of these co-crystals, **19·21**[36] and **20**$_2$**·21**,[37] despite the difference in stoichiometry, are respectively sustained by similar C–I···N (d(I···N) = 2.93 Å, R_{XB} = 0.830, <(C–I···N) = 174.3 or 175.0°) and C–Br···N (d(Br···N) = 3.00 Å, R_{XB} = 0.882, <(C–Br···N) = 175.3°) halogen bonds and C–H···N hydrogen bonds (Figure 3.2). Some two decades later, Reddy *et al.* reported a similar co-crystal using carbon tetrabromide **22** as the halogen bond donor.[38] In **22·21**, both the carbon tetrabromide and the hexamethylenetetramine are fully engaged as donor and acceptor, respectively, forming four C–Br···N halogen bonds (d(Br···N) = 2.61(2) Å, R_{XB} = 0.768, <(C–Br···N) = 180°) each. Even more exciting, the crystal structure reveals the formation of a diamondoid network with alternating molecules of **22** and **21** as the nodes (Figure 3.2). Interestingly, a similar, albeit distorted, diamondoid network is also observed in the structure of **19·21**, with each molecule of iodoform **19** forming three C–I···N halogen bonds and one C–H···N hydrogen bond instead of the four C–Br···N halogen bonds observed in **22·21**. On the other hand, co-crystals **23·21**$_2$ and **23·22·21**, described later by Reddy *et al.*,[39] were found to be sustained solely by C–H···N hydrogen bonds, suggesting that the C–Br···N halogen bonds can easily succumb to weak hydrogen bonds and other packing interactions. Finally, in contrast to **22·21**, Kochi and co-workers reported co-crystals **22·24**[40] and **22·25**,[41] where each molecule of **22** forms two C–Br···N halogen bonds only, giving rise to 1-D chains, and **22·26**,[40] where **22** forms only one such halogen bond, resulting in a discrete 1 : 1 molecular pair (Figure 3.2). In all these co-crystals, the halogen atoms not involved in C–X···N (X = Br/I) halogen bonds were found to engage in C–X···X (X = Br/I) type II halogen bonds.

3.4 Haloalkene Donors

The examples of simple halocarbon donors are not limited only to saturated donors. In fact, unsaturated halocarbons have also been used as the halogen bond donors from the very first studies. The simplest such compounds, *e.g.* carbon bromides and iodides, have already been studied by Hassel and co-workers. They described the two co-crystals of tetrabromoethene **27** and

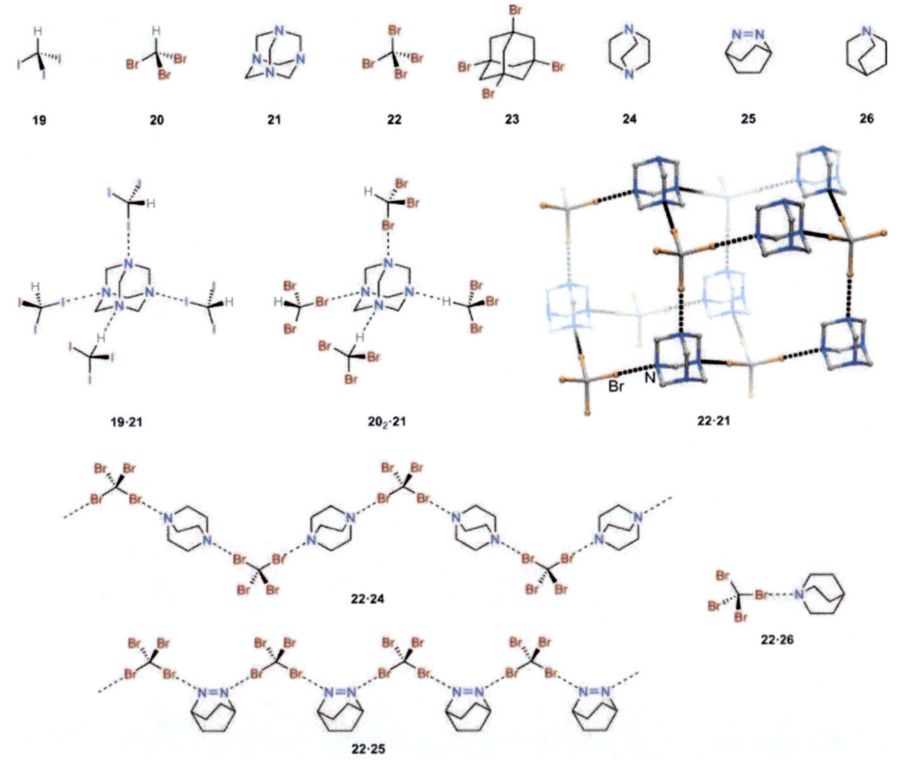

Figure 3.2 Halogen bonding around **21** in the co-crystals **19·21** and $\mathbf{20_2{\cdot}21}$. A fragment of the diamondoid structure of **22·21**. Halogen bonding in the co-crystals **22·24**, **22·25** and **22·26**. Halogen bonds are shown as black dashed lines.

tetraiodoethene **28**, respectively, with pyrazine **29**.[42] Both of these co-crystals, **27·29** and **28·29**, were found to consist of 1-D chains respectively sustained by C–Br···N (d(Br···N) = 3.019(15) Å, R_{XB} = 0.888, <(C–Br···N) = 174.8°) and C–I···N (d(I···N) = 2.98(3) Å, R_{XB} = 0.844, <(C–I···N) = 175.2°) halogen bonds (Scheme 3.6). The overall structures, however, are not identical due to different packing of these chains, respectively dictated by C–X···X (X = Br/I) type II halogen bonds. Similar halogen bonding synthons are also observed in the co-crystal of hexaiodobutadiene **30** and 1,7-phenanthroline **31**,[43] once again forming 1-D chains (Scheme 3.6), as well as in the co-crystal of the exotic tetrabromobutatriene **32** with phenazine **13**,[44] where it gives rise to a 1 : 2 discrete assembly (Scheme 3.6).

3.5 1-Haloalkyne Donors

However, in this comparison of simple halocarbons, 1-haloalkynes, *i.e.* those with the donor halogen atom attached to a terminal sp-hybridized carbon, have possibly been the most widely used ones. A computational study by Zou

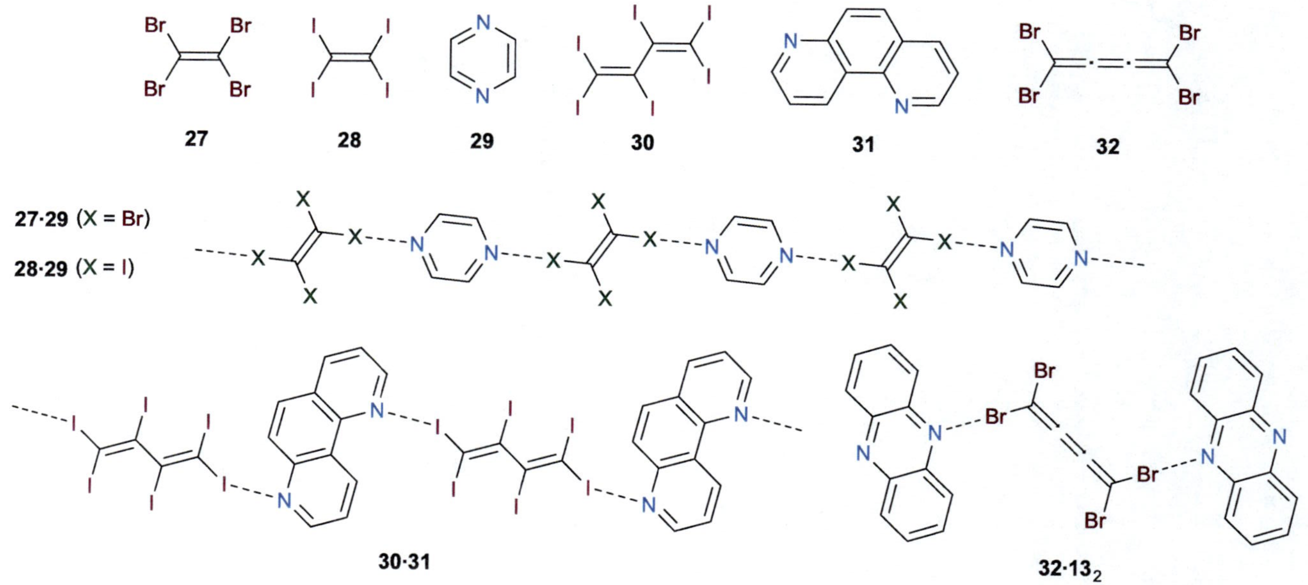

Scheme 3.6 Halogen-bonded 1-D chains in the co-crystals **27·29**, **28·29** and **30·31**. Discrete halogen-bonded assembly in the co-crystal **32·13**$_2$.

et al. suggested that the strength of a halogen bond with an ammonium (**2**) molecule as the acceptor should increase going from an sp^3- to an sp-bound halogen (X): C(sp^3)–X < C(sp^2)–X < C(sp)–X.[45] This can also be seen *e.g.* when comparing the structure of a co-crystal consisting of diiodoacetylene **33** and pyrazine **29**, **33·29**[46] (Scheme 3.7), with that of tetraiodoethene **28** with the same acceptor, **28·29**[42] (Scheme 3.6). Both structures comprise 1-D chains sustained by C–I···N halogen bonds, with d(I···N) = 2.832(7) Å, R_{XB} = 0.802, <(C–I···N) = 175.3(3)° in the case of **33·29** and d(I···N) = 2.98 Å, R_{XB} = 0.844, <(C–I···N) = 175.2° in the case of **28·29**, respectively, suggesting that the C–I···N halogen bond is significantly stronger with the sp-donor **33** than with the sp^2-donor **28**. In addition to that, the work of Brammer and co-workers also allows for the comparison between the co-crystals comprising ditopic acceptors based on aromatic sp^2 nitrogen, **33·29**, and sp^3 nitrogen, **33·24** (Scheme 3.7).[46] Based on the halogen bond parameters for **33·29** (see above) and **33·24** (d(I···N) = 2.715(3) Å, R_{XB} = 0.769, <(C–I···N) = 178.2(2)°; d(I···N) = 2.719(4) Å, R_{XB} = 0.770, <(C–I···N) = 178.7(2)°), the sp^3 nitrogen gives rise to stronger C–I···N halogen bonds than the aromatic sp^2 nitrogen. Besides these two, sp-hybridized nitrogen can also act as a halogen bond acceptor in the form of a nitrile, as established in the work of Goroff and co-workers, and illustrated by an example of the co-crystal **34·35** (Scheme 3.7; d(I···N) = 3.002(12) Å, R_{XB} = 0.850, <(C–I···N) = 174.8(4)°; d(I···N) = 3.073(10) Å, R_{XB} = 0.871, <(C–I···N) = 174.4(4)°).[47] With R_{XB} values of 0.85 or more, these halogen bonds are weaker than the ones in either **33·29** or **33·24**.

As could be expected, diiodoacetylene **33** also readily forms co-crystals with oxygen-based acceptors. For example, in the structure of a co-crystal with *N,N*-dimethylformamide **36**, **33·36** (Scheme 3.7), the **36** acts as a bifurcated acceptor through the oxygen atom of the amide group (d(I···O) = 2.834(4) Å, R_{XB} = 0.810, <(C–I···O) = 170.7(2)° and d(I···O) = 2.888(4) Å, R_{XB} = 0.825, <(C–I···O) = 173.0(2)°), forming a zig-zag chain motif. Still, the halogen bonds deviate significantly from the angles expected with the oxygen lone pair directions in mind (<(C=O···I) = 125.1(4)° and 145.4(4)°, <(I···O···I) = 89.4(1)°). Next, the 1,4-dioxane **4** also readily forms a 1 : 1 co-crystal **33·4**.[48] Here, however, **4** only engages in halogen bonding through one of the oxygen lone pairs (d(I···O) = 2.65 Å, R_{XB} = 0.757, <(C–I···O) = 176°), namely the equatorial one, to form 1-D chains (Scheme 3.7).

An obvious advantage of the 1-haloalkyne halogen bond donors is that such functionality can be easily introduced to almost any kind of scaffolding, as it only requires three additional atoms at best (–C≡C–X, X = I/Br/Cl) and the necessary chemistry is well developed. Batsanov and Howard showed how, even when attached to a plain phenyl ring such as in **37**, exceptionally short (and presumably rather strong) C–I···N halogen bonds (d(I···N) = 2.712(2) Å, R_{XB} = 0.768, <(C–I···N) = 177.84(10)°) can be obtained using an iodoethynyl donor group, here in combination with morpholine **38** as the acceptor through its amine nitrogen (Scheme 3.8).[49] With the halogen bond donor approaching equatorially, the NH hydrogen atom is oriented axially and forms a hydrogen bond with the neighbouring molecule

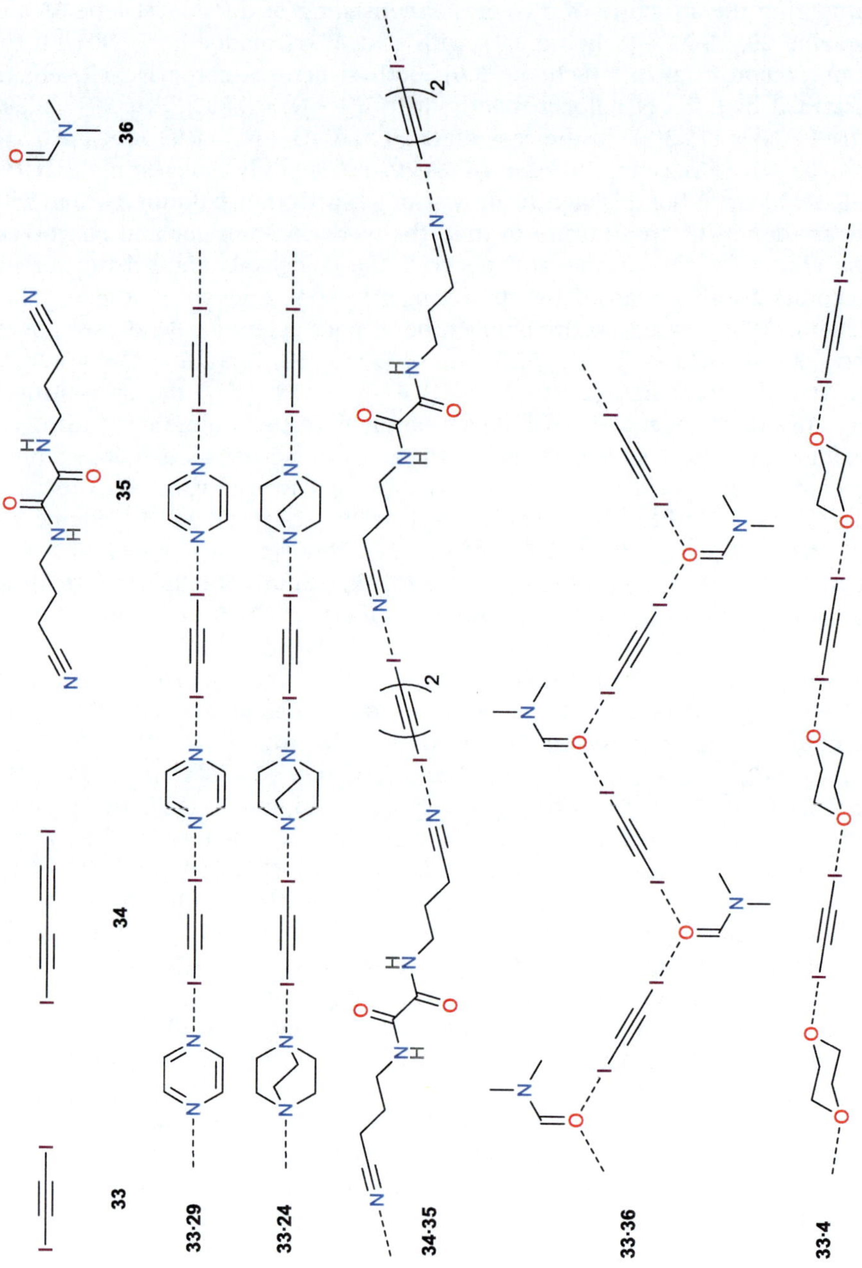

Scheme 3.7 Halogen-bonded 1-D chains in the co-crystals **33·29**, **33·24**, **34·35**, **33·36** and **33·4**.

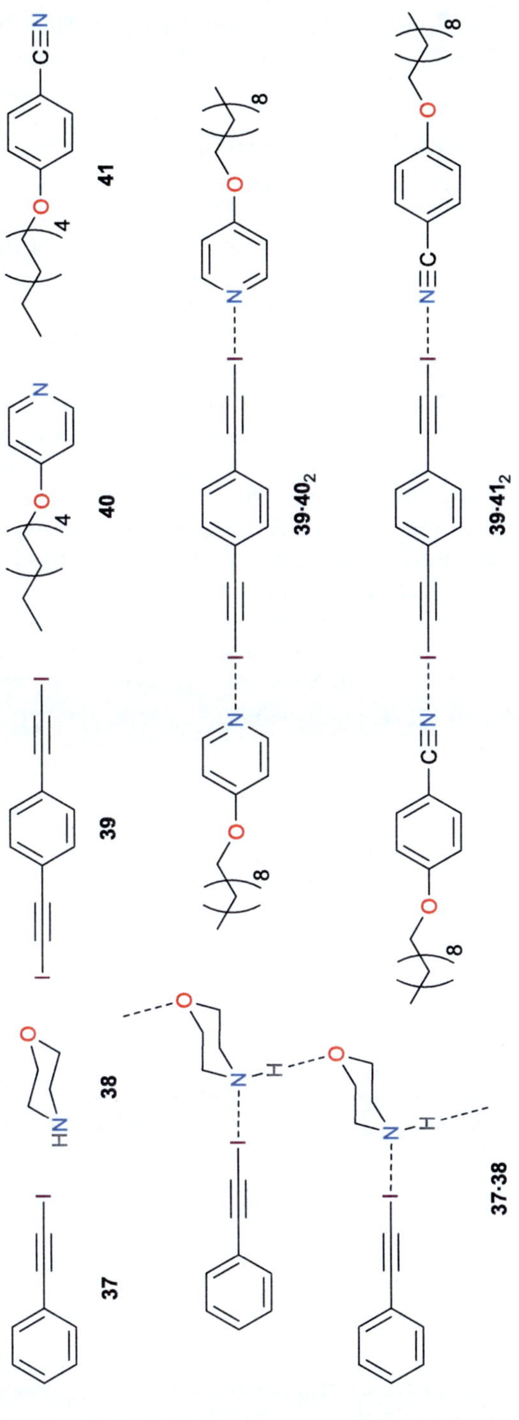

Scheme 3.8 Halogen-bonded discrete assemblies in the co-crystals **37·38**, **39·40$_2$** and **39·41$_2$**.

of **38** through its oxygen atom, which on the other hand does not participate in halogen bonding. Another convenient aspect is that more than one iodoethynyl group can be attached to the scaffold to obtain di- or polytopic donors. Thus, Serrano and co-workers reported the preparation of two different halogen bond donors 1,4-bis(iodoethynyl)benzene **39** and 1,3-bis-(iodoethynyl)benzene, which were then combined with a number of different donors to obtain co-crystals exhibiting interesting liquid crystalline properties.[50] Crystal structures were reported of two co-crystals **39·40**$_2$ and **39·41**$_2$ (Scheme 3.8), both of which show the expected 2 : 1 assemblies sustained by C–I···N halogen bonds of **39** with pyridine of **40** (d(I···N) = 2.757(15) Å, R_{XB} = 0.781, <(C–I···N) = 172.8(6)°) or the cyanide group of **41** (d(I···N) = 2.946(16) Å, R_{XB} = 0.835, <(C–I···N) = 176.4(6)°). Besides showing the practicality of appending iodoethynyl groups to introduce halogen-bonding functionality, it again demonstrates the convenience of switching from a pyridine- to a cyano-acceptor as means of adjusting the strength of the halogen bond.

Reporting the structures of three co-crystals based on different iodoethynylbenzene-type donors, Bosch found what was the shortest C–I···N halogen bond of its kind at the time (d(I···N) = 2.622(3) Å, R_{XB} = 0.743, <(C–I···N) = 180.00(1)°) in the co-crystal **42·45** (Scheme 3.9), which contained four chemically equivalent but symmetrically non-equivalent halogen bonds in total (with the parameters of the other three fairly similar: d(I···N) = 2.675(2) Å, R_{XB} = 0.758, <(C–I···N) = 174.86(11)°; d(I···N) = 2.700(2) Å, R_{XB} = 0.765, <(C–I···N) = 176.37(12)° and d(I···N) = 2.705(3) Å, R_{XB} = 0.766,

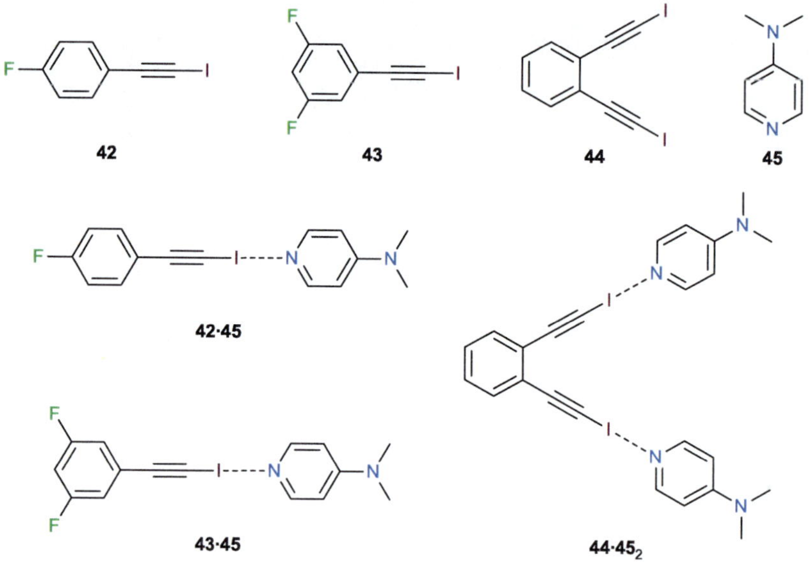

Scheme 3.9 Halogen-bonded discrete assemblies in the co-crystals **42·45**, **43·45** and **44·45**$_2$.

<(C–I···N) = 174.30(12)°).[51] The electrostatic potential calculations suggested that the **43** should form stronger halogen bonds due to the presence of an additional fluorine substituent on the benzene ring and, as a result, a higher positive potential on the iodine corresponding to a deeper σ-hole. This, however, did not result in shorter halogen bonds in **43·45** (Scheme 3.9; d(I···N) = 2.680(3) Å, R_{XB} = 0.759, <(C–I···N) = 175.18(12)°). Even **44·45$_2$** (Scheme 3.9), the co-crystal of the non-fluorinated 1,2-bis(iodoethynyl)benzene **42** with **45**, exhibited slightly shorter halogen bonds (d(I···N) = 2.661(2) Å, R_{XB} = 0.754, <(C–I···N) = 175.10(8)° and d(I···N) = 2.666(2) Å, R_{XB} = 0.755, <(C–I···N) = 174.21(8)°).

This 'double activation' (the term can be attributed to Aakeröy *et al.*[52]) approach of Bosch[51] sought to use an iodoethynyl donor group attached to a benzene ring activated with one (**42**) or two (**43**) atoms of fluorine in order to create a donor that would form stronger halogen bonds. Unfortunately, this approach could not be convincingly validated, possibly because of the limited number of the studied co-crystals or the relatively unreliable correlation between the halogen bond strength and its length as measured from the crystal structures.

A similar but broader study was subsequently undertaken by Aakeröy *et al.*, where the 'double activation' strategy was employed to design a number of halogen bond donors, this time by attaching a haloethynyl donor group to a benzene ring activated by one or two nitro groups.[52] The electrostatic potential calculations (Scheme 3.10) suggested that the iodoethynylnitrobenzene donors **49–51** should indeed be better halogen bond donors than **37**. In case of mono-nitro donors **49** and **50**, the maximum potential was found to be 20 kJ mol^{-1} higher compared to a classical halogen bond donor such as pentafluoroiodobenzene **48**, while with the di-nitro donor **51** the maximum potential was over 40 kJ mol^{-1} higher than in **48**! Interestingly, a similar difference was observed between **47** and **51**, once again illustrating the strong activating effect of the ethynyl group.

The co-crystallization experiments were conducted with a number of potential acceptors, some of which are shown in Scheme 3.11. They turned out a success, as the iodoethynyl-based donors **49–51** formed co-crystals in all cases, resulting in a 100% supramolecular yield,[53] while the corresponding

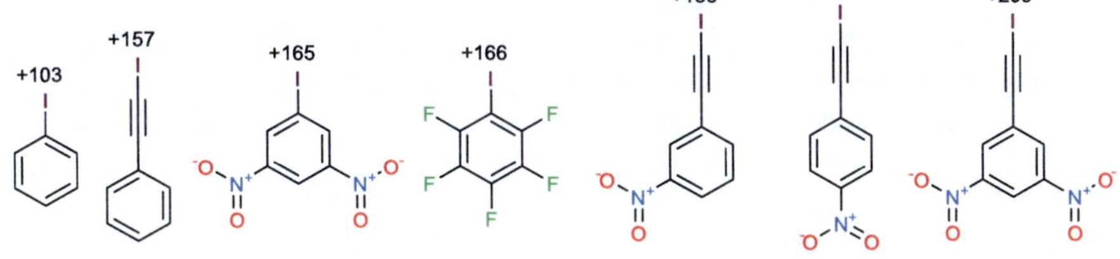

Scheme 3.10 Calculated electrostatic potential values (in kJ mol^{-1}) of the halogen bond donor atoms in **46, 37, 47–51**.

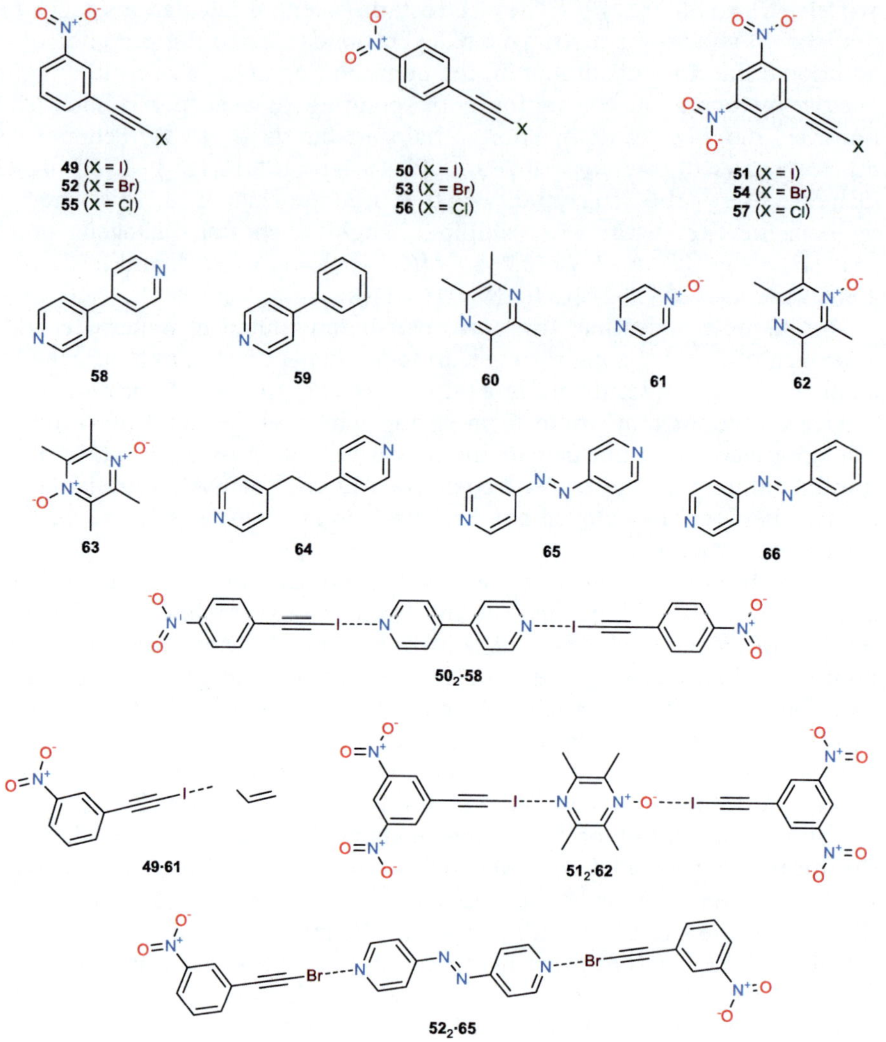

Scheme 3.11 Halogen-bonded discrete assemblies in the co-crystals **50$_2$·58**, **49·61**, **51$_2$·62** and **52$_2$·65**.

bromoethynyl-donors **52–54** yielded a more modest 60% success rate. Remarkably, the halogen bond lengths in the co-crystals with **49–51** were compared with those of different donors with the same acceptors and were found to be consistently shorter, suggesting that the **49–51** are very strongly activated and are among the strongest halogen bond donors.

Another interesting aspect of the iodoethynyl donor group is its resemblance to the unsubstituted ethynyl group. This idea was experimentally examined by Aakeröy *et al.*,[54] who prepared an iodoethynyl-based halogen bond donor **67** and an ethynyl-based hydrogen bond donor **68** (Figure 3.3).

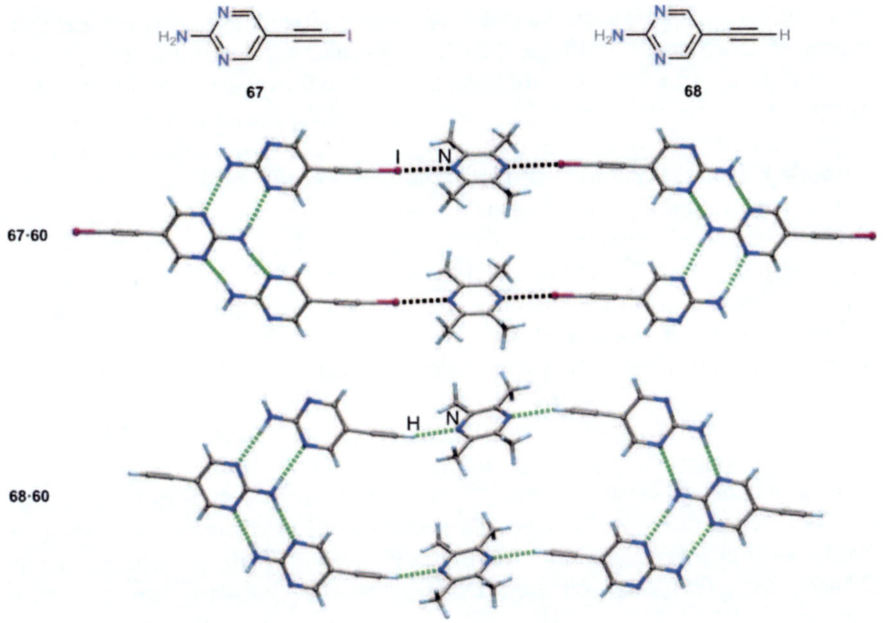

Figure 3.3 Equivalence of halogen and hydrogen bonding in the co-crystals of **67** or **68** with **60**. Halogen and hydrogen bonds are shown as black and green dashed lines, respectively.

The two were successfully co-crystallized with either tetramethylpyrazine **60** or 1,2-bis(4-pyridyl)ethylene **11** to yield 1 : 1 co-crystals **67·60**, **68·60**, **67·11** and **68·11**. The structures containing the same hydrogen/halogen bond acceptor were found to be remarkably similar, as can be seen from the crystal structures of **67·60** and **68·60** (Figure 3.3). Comparing the halogen bonding in **67·60** (d(I···N) = 2.897(2) Å, R_{XB} = 0.821, <(C–I···N) = 176.76(11)°) and the hydrogen bonding in **68·60** (d(H···N) = 2.414(16) Å, R_{HB} = 0.878, <(C–H···N) = 166.2(14)°) shows that the two interactions are remarkably similar, but that the halogen bond is also more linear, as expected. The same is true for the other pair of co-crystals, **67·11** and **68·11**, suggesting that the ethynyl/iodoethynyl equivalence is an important property of such systems, relevant both in crystal engineering and supramolecular chemistry in general.

3.6 Aryl Halide Donors

Several elegant studies have employed simple aromatic donors where the bromine or iodine atoms are attached to an unsubstituted benzene ring. While these might seem insufficiently polarized to engage in halogen bonding, they have been shown to form halogen bonds with a few different acceptors. For example, Allen *et al.* showed that the 1,4-diiodobenzene **69** readily forms co-crystals with 1,4-dinitrobenzene **70**, forming 1-D chains sustained by C–I···O halogen bonds with the nitro group as an acceptor

(Scheme 3.12).[55] This type of C–I···O(nitro) halogen bonds often exhibit a degree of bifurcation, with the two C–I···O distances similar or even equal (Scheme 3.12). In **69·70**, the distances to the two oxygen atoms of the nitro group are exactly the same due to the crystallographic symmetry requirements (d(I···O) = 3.453(3) Å, R_{XB} = 0.987, <(C–I···O) = 161.32(7)°). While seemingly weak, the C–I···O(nitro) halogen bonds have since been successfully exploited by Tothadi and Desiraju to construct more co-crystals where this recognition motif works in parallel to a hydrogen bonding one (Scheme 3.12).[56] Thus in the co-crystal **69·71₂**, the molecules of 4-nitrobenzamide **71** were found to interact through an amide–amide hydrogen bonding homosynthon, forming dimers that were subsequently linked through C–I···O(nitro) halogen bonds (d(I···O) = 3.340(9) Å, R_{XB} = 0.954, <(C–I···O) = 162.4(3)° and d(I···O) = 3.406(9) Å, R_{XB} = 0.973, <(C–I···O) = 160.0(3)°) to form infinite 1-D chains (Scheme 3.12).

In the next step, a similar system was examined where the fumaric acid **72** was combined with **69** and **71** to give a three-component system, a ternary co-crystal **69·71$_2$·72**. Here, **72** inserts itself into the hydrogen-bonded amide homodimeric motif, leaving the C–I···O(nitro) halogen bonds between **69** and **71** largely intact. Still, the C–I···O(nitro) motif is fairly asymmetric (d(I···O) = 3.271(2) Å, R_{XB} = 0.935, <(C–I···O) = 166.08(9)° and d(I···O) = 3.576(3) Å, R_{XB} = 1.022, <(C–I···O) = 157.42(9)°), with one of the

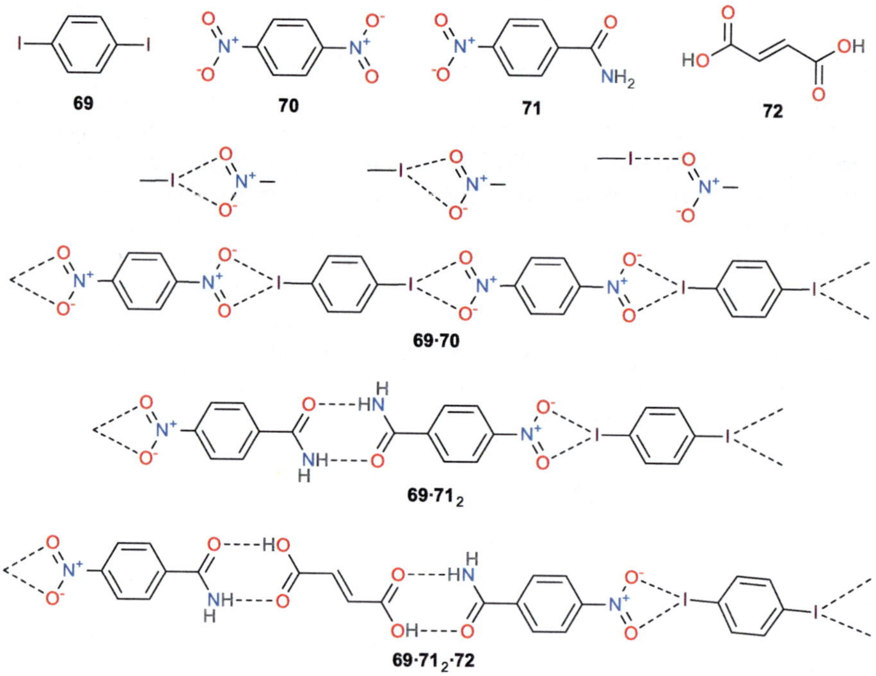

Scheme 3.12 Halogen (and hydrogen) bonded 1-D chains in the co-crystals **69·70**, **69·71$_2$** and **69·71$_2$·72**.

contact distances larger than the sum of the van der Waals radii of oxygen and iodine! Moreover, Tothadi and Desiraju succeeded in obtaining a number of similar ternary co-crystals, *e.g.* one using a 1-bromo-4-iodo-benzene instead of **69**,[56] and a few more by combining **70** with 4-bromo/iodobenzamide and different dicarboxylic acids,[57] demonstrating among other things a surprising level of robustness of the C–I···O(nitro) halogen bonds[58] and the modularity of their crystal engineering strategy.

The same donor (**69**) was also successfully used by Pennington and co-workers,[28] and Cinčić *et al.*,[59] in co-crystallization with more traditional halogen bond acceptors such as 1,2-bis(4-pyridyl)ethylene **11**, hexamethyl-enetetramine **21**, 1,4-diazabicyclo[2.2.2]octane **24** and 4,4′-bipyridine **58**. This yielded co-crystals **69·11**, **69·21**, **69·24** and **69·58** (Scheme 3.13, Table 3.1), all of which contain 1-D halogen-bonded chains, and in the case of **69·21** a pronouncedly zig-zag one. Based on the R_{XB} values, these halogen bonds seem significantly stronger than the C–I···O(nitro) halogen bonds in the co-crystals comprising **72** and nitro-acceptors.

3.7 Perfluoroaromatic Donors

The most popular halogen bond donors, however, have been those with the donor halogen atom(s), usually iodine or bromine, attached to a (per)-fluorinated aromatic or aliphatic backbone. Conveniently, Pennington and co-workers[28] and Syssa-Magalé *et al.*[60] also provided the necessary data on the systems comprising the fluorinated analogue of **69**, tetrafluoro-1,4-diiodobenzene **73**, enabling a direct comparison of the two donors. The key parameters of the C–I···N bonds in the two sets of co-crystals of **69** or **73** with **11**, **21**, **24** and **58** are listed in Table 3.1. It is interesting to note the systematic difference in the R_{XB} values between the two sets of co-crystals, with those in the co-crystals of **73** being around 0.04–0.05 (4–5%) lower than in the co-crystals of **69**, coming from the stronger donor ability of **73** due to the activation by the fluorine atoms.

A very elegant work illustrating the effect of the fluorination was presented by Bruce and co-workers, where a number of co-crystals between iodobenzenes with various degrees of fluorination (**74–80** and **48**) and 4-dimethylaminopyridine **45** were examined (Scheme 3.14, Table 3.2).[61] A powerful message is delivered by a simple plot of R_{XB} values against the number of fluorine atoms on the benzene ring of the donor (Figure 3.4), which shows a clear correlation between the two, with the R_{XB} values decreasing as the number of fluorine atoms of the donor increases. On the other hand, the <(C–I···N) angle does not show any clear correlation with the degree of fluorination. Still, the observed link between the degree of fluorination and R_{XB} values nicely demonstrates the activating effect of the fluorine substitution on the halogen bond donors.

A study by Aakeröy *et al.* compared the co-crystallization performance of 6 carefully chosen halogen bond donors (Scheme 3.15) with 21 different acceptors, including pyridines, pyridine-*N*-oxides, pyrazoles and

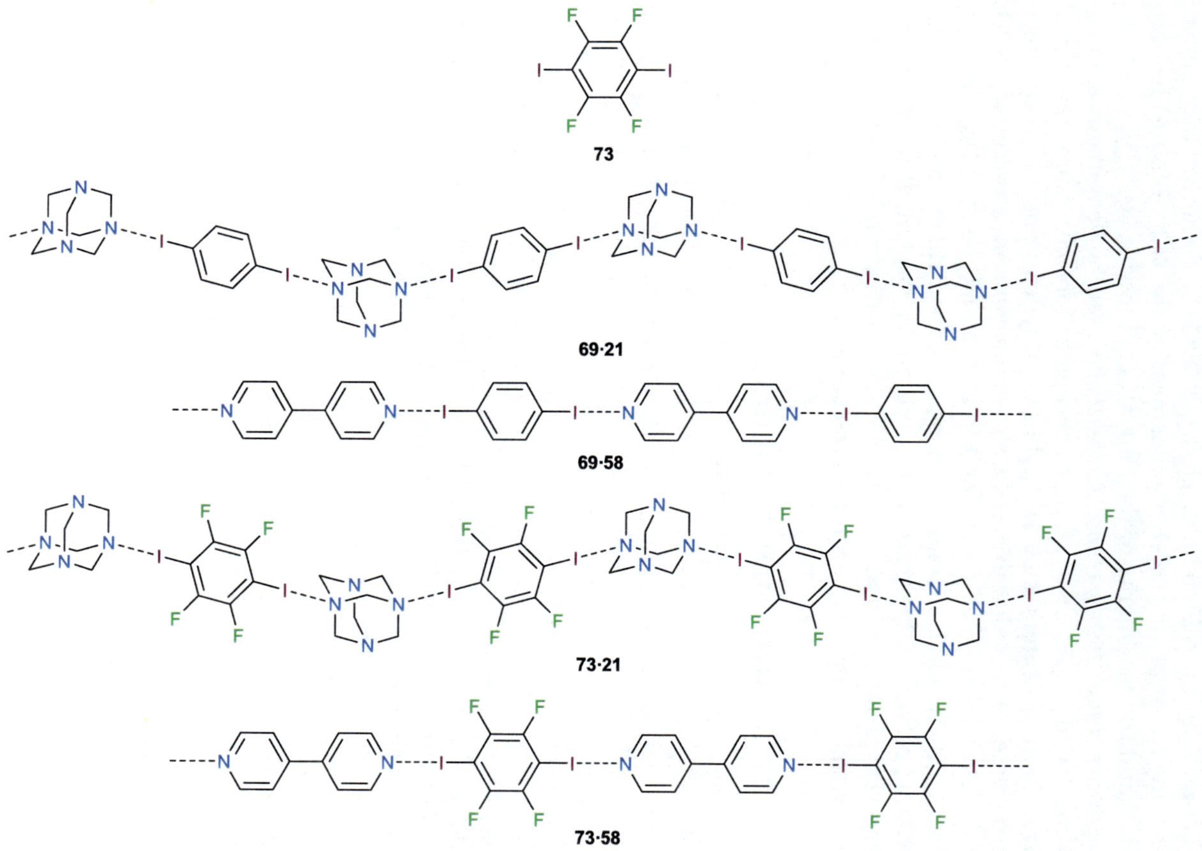

Scheme 3.13 Halogen-bonded 1-D chains in the co-crystals **69·21**, **69·58**, **73·21** and **73·58**.

Table 3.1 Halogen bond parameters in the co-crystals of **69** or **73** with **11**, **21**, **24** and **58**.

	69·11	**69·21**	**69·24**		**69·58**
d(I···N)/Å	2.996(5)	3.010(4)	2.928(5)	2.959(5)	3.032(3)
R_{XB}	0.849	0.853	0.829	0.838	0.859
<(C–I···N)/°	176.91(10)	173.21(15)	175.4(2)	176.3(2)	176.05(11)
	73·11	**73·21**	**73·24**		**73·58**
d(I···N)/Å	2.811(6)	2.845(3)	2.764(3)	2.774(3)	2.851(4)
R_{XB}	0.796	0.806	0.783	0.786	0.808
<(C–I···N)/°	179.3(2)	169.05(14)	173.21(12)	174.08(11)	177.27(12)

benzimidazoles.[62] The co-crystallization was conducted by the efficient solid-state grinding method and the successful formation of co-crystals was detected by means of IR spectroscopy. For each particular donor **69, 73, 81–84,** the supramolecular yield was calculated as the percentage of successful co-crystallizations. Interestingly, the obtained supramolecular yields were found to follow the same trend as the electrostatic potentials of the donors (Figure 3.5), although the correlation is not necessarily linear. It is interesting to note again that the ethynyl group is just as successful (if not slightly better) as a perfluorinated benzene ring at activating the iodine or bromine atom. Moreover, the activation is shown to dramatically enhance the success of co-crystallization, especially so in the case of bromo-donors. Remarkably, activated bromo-donors could easily outperform a non-activated iodo-donor.

Another interesting aspect of the perfluoroaromatic halogen bond donors is their rigidity and the possibility of controlling the directionality of the halogen bonding through different substitution patterns. Thus, the tetrafluoro-1,2-diiodobenzene **88** (Scheme 3.16) would be expected to form halogen bonds with a 60° angle between them. The trifluoro-1,3,5-triiodobenzene **89** (Scheme 3.16) would, on the other hand, be expected to form halogen bonds at a 120° angle. Aakeröy *et al.* decided to take advantage of this by combining the two halogen bond donors **88** and **89** with a number of pyridine-*N*-oxide acceptors **85–87** and **63.**[63] Assuming that the pyridine-*N*-oxides would act as bifurcated acceptors, forming two C–I···O halogen bonds at a *ca.* 120° angle, combining these *N*-oxides with **88** and **89** was expected to give rise to supramolecular rhombi or hexagons. Indeed, a number of co-crystals were obtained, some of which exhibited the planned motifs. For example, the structure of **88·87** revealed the formation of discrete $\mathbf{88_2{\cdot}87_2}$ rhombi sustained by four C–I···O halogen bonds (Scheme 3.16). The structure of $\mathbf{89_2{\cdot}63}$, on the other hand, was found to consist of fused hexagons, with the molecule of **63** shared between the neighbouring hexagons (Scheme 3.16).

Interestingly, in $\mathbf{89_2{\cdot}63}$ only two iodine atoms of **89** are involved in the C–I···O halogen bonds with **63**, with the remaining iodine only forming a weak C–I···I interaction. This behaviour was investigated by van der Boom and co-workers, both experimentally and computationally.[64] First, they described the structures of three co-crystals of **89** with different bipyridyl

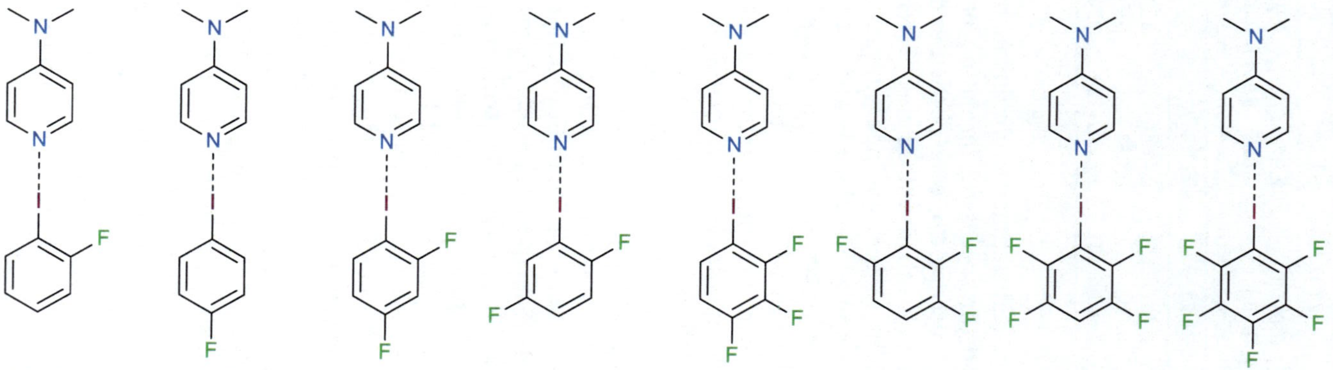

74·45 75·45 76·45 77·45 78·45 79·45 80·45 48·45

Scheme 3.14 Discrete halogen-bonded assemblies in the co-crystals **74·45–80·45** and **48·45**.

Table 3.2 Halogen bond parameters in co-crystals of **76–82** and **49** with **45**.

	74·45	**75·45**	**76·45**	**77·45**
d(I···N)/Å	2.9092(15)	2.9257(16)	2.880(4)	2.8481(15)
R_{XB}	0.824	0.829	0.816	0.809
<(C–I···N)/°	174.21(5)	175.48(6)	174.70(14)	177.83(5)
	78·45	**79·45**	**80·45**	**48·45**
d(I···N)/Å	2.817(4)	2.872(2)	2.702(2)	2.693(3)
R_{XB}	0.798	0.814	0.765	0.763
<(C–I···N)/°	175.55(13)	172.05(8)	171.49(7)	178.43(12)

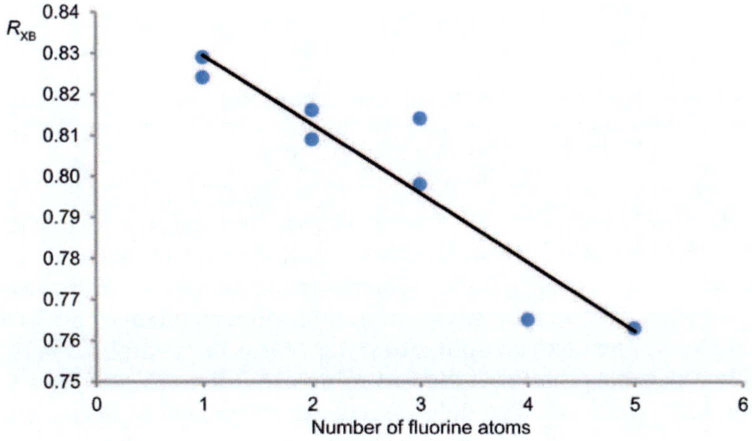

Figure 3.4 R_{XB} values for the C–I···N halogen bonds in the co-crystals **74·45–80·45** and **48·45** plotted as a function of the number of fluorines attached to the iodobenzene ring of the donor.

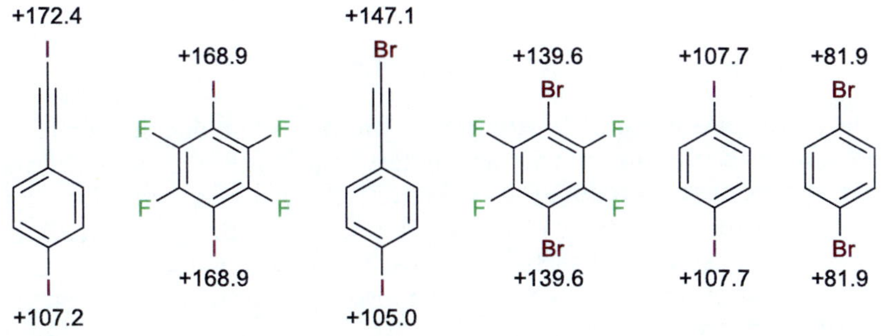

81 **73** **82** **83** **69** **84**

Scheme 3.15 Halogen bond donors **69**, **73** and **81–84**, ranked from left to right according to the calculated electrostatic potential values (in kJ mol^{-1}) of the donor atoms.

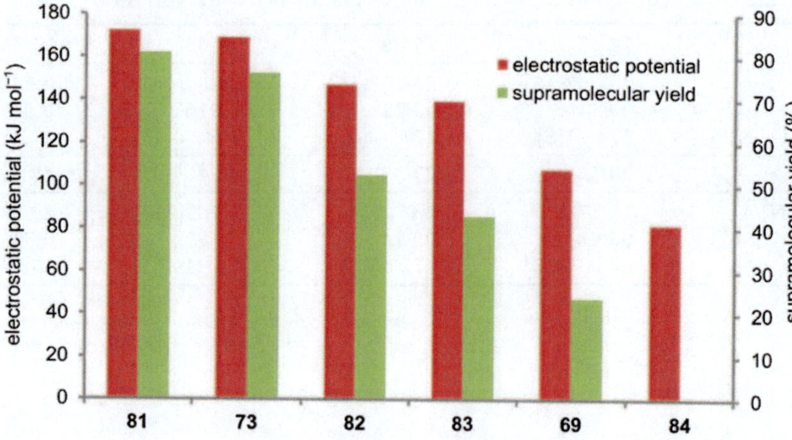

Figure 3.5 Comparison of the calculated electrostatic potentials of the halogen bond donors **69**, **73** and **81–84**, and the respective supramolecular yields as obtained in the co-crystallization experiments.

acceptors, all of which show **89** acting only as a ditopic donor (two of them are shown in Scheme 3.17). Next, through calculations, they showed that each successive C–I···N halogen bond formed by the **89** slightly lowers the binding energy for the next acceptor, with a more pronounced change going from the 1 : 2 to 1 : 3 complex than on going from the 1 : 1 to 1 : 2 complex. On the other hand, they also suggested that the 1 : 2 complexes might pack more efficiently than the 1 : 3 complexes. The delicate balance of these two effects could then dictate the outcome of the crystallization. Interestingly, Karadakov, Bruce and co-workers later described a 1 : 3 co-crystal of **89** with 4-dimethylaminopyridine **45**, $\mathbf{89}\cdot\mathbf{45}_3$.[65] While it might be tempting to conclude that this is a consequence of the higher Lewis basicity of **45**, the possibility that the formation of a 1 : 3 co-crystal is facilitated by more favourable packing and weak interactions formed by **45** cannot be entirely dismissed.

3.8 Perfluoroaliphatic Donors

Similarly to perfluorinated aromatics, bromo- or iodo-substituted perfluoroalkanes represent another widely used class of halogen-bond donors, with 1,ω-diiodoperfluoroalkanes as the most typical representatives. Based on the calculated molecular electrostatic potentials, the perfluoroaliphatic iodides were reasonably expected to be just as effective halogen bond donors as the perfluoroaromatic iodides.[66] Moreover, perfluoroaliphatic backbone can offer different metrics (chain length), conformational flexibility and, unlike the perfluoroaromatic backbone, no particular stacking preferences due to the lack of an aromatic π-system.

In fact, the halogen bonding with 1,ω-diiodoperfluoroalkanes was reported before that with perfluoroaromatic iodides. In 1998, Resnati and co-workers

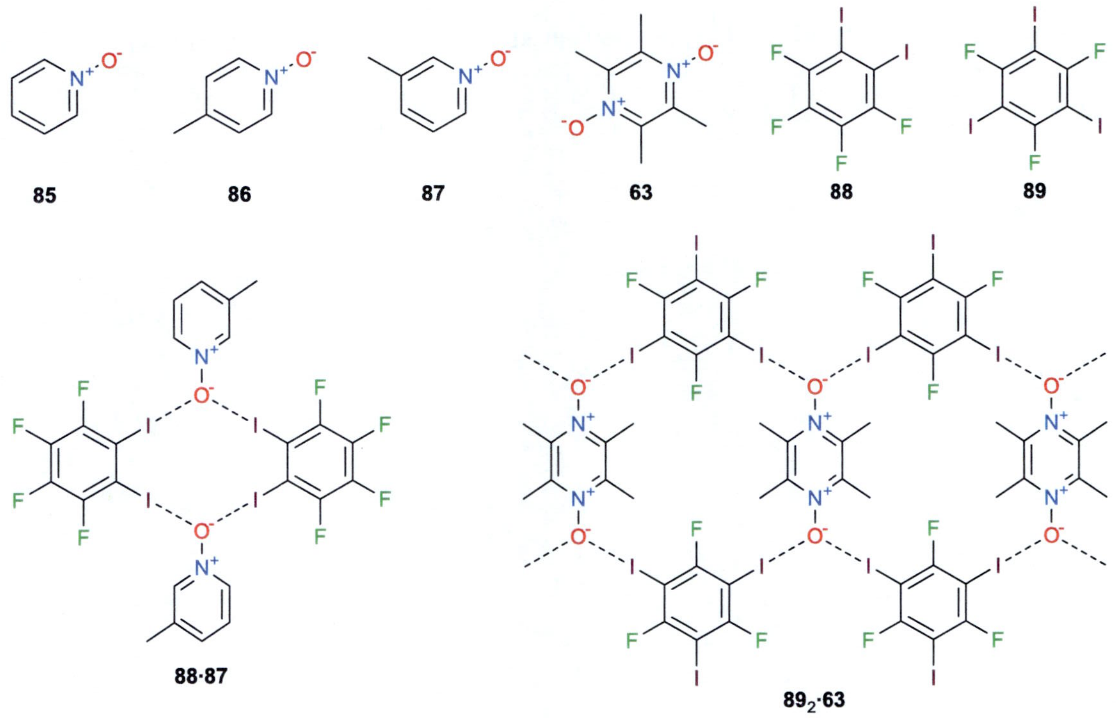

Scheme 3.16 Halogen-bonded molecular rhombus in the structure of **88·87** and the fused hexagons in the structure of **89_2·63**.

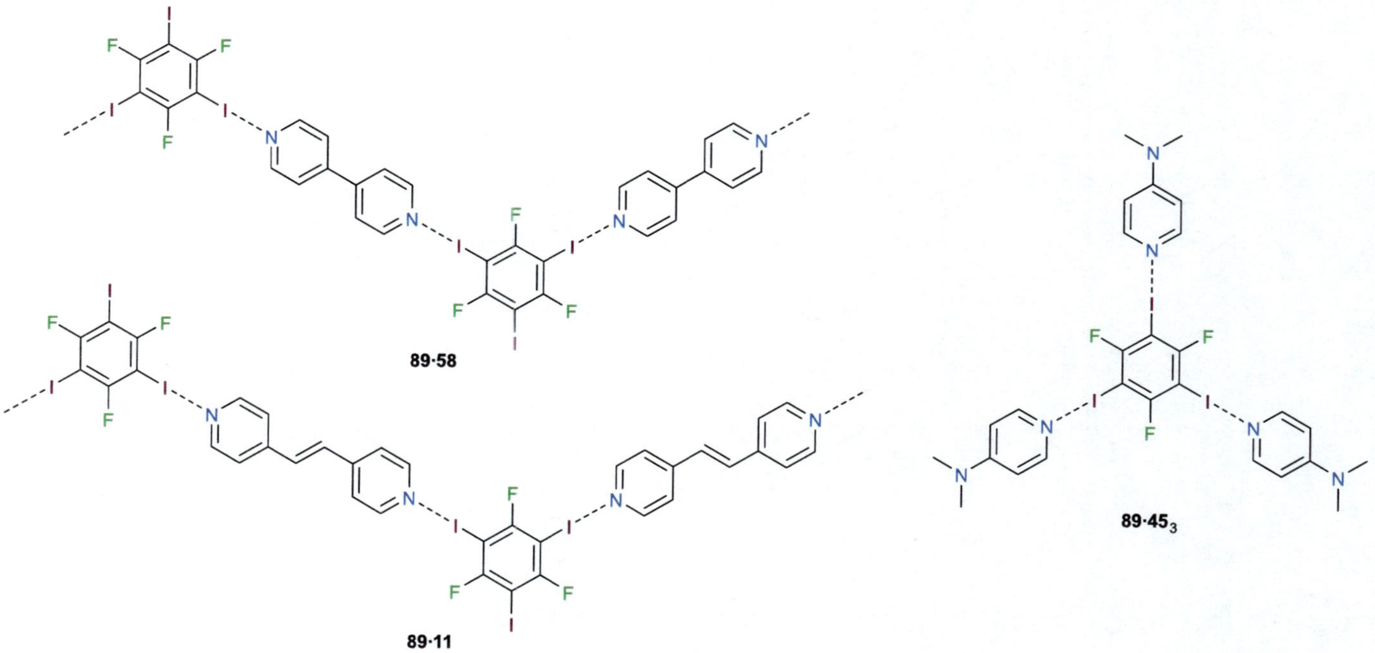

Scheme 3.17 Halogen-bonded 1-D chains in the co-crystals **89·58** and **89·11** and the discrete 1 : 3 assembly in the co-crystal **89·45$_3$**.

reported their work on self-assembly between perfluorocarbons and hydrocarbons mediated by nitrogen–iodine interactions.[3] Specifically, they combined the liquid tetrafluoro-1,2-diiodoethane **90** (b.p. 112–113 °C) as a donor with the liquid (b.p. 121 °C) *N,N,N′,N′*-tetramethylethylenediamine **91** in chloroform solution, obtaining a solid, 1 : 1 co-crystal comprising 1-D chains **90·91** propagated by C–I···N halogen bonds (Scheme 3.18, 2.797(7) Å, $R_{XB}=0.792$, <(C–I···N) = 170.6(8)°). Similarly, using the solid 2.2.2-cryptand **92** as an acceptor in place of **91** resulted again in a 1 : 1 co-crystal **90·92** featuring an analogous 1-D chain sustained by C–I···N halogen bonds (Scheme 3.18, 2.86(2) Å, $R_{XB}=0.810$, <(C–I···N) = 171.0(5)°). The structures of the two co-crystals were analysed by single crystal X-ray diffraction. Moreover, using the longer homologues of **90** (of 4, 6 or 8 carbon atoms in length) with either **91** or **92** also yielded 1 : 1 co-crystals, which were reasonably assumed to be sustained by C–I···N halogen bonds. This work established the use of 1,ω-diiodoperfluoroalkanes as halogen bond donors.

Complementary results were soon published by Zhu and co-workers, who combined the 1,ω-diiodoperfluoroalkanes with oxygen-based halogen bond acceptors.[67] Interestingly, mixing of **90** with 1,4-dioxane **4**, both liquids at room temperature, in chloroform, again yielded a solid, found to be a 1 : 1 co-crystal **90·4**. Unsurprisingly, the structure was again found to feature 1-D halogen-bonded chains propagated by C–I···O halogen bonds (Scheme 3.19, 2.81 Å, $R_{XB}=0.803$, <(C–I···O) = 163.3°). Despite the presence of two lone pairs on the sp^3 oxygen in **4**, each oxygen atom was found to only form one halogen bond. This was, however, not the case with **93·94**, a 1 : 1 co-crystal of octafluoro-1,4-diiodobutane **93** and hexamethylphosphoramide **94** (again, both liquids at room temperature). Its crystal structure showed the formation of zig-zag chains where each sp^2 oxygen atom of **94** forms two C–I···O halogen bonds (Scheme 3.19, 2.84(1) Å, $R_{XB}=0.811$, <(C–I···O) = 170.6(9)° and 2.86(1) Å, $R_{XB}=0.817$, <(C–I···O) = 177.0(7)°; <(I···O···I) = 88.7(3)°). Here, both sp^3 and sp^2 oxygen were demonstrated to be potential halogen bond acceptors, acting as mono- or bidentate acceptors, respectively.

Metrangolo, Resnati and co-workers were also successful in obtaining halogen-bonded co-crystals using 1-iodoperfluoroalkane and 1,ω-dibromoperfluoroalkane donors. In the case of the co-crystal $\mathbf{96_2{\cdot}91}$, two liquid components, 1-iodoperfluoroheptane **96** and *N,N,N′,N′*-tetramethyl-*p*-phenylenediamine **95**, co-crystallized to give a solid consisting of discrete $\mathbf{96_2{\cdot}91}$ assemblies sustained by two C–I···N halogen bonds (Scheme 3.20).[68] Remarkably, despite the absence of any extended halogen-bonded chains or networks, the co-crystal, whose components' melting points were both under 0 °C, exhibited a melting point of 52–53 °C.

Metrangolo, Resnati and co-workers also reported the co-crystallization of 1,8-dibromoperfluorooctane **97** and **95**.[69] The resulting co-crystal **97·95** was found to exhibit, expectedly, 1-D halogen-bonded chains sustained by C–Br···N halogen bonds, showing how the bromoperfluoroalkanes can also be employed as halogen bond donors (Scheme 3.20).

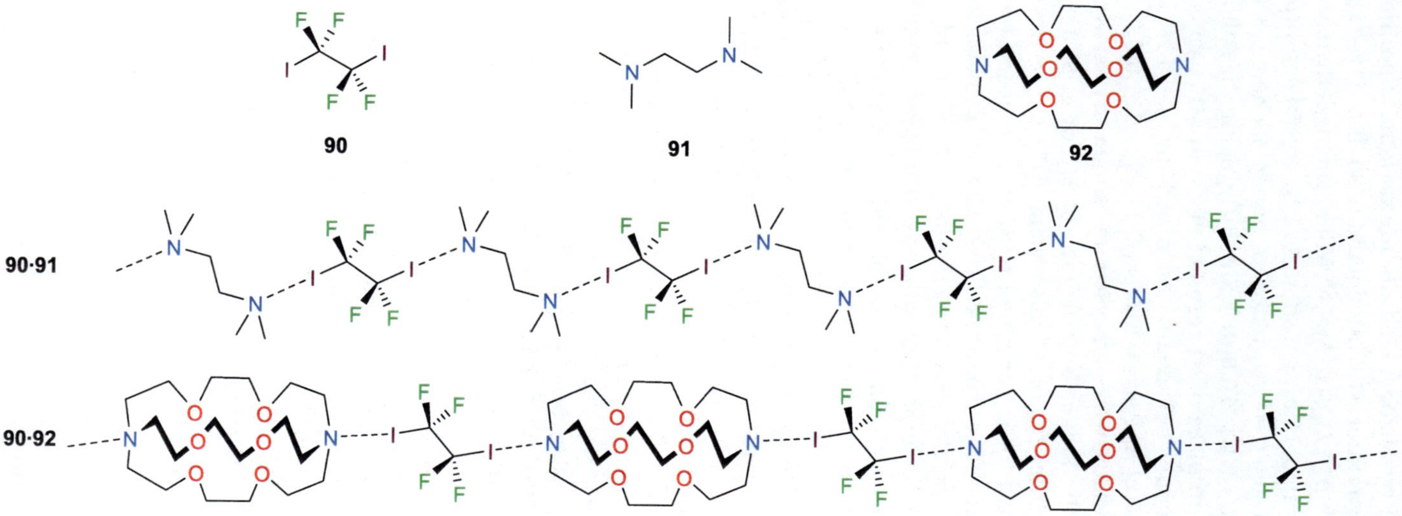

Scheme 3.18 Halogen-bonded chains in **90·91** and **90·92**.

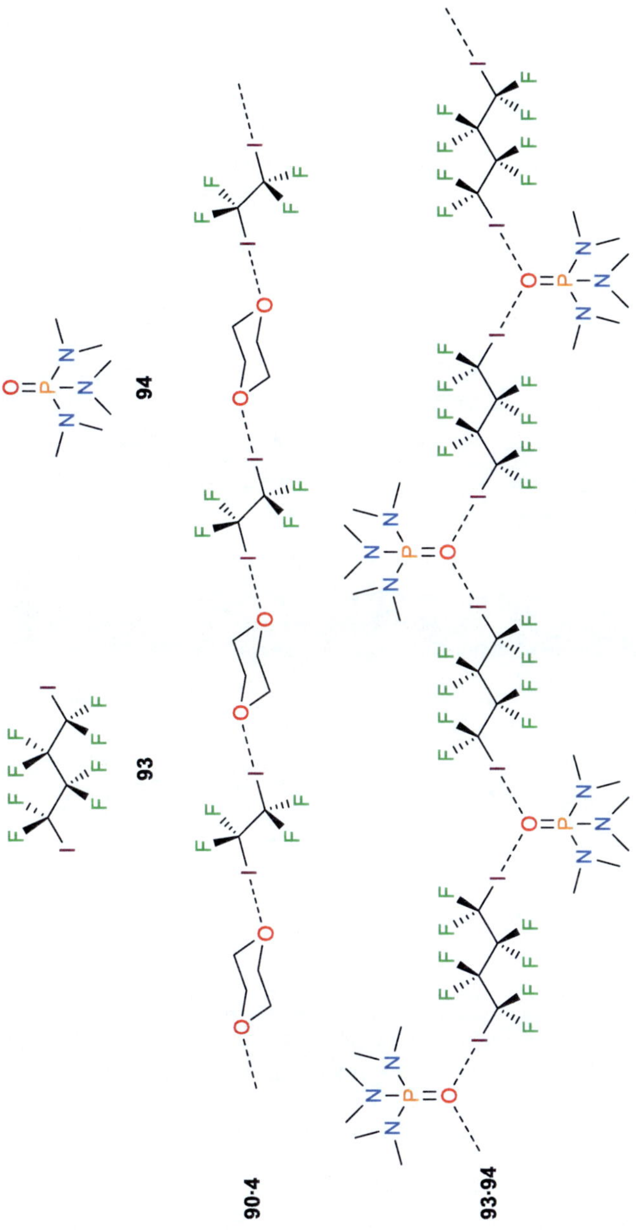

Scheme 3.19 Halogen-bonded 1-D chains in **90·4** and **93·94**.

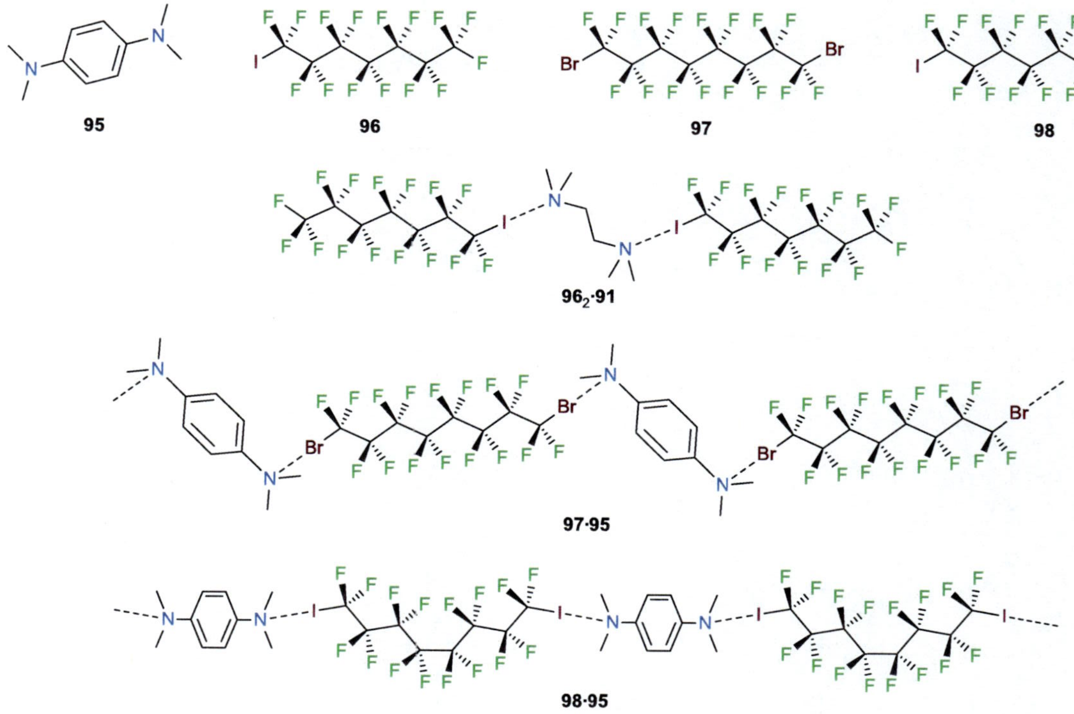

Scheme 3.20 Halogen-bonded 2 : 1 discrete assembly in the co-crystal $\mathbf{96_2{\cdot}91}$ and the 1-D halogen-bonded chains in the co-crystals **97·95** and **98·95**.

Next, using the iodo-analogue **98** in place of **97** yielded upon co-crystallization with **95** a 1 : 1 co-crystal **98·95**.[70] Similarly to **97·95**, the crystal structure exhibits 1-D chains formed through C–I···N halogen bonding. However, while the molecule of **97** appears in an expected *all-trans* conformation in **97·95**, in **98·95** the donor **98** unexpectedly assumes a *gauche* conformation (Scheme 3.20), illustrating the conformational flexibility of the perfluoroalkanes as a distinctive feature in comparison to *e.g.* perfluoroaromatic halogen bond donors.

As already mentioned above, one specific feature of the perfluoroalkane halogen bond donors is also the availability of donors with different lengths. This can be used to prepare co-crystals with desired metrics, as the donor length is often successfully reflected in the metrics of the crystal structure. Metrangolo, Resnati and co-workers demonstrated this using two different bispyridyl donors, 4,4′-bipyridine **58** and 1,2-bis(4-pyridyl)ethylene **11**, which were co-crystallized with a range of even-chain 1,ω-diiodoperfluoroalkanes from tetrafluoro-1,2-diiodoethane to icosafluoro-1,10-diiododecane (**90**, **93**, **99**, **98** and **100**).[71] In all structures, the hydrocarbon part (acceptor) and the fluorocarbon parts (donor) were found to be segregated, each forming a separate layer (Figure 3.6). For both co-crystal series, the chain pitch length was found to change linearly with the number of CF_2 groups of the donor.

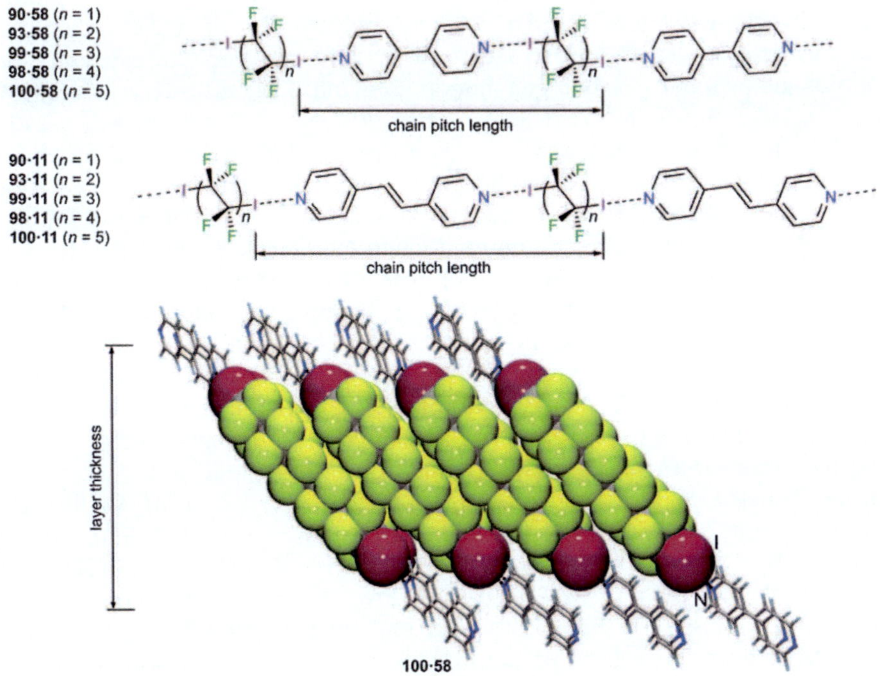

Figure 3.6 Halogen-bonded 1-D chains in co-crystals of **90**, **93**, **99**, **98** and **100** with **58** or **11**. Packing of **100·58**, showing the segregation of hydrocarbon and fluorocarbon components, as also observed in other co-crystals. Halogen bonds are shown as black dashed lines.

Similarly, in all structures the fluorocarbon layer thickness was also found to exhibit a linear dependence on the number of CF_2 groups in the donor.

A particular property of fluorocarbons, of great practical importance, is their low melting and boiling points as well as high volatility. This is especially so in the case of perfluoroalkanes. For example, tetrafluoro-1,2-diiodoethane **90** exhibits a melting point of −21 °C and a boiling point of 112–113 °C, while the octafluoro-1,4-diiodobutane **93** has a melting point of −9 °C and a boiling point of 150 °C. Dodecafluoro-1,6-diiodohexane **99** is barely solid at room temperature, with a melting point of 25–30 °C. Aakeröy *et al.* studied the co-crystallization of **90**, **93** and **99** with a number of halogen bond acceptors.[72] They showed that the co-crystallization conferred thermal stability, stability towards moisture and prolonged stability at ambient conditions. Moreover, the 1,ω-diiodoperfluoroalkanes could be easily isolated from the co-crystals by simple solvent extraction.

3.9 Nitroaryl Donors

In search for new halogen bond donors, Raatikainen and Rissanen probed the behaviour of 1-iodo-3,5-dinitrobenzene **47**, expecting that the two nitro-groups could significantly activate the iodine substituent and make **47** a fairly strong halogen bond donor.[73] Co-crystallization of **47** and 1,4-diazabicyclo[2.2.2]octane **24** yielded two different polymorphs of a 2 : 1 co-crystal, $\mathbf{47_2{\cdot}24}$**-I** and $\mathbf{47_2{\cdot}24}$**-II**. The structure of $\mathbf{47_2{\cdot}24}$**-I** revealed the expected formation of two C–I···N halogen bonds (Scheme 3.21; $d(\text{I}\cdots\text{N}) = 2.765(9)$ Å, $R_{XB} = 0.783$, $<(\text{C–I}\cdots\text{N}) = 176.32(19)°$ and $d(\text{I}\cdots\text{N}) = 2.734(9)$ Å, $R_{XB} = 0.775$, $<(\text{C–I}\cdots\text{N}) = 175.77(19)°$), resulting in a discrete halogen-bonded assembly $\mathbf{47_2{\cdot}24}$**-I**. The structure of $\mathbf{47_2{\cdot}24}$**-II**, on the other hand, showed the formation of one C–I···N ($d(\text{I}\cdots\text{N}) = 2.762(3)$ Å, $R_{XB} = 0.782$, $<(\text{C–I}\cdots\text{N}) = 174.39(12)°$) and one C–I···O halogen bond (Scheme 3.21; $d(\text{I}\cdots\text{O}) = 3.051(3)$ Å, $R_{XB} = 0.872$, $<(\text{C–I}\cdots\text{O}) = 166.80(13)°$). The two polymorphs were obtained from the same sample with $\mathbf{47_2{\cdot}24}$**-II** crystallizing first, followed by $\mathbf{47_2{\cdot}24}$**-I**, suggesting that $\mathbf{47_2{\cdot}24}$**-I** is the more stable polymorph. This is in good agreement with their structures, as the more stable $\mathbf{47_2{\cdot}24}$**-I** shows the presence of two strong C–I···N halogen bonds, whereas the less stable $\mathbf{47_2{\cdot}24}$**-II** only has one strong C–I···N and one weaker C–I···O halogen bond.

Co-crystallization of **47** with 4,4′-bipyridine **58** yielded a 1 : 1 co-crystal **47·58** (Scheme 3.21). The structure showed the molecule of **47** forming one (expected) C–I···N halogen bond ($d(\text{I}\cdots\text{N}) = 2.874(2)$ Å, $R_{XB} = 0.814$, $<(\text{C–I}\cdots\text{N}) = 172.02(6)°$), as well as one C–H···N ($d(\text{C}\cdots\text{N}) = 3.444(3)$ Å, $<(\text{C–H}\cdots\text{N}) = 173°$) and two C–H···O hydrogen bonds ($d(\text{C}\cdots\text{O}) = 3.193(3)$ and 3.303(2) Å, $<(\text{C–H}\cdots\text{O}) = 126°$ and 134°, respectively).

Matzger and co-workers prepared and investigated co-crystal formation with 1,3,5-triiodo-2,4,6-trinitrobenzene **101**,[74] another halogen bond donor activated by nitro-groups, but also an energetic material.[75] As a first step in the study, electrostatic potential calculations were conducted, showing the sigma holes on the iodine atoms of **101** were the largest of all the halogen

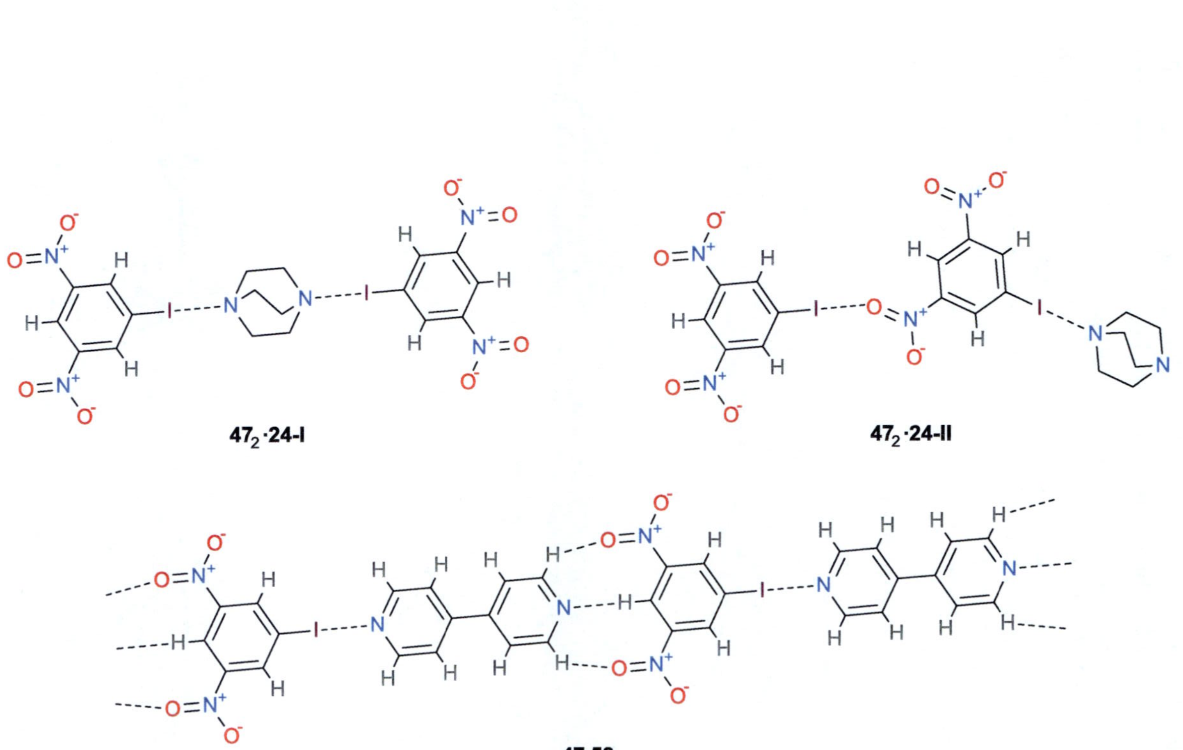

$\mathbf{47_2 \cdot 24}$**-I**

$\mathbf{47_2 \cdot 24}$**-II**

47·58

Scheme 3.21 Halogen-bonded 2 : 1 discrete assemblies in the two polymorphs of the co-crystal $\mathbf{47_2}$·**24** and the 1-D halogen and hydrogen-bonded chain in the co-crystal **47**·**58**.

bond donors investigated up to that point, surpassing *e.g.* *N*-iodosuccinimide **102** (Figure 3.7) and 1,3,5-trifluoro-2,4,6-triiodobenzene **89**. Next, **101** was successfully co-crystallized with a number of different halogen bond acceptors, *e.g.* 4,4′-bipyridine **58**, pyridine-*N*-oxide **85**, 1,4-dithiane **14** and even benzene **103**, to name but a few. The structures of these co-crystals revealed the presence of very strong halogen bonds between **101** and acceptor molecules (Table 3.3), but also the tendency towards narcissistic self-association of **101** through C–I···O(nitro) halogen bonds, forming chains (**101·58** and **101·14**, Figure 3.7) or tapes (**101·103**, Figure 3.7). The observed tendency of **101** to form C–I···O(nitro) halogen bonds was also used by Matzger and co-workers to co-crystallize **101** or its bromo analogue 1,3,5-tribromo-2,4,6-trinitrobenzene **104** with 1,3,5-trinitrobenzene **105** (Figure 3.7), yielding a pair of isostructural energetic co-crystals sustained by C–I···O(nitro) or C–Br···O(nitro) halogen bonds.[76]

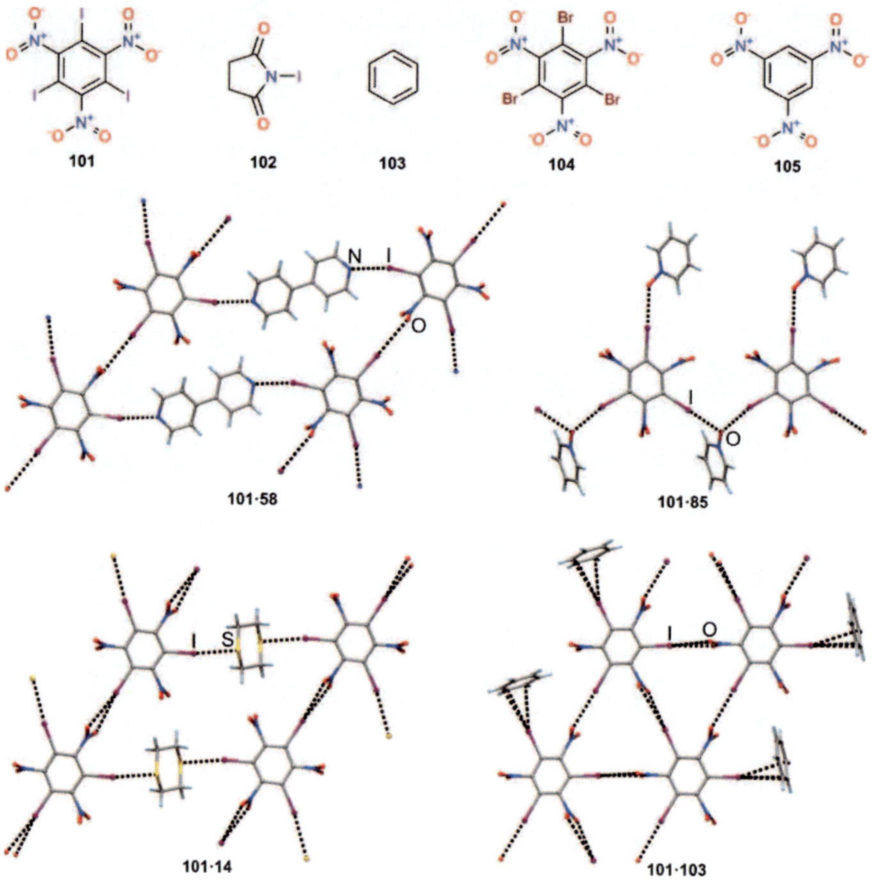

Figure 3.7 Details from the structures of the co-crystals of **101** with different acceptors. Halogen bonds are shown as black dashed lines.

Table 3.3 Halogen bond parameters in co-crystals of **101** with **58**, **85**, **14** and **103**.

	101·58 (X=N)	**101·85** (X=O)	**101·14** (X=S)	**101·103** (X=C)
d(I···X)/Å	2.705(3)	2.817(14)	3.1258(3)	3.369(13)
R_{XB}	0.766	0.805	0.827	0.915
<(C–I···X)/°	169.53(8)	177.5(6)	171.10(3)	179.6(3)
d(I···X)/Å	2.765(2)	2.776(15)	—	3.481(12)
R_{XB}	0.783	0.793	—	0.946
<(C–I···X)/°	166.16(7)	177.6(6)	—	156.1(4)
d(I···X)/Å	—	2.710(15)	—	3.477(13)
R_{XB}	—	0.774	—	0.945
<(C–I···X)/°	—	169.4(6)	—	157.4(3)

3.10 *N*-Haloimide Donors

In contrast to the wealth of data gathered on the halogen bonding involving the (pseudo)dihalogens and various halocarbons, halogen bond donors in which the halogen atoms are bound to nitrogen, such as *N*-haloimides, have only recently started to receive more attention. The first structural study of a halogen-bonded co-crystal with *N*-bromosuccinimide **106** was reported by Crowston *et al.*, who described $\mathbf{106_2{\cdot}24}$, a co-crystal of **106** with 1,4-diazabicyclo[2.2.2]octane **24**.[77] The crystal structure revealed discrete 2:1 assemblies sustained by N–Br···N halogen bonds (Scheme 3.22; Table 3.4). A striking feature is the R_{XB} lower than 0.7, which indicates a very strong halogen bond. This agrees well with another observation, namely that the N–Br bond in **106** (*d*(N–Br) = 1.945(4) Å) is significantly elongated compared to that observed in the crystal structure of pure **106** (*d*(N–Br) = 1.817(7) Å).[78] Still, the halogen bonding was neither widely recognized under that name nor systematically studied at the time.

The potential of the *N*-haloimides has thus remained largely unexploited until Raatikainen and Rissanen reported the crystal structures of the co-crystals of *N*-iodosuccinimide **102** and *N*-bromosuccinimide **106** with nitrogen-based acceptors such as hexamethylenetetramine **21**, 1,4-diazabicyclo[2.2.2]octane **24** and 1,3,5-triazine **107** (Scheme 3.22).[79] Interestingly, they described the structure of $\mathbf{106_2{\cdot}24}{\cdot}(CHCl_3)_2$, a chloroform solvate of $\mathbf{106_2{\cdot}24}$, whose halogen bond parameters are very close to those of the non-solvated $\mathbf{106_2{\cdot}24}$ (Table 3.4), suggesting that the solvation and different packing do not significantly affect the halogen bonds. Moreover, it also showed the significant elongation of N–Br bonds compared to the pure **106** (Table 3.4). This, however, is in contrast with the behaviour of **102**, which shows no significant elongation of the N–I bond in *e.g.* $\mathbf{102_2{\cdot}107}$ compared to the pure **102**. Still, it might be due to the presence of fairly short N–I···O halogen bonds in pure **102** (*d*(I···O) = 2.580(6) Å, R_{XB} = 0.737, <(N–I···O) = 175.71(19)°) making the N–I bond longer than expected already in the pure **102**. The structure of $\mathbf{102_4{\cdot}21}$ supports this idea, as the

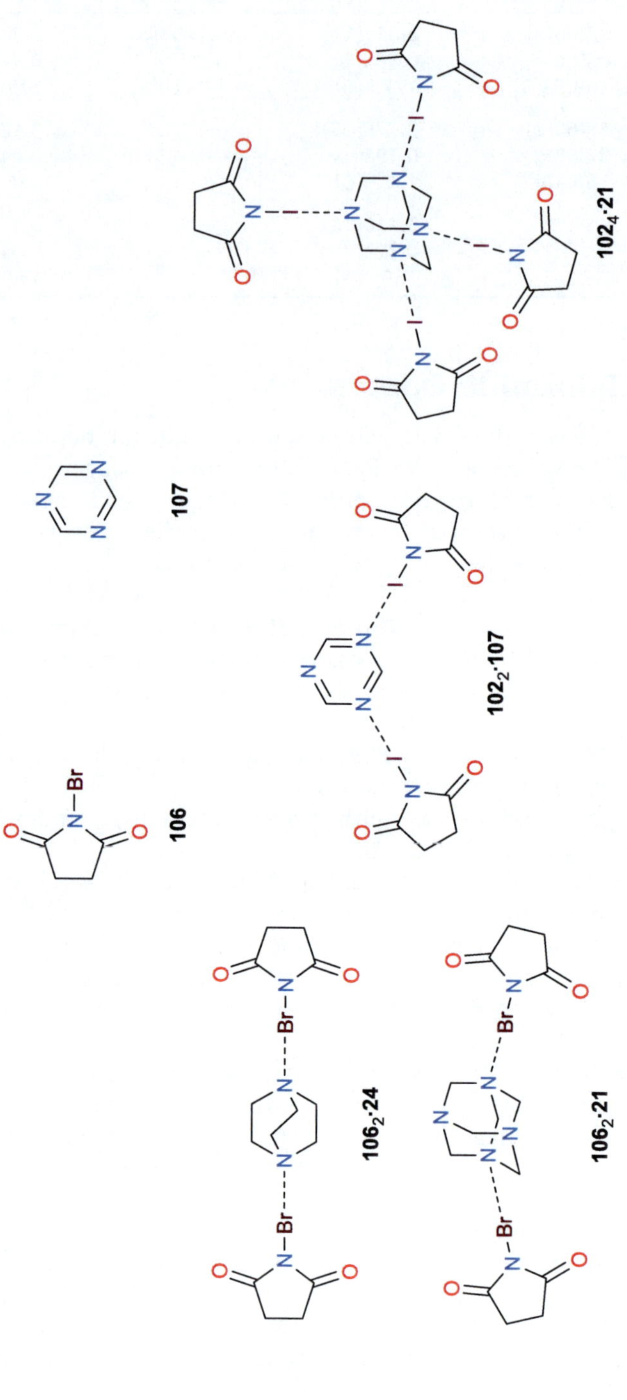

Scheme 3.22 Discrete halogen-bonded assemblies in the co-crystals of *N*-iodosuccinimide **102** and *N*-bromosuccinimide **106**.

Table 3.4 Halogen bond parameters in co-crystals **106**$_2$·**24**, **106**$_2$·**21**, **102**$_2$·**107** and **102**$_4$·**21**.

	106$_2$·**24** (X = Br)	**106**$_2$·**24**·$(CHCl_3)_2$ (X = Br)	**106**$_2$·**21** (X = Br)	**102**$_2$·**107** (X = I)	**102**$_4$·**21**·(guest) (X = I)
d(X···N)/Å	2.332(4)	2.3642(19)	2.433(3)	2.5689(18)	2.550(8)
R_{XB}	0.686	0.700	0.716	0.728	0.722
<(N–X···N)/°	175.1(2)	175.95(8)	172.04(12)	177.70(7)	175.2(3)
d(N–X)/Å	1.945(4)	1.9356(19)	1.917(3)	2.0791(18)	2.121(9)
	106 (X = Br)			**102** (X = I)	
d(X···N)/Å	—	2.3468(19)	2.414(3)	—	—
R_{XB}	—	0.690	0.710	—	—
<(N–X···N)/°	—	175.78(8)	173.63(12)	—	—
d(N–X)/Å	1.817(7)	1.9334(19)	1.925(3)	2.060(6)	—

presence of a better halogen bond acceptor **21** (compared to **107** or the carbonyl oxygen in **102**) makes the N–I bond somewhat (more) elongated. These structures showed that *N*-haloimides such as **102** and **106** can form very short ($R_{XB} \approx 0.7$) halogen bonds with different acceptors.

Interestingly, the crystal structure of **102**$_4$·**21** was shown to exhibit significant voids filled with disordered acetonitrile, which could not be modelled. Further experiments revealed that **102**$_4$·**21** could form three different frameworks, all consisting of **102**$_4$·**21** tetrahedra (Scheme 3.21), depending on the type of the trapped solvent (Figure 3.8). Altogether, six different structures were obtained with six different solvents, exhibiting three different frameworks. Pairs of structures **102**$_4$·**21**·$(CH_3NO_2)_4$ and **102**$_4$·**21** (acetonitrile solvate), **102**$_4$·**21**·$(CH_2Cl_2)_4$ and **102**$_4$·**21**·$(CCl_4)_2$, as well as **102**$_4$·**21**·$(CHCl_3)_2$ and **102**$_4$·**21**·$(C_6H_5CH_3)_2$ were found to exhibit the same framework structures (all three framework types are represented in Figure 3.8). Among these six structures, only minimal changes in the halogen bonding parameters can be observed, with d(I···N) varying between 2.486(3) and 2.586(3) Å and d(N–I) between 2.086(3) and 2.121(9) Å. Molecules of CH_2Cl_2 in **102**$_4$·**21**·$(CH_2Cl_2)_4$ could be exchanged for CCl_4, either by soaking in liquid CCl_4 or exposure to its vapours, to give **102**$_4$·**21**·$(CCl_4)_2$ in a single-crystal-to-single-crystal manner.

Stilinović *et al.* examined very thoroughly the halogen bonding between *N*-halosuccinimides (*N*-iodosuccinimide **102**, *N*-bromosuccinimide **106**, *N*-chlorosuccinimide **108**) and various 4-substituted pyridine acceptors (**7**, **12**, **17**, **45**, **110–112**) in solid state, solution and computationally (Scheme 3.23).[80] This choice of donors and acceptors also enabled the comparison between the hydrogen bonding with parent succinimide and the halogen bonding with *N*-halosuccinimides. The N–X···N (X = I/Br/Cl) halogen bonds in the studied systems were found to be almost entirely of electrostatic nature, with energies between −60 and −20 kJ mol^{-1}. Moreover, their energies and geometric parameters were shown to correlate well both with the Hammett constant for different substituents on the pyridine

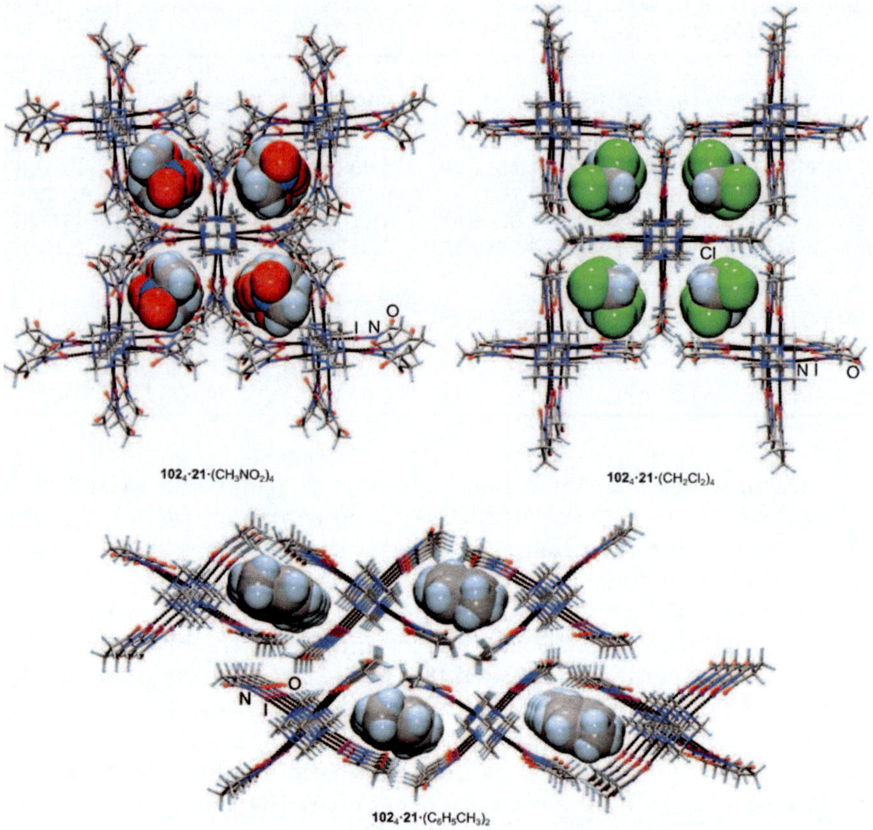

Figure 3.8 Packing in three different solvates of **102_4·21** with three different framework structures, namely **102_4·21**·$(CH_3NO_2)_4$, **102_4·21**·$(CH_2Cl_2)_4$ and **102_4·21**·$(C_6H_5CH_3)_2$. Halogen bonds are shown as black dashed lines.

ring (Scheme 3.23) and the calculated atomic charges on the pyridine nitrogen and the donor halogen. With **102** as the donor, co-crystals were successfully obtained with all the acceptors and the halogen bond lengths *d*(I···N) were found to be in the range from 2.403(5) to 2.500(2) Å, corresponding to R_{XB} values between 0.681 and 0.708. These halogen bonds were found to be significantly affected by their supramolecular environment, best illustrated by the example of **102·110**, where the two symmetrically independent complexes exhibited halogen bonds with lengths *d*(I···N) of 2.427(6) and 2.488(6), *i.e.* a difference of 0.061(6). The whole series of co-crystals was also successfully obtained with **106** as the donor, exhibiting halogen bonds with lengths *d*(Br···N) between 2.418(2) and 2.597(3) Å, *i.e.* with R_{XB} values between 0.711 and 0.764. Interestingly, the strength of the hydrogen bonding of pyridines **7**, **12**, **17**, **45**, **110–112** with **109** was found to be between that of the halogen bonding with **102** and **106**, based on the gas-phase computational and solution studies. On the other hand, the

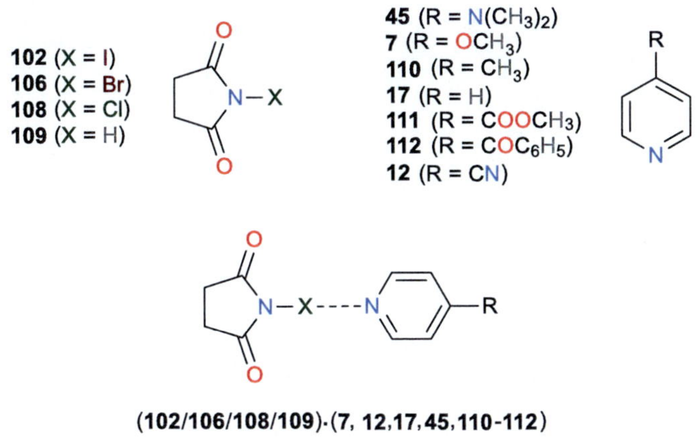

Scheme 3.23 Halogen- or hydrogen-bonded 1 : 1 discrete assemblies formed between the succinimides **102**, **106**, **108**, **109** and 4-substituted pyridines **7**, **12**, **17**, **45**, **110–112**.

hydrogen-bonded co-crystals with **109** were much more difficult to obtain in the solid state. Thus, while all the possible donor/acceptor combinations involving **102** and **106** yielded co-crystals, only two co-crystals were obtained with **109**. This was suggested to arise due to the differences between the hydrogen bonding of **109** and halogen bonding of **102** and **106**, such as the charge distribution around the donor atom and the possible formation of hydrogen-bonded dimers of **109**; similar halogen-bonded dimers of **102** or **106** cannot be formed due to steric reasons.

Dolenc and Modec examined the halogen bonding behaviour of another type of *N*-haloimides–*N*-halosaccharins, represented by *N*-iodosaccharin **113** and *N*-bromosaccharin **114**.[81] Starting with **113**, two co-crystals/solvates were formed respectively with water **115** and tetrahydrofuran **116**, both exhibiting fairly strong, albeit expected N–I···O halogen bonds with $R_{XB} \approx 0.7$ (Scheme 3.24, Table 3.5). Pyridine acceptors **17** and **29** respectively gave 1 : 1 and 2 : 1 co-crystals **113·17** and **113$_2$·29**, both of which exhibited expected N–I···N halogen bonds with R_{XB} values as low as 0.646 in the case of **113·17** (Scheme 3.24, Table 3.5). Remarkably, formal N–I covalent and (N–)I···N halogen bonds are almost indistinguishable in **113·17**, with d(N–I) = 2.254(11) Å and d(I···N) = 2.279(11) Å, respectively, due to elongation of the N–I covalent bond and the very short I···N halogen bond. Finally, co-crystallization of **114** and **29** successfully yielded **114$_2$·29**, with a fairly short N–Br···N halogen bond (R_{XB} = 0.709, Table 3.5).

Fourmigué and Espinosa undertook a comparative investigation of the halogen bonding in the co-crystals of *N*-iodosuccinimide **102** or *N*-iodosaccharin **113** with 4-dimethylaminopyridine **45**, 4-methylpyridine **110** or pyridine **17** (Scheme 3.25).[82] All combinations yielded 1 : 1 co-crystals with fairly short halogen bonds (d(I···N) = 2.218(2)–2.493(8) Å, R_{XB} = 0.628–0.706). However, in the case of **113·45** (structure determined at T = 150 K), the length of the halogen

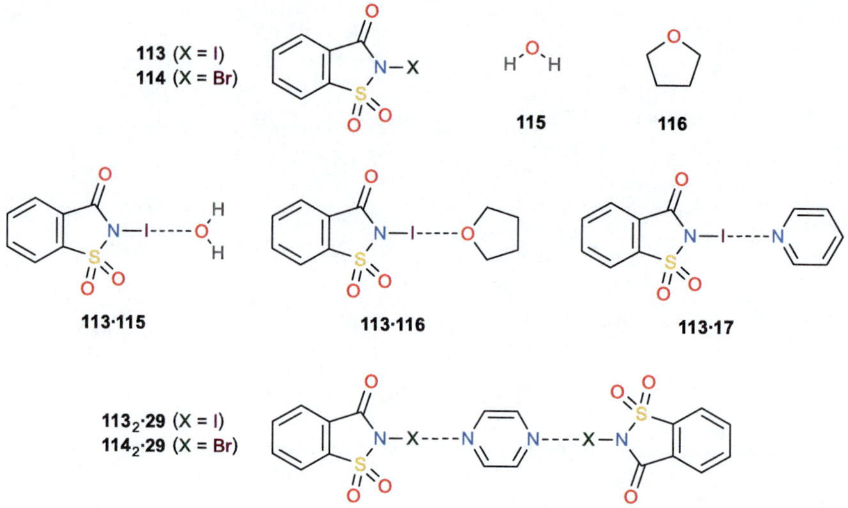

Scheme 3.24 Halogen-bonded discrete assemblies in the co-crystals of *N*-iodosaccharin **113** or *N*-bromosaccharin **114** with water **115**, tetrahydrofurane **116**, pyridine **17** or pyrazine **29**.

Table 3.5 Halogen bond parameters in co-crystals **113·115**, **113·116**, **113·17**, **113_2·29** and **114_2·29**.

	113·115 (X = I, Y = O)	**113·116** (X = I, Y = O)	**113·17** (X = I, Y = N)	**113_2·29** (X = I, Y = N)	**114_2·29** (X = Br, Y = N)
d(X···Y)/Å	2.443(2)	2.512(2)	2.279(11)	2.423(2)	2.410(1)
R_{XB}	0.698	0.718	0.646	0.686	0.709
∠(N–X···Y)/°	177.68(7)	178.51(9)	174.5(4)	175.26(6)	175.22(6)
d(N–X)/Å	2.096(2)	2.073(2)	2.254(11)	2.142(2)	1.906(1)

bond (*d*(I···N) = 2.218(2) Å, R_{XB} = 0.628) was found to be less than the length of the corresponding covalent bond (*d*(N–I) = 2.292(2) Å). This could be interpreted as effectively resulting from a transfer of iodine atom from **113** to **45**, formally giving a salt, analogously to proton transfer (Scheme 3.25). Computational studies showed that the supramolecular environment in the crystal is crucial in stabilizing the structure where the iodine is shifted towards the acceptor along the N–I···N halogen bond.

Puttreddy *et al.* examined the halogen bonding of *N*-iodosaccharin **113** or *N*-iodosuccinimide **102** with *N*-oxides **85, 117, 118** and **119** in solid state and in solution. Two series of 1 : 1 co-crystals were successfully obtained with either **113** or **102** and **85, 117** and **118**, all sustained by very short N–I···O halogen bonds (Scheme 3.26; *d*(I···N) = 2.316(3)–2.328(8) Å, R_{XB} = 0.662–0.665 for the co-crystals with **113**; *d*(I···N) = 2.383(2)–2.453(2) Å, R_{XB} = 0.681–0.701 for the co-crystals with **102**). No co-crystals were obtained with the electron-poor acceptor

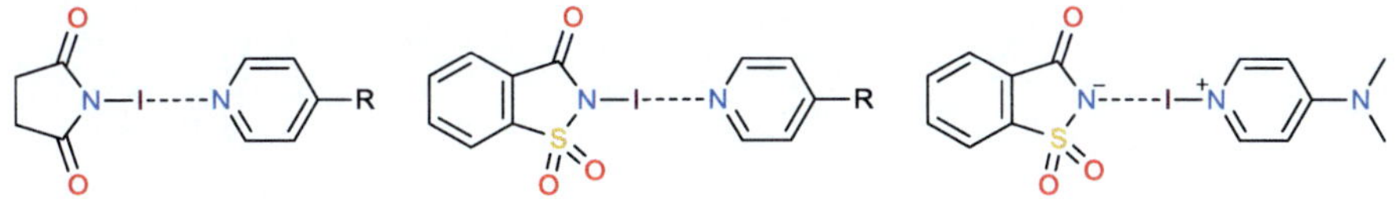

102·45 (R = N(CH3)2)
102·110 (R = CH3)
102·17 (R = H)

113·110 (R = CH3)
113·17 (R = H)

113·45

Scheme 3.25 Halogen- or hydrogen-bonded 1 : 1 discrete assemblies formed between *N*-iodosuccinimide **102** or *N*-iodosaccharin **113** and 4-substituted pyridines **17, 45** or **110.**

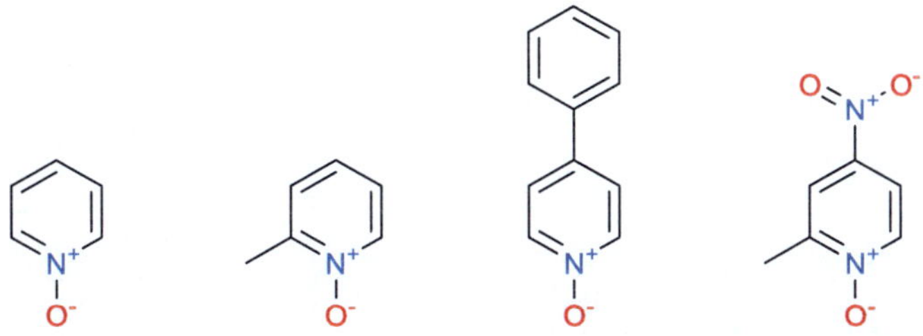

85 **117** **118** **119**

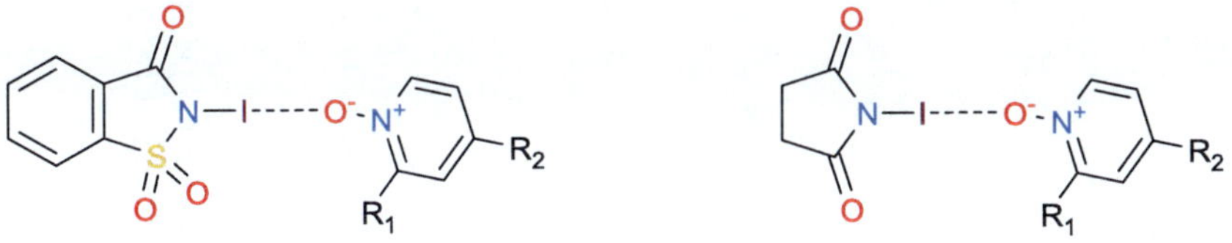

113·85 ($R_1 = R_2 = H$)
113·117 ($R_1 = CH_3$, $R_2 = H$)
113·118 ($R_1 = H$, $R_2 = C_6H_5$)

102·85 ($R_1 = R_2 = H$)
102·117 ($R_1 = CH_3$, $R_2 = H$)
102·118 ($R_1 = H$, $R_2 = C_6H_5$)

Scheme 3.26 Halogen-bonded 1 : 1 discrete assemblies in the co-crystals of *N*-iodosaccharin **113** or *N*-iodosuccinimide **102** with *N*-oxides **85, 117** or **118.**

119. Remarkably, the 1 : 1 complexation was also found to persist in solution, with association constants as high as ~16 000 M^{-1} in case of **113·117** in chloroform, roughly 20 times that of **102·117.** This nicely illustrates how the strength of an interaction cannot be fully appreciated solely by examining the geometrical parameters extracted from the crystal structures.

A combined *N*-haloimide and *N*-haloamide ditopic donor *N,N′*-diiododimethylhydantoin **120** (Scheme 3.27) was used by Fourmigué and co-workers to address the question of (de)activation of polytopic halogen bond donors upon engaging in halogen bonding, which often causes them to remain unsaturated.[83] Computation of the electrostatic potential surface for **120** revealed the amide-bound iodine atom to have a slightly

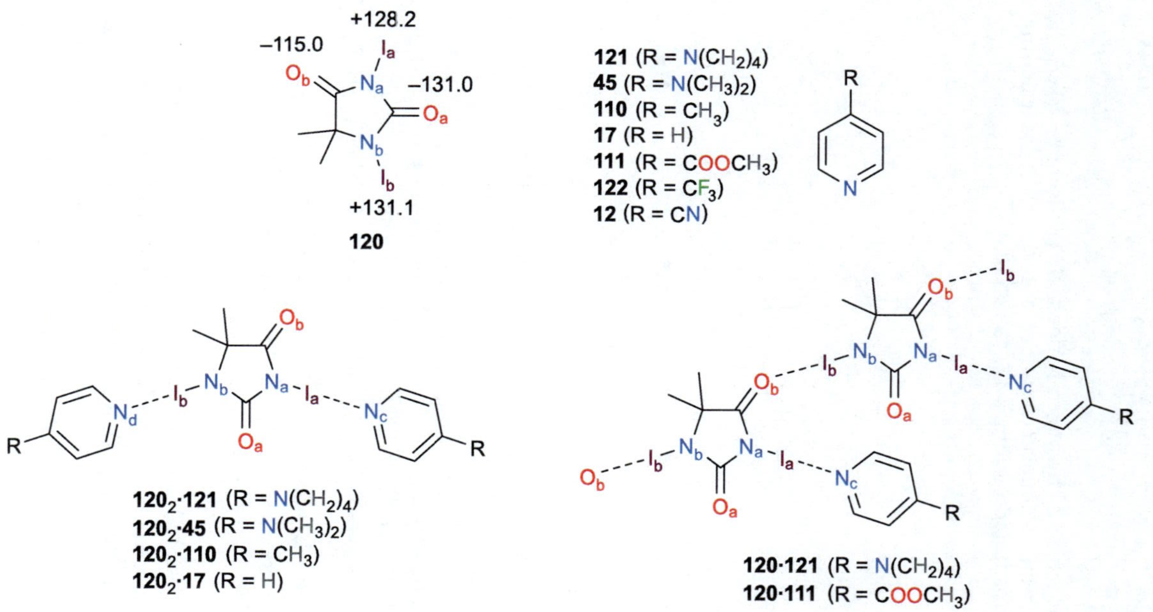

Scheme 3.27 Calculated electrostatic potential values (in kJ mol^{-1}) at the iodine and oxygen atoms in *N*,*N*′-diiododimethylhydantoin **120**. Halogen-bonded 2 : 1 and 1 : 1 discrete assemblies formed between the **120** and 4-substituted pyridines **17, 45, 110, 111** and **121**.

higher positive potential (131.1 kJ mol^{-1}) than the imide-bound iodine (128.2 kJ mol^{-1}). While seemingly counterintuitive, it was proposed to be a consequence of the molecular dipole moment of **120**, roughly parallel but opposite in orientation to the N_a–I_a bond (Scheme 3.27). Nevertheless, crystallization attempts with different pyridines yielded a number of 2 : 1 co-crystals (**120$_2$·121**, **120$_2$·45**, **120$_2$·110**, **120$_2$·17**) and four 1 : 1 co-crystals (**120·121**, **120·111**, **120·122**, **120·12**). In 2 : 1 co-crystals, both iodine atoms of **121** are engaged in halogen bonding with pyridine (Scheme 3.27), with $I_a \cdots N_c$ halogen bonds ($d(I \cdots N) = 2.349(6)$–$2.457(5)$ Å, $R_{XB} = 0.665$–0.696) shorter than the $I_b \cdots N_d$ halogen bonds ($d(I \cdots N) = 2.438(12)$–$2.532(10)$ Å, $R_{XB} = 0.691$–0.717). On the other hand, only two structures of the 1 : 1 co-crystals were successfully elucidated (**120·121**, **120·111**; Scheme 3.27), both exhibiting $I_a \cdots N_c$ halogen bonds between the imide iodine and the pyridine ($d(I \cdots N) = 2.294(4)$ and $2.398(6)$ Å, $R_{XB} = 0.650$ and 0.679) and the $I_b \cdots O_b$ halogen bonds between the amide iodine and one of the oxygen atoms ($d(I \cdots O) = 2.692(3)$ and $2.720(4)$ Å, $R_{XB} = 0.769$ and 0.777). Comparing the two co-crystals of **120** and **121** with different stoichiometries showed that the $I_a \cdots N_c$ halogen bond is significantly shorter in **120·121** (2.294(4) Å) than in **120$_2$·121** (2.392(10) and 2.385(12) Å), suggesting that the first halogen bond is preferably formed with the imide iodine I_a, that its formation significantly deactivates the amide iodine site and that, in the 1 : 1 complex, the oxygen O_b can successfully compete with the pyridine nitrogen. Moreover, the formation of the second halogen bond at the I_b site weakens the first one at the I_a site, as witnessed by the elongation of the $I_a \cdots N_c$ halogen bond in the 2 : 1 co-crystal compared to the 1 : 1 co-crystal. All these conclusions were also in good agreement with the computational studies of the 1:1 and 2:1 assemblies of **120** with the same pyridines.

3.11 Structural Equivalence of Donors and Acceptors

One obvious, attractive aspect of the halogen bonding is the potential for facile tuning of its strength by substitution of *e.g.* iodine with bromine or vice versa. Such a study was reported by Cavallo, Terraneo and co-workers, who examined a family of isostructural co-crystals comprising hexamethylphosphoramide **94** and four different tetrafluoro-1,2-dihaloethanes, namely the 1,2-diiodo **90**, 1-bromo-2-iodo- **123**, 1,2-dibromo- **124** and 1-chloro-2-iodo- **125** (Scheme 3.28).[84] Given the fundamental similarity of these molecules, which only differ in the halogen bond donor atom(s), the isostructurality of **90·94**, **123·94**, **124·94** and **125·94** is not entirely surprising. Moreover, it provided an opportunity to compare the effect of donor atom(s) on the strengths of C–X···O (X = I, Br, Cl) halogen bonds and the resulting bulk properties of the co-crystals. Expectedly, the melting points were found to decrease from **90·94** (90 °C) with two C–I···O halogen bonds, **123·94** (28 °C) with one C–I···O and one C–Br···O halogen bond, through **124·94** (14 °C) with two C–Br···O halogen bonds to **125·94** (−12 °C) with one C–I···O and one C–Cl···O halogen bond. The IR spectroscopy of the co-crystals

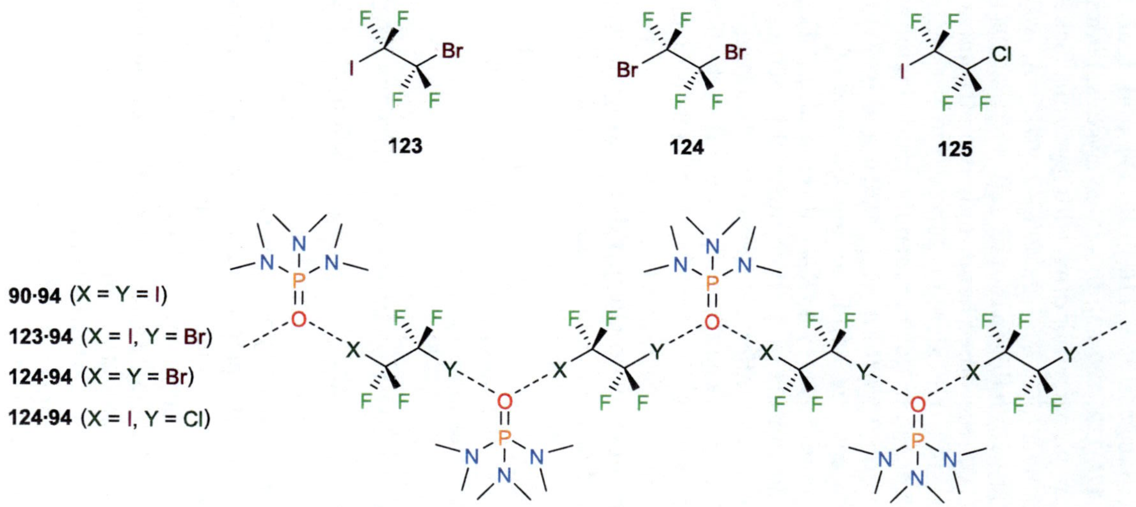

Scheme 3.28 Halogen-bonded chains in **90·94**, **123·94**, **124·94** and **125·94**.

showed the expected red shift of the P=O stretching band: from 1202 cm^{-1} in liquid **94** to 1192 cm^{-1} in **90·94**, 1197 cm^{-1} in **123·94** and **125·94** and 1199 cm^{-1} in **124·94**.

Cinčić *et al.* examined two pairs of co-crystals of phenazine **13** and acridine **126** with tetrafluoro-1,4-diiodobenzene **73** and 1,4-dibromotetrafluorobenzene **83**. The co-crystals **73·13**$_2$ (d(I···N) = 2.970(3) Å, $R_{XB} = 0.841$, $<$(C–I···N) = 175.45(11)°) and **83·13**$_2$ (d(Br···N) = 2.985(2) Å, $R_{XB} = 0.878$, $<$(C–Br···N) = 172.59(8)°) could be obtained by solution or solid-state grinding co-crystallization and were indeed found to be isostructural, in both cases forming 1 : 2 assemblies through C–I···N or C–Br···N halogen bonds, respectively. However, with **13** only acting as a single halogen bond acceptor, the nitrogen atom not involved in halogen bonding was found to form a C–H···N hydrogen bond to another molecule of **13**, with two such hydrogen bonds connecting the **73·13**$_2$/**83·13**$_2$ into a 1-D chain (Scheme 3.29). Using **126** instead of **13** again yielded two co-crystals **73·126**$_2$ (d(I···N) = 2.971(2) Å, $R_{XB} = 0.842$, $<$(C–I···N) = 176.55(8)°) and **83·126**$_2$ (d(Br···N) = 3.075(7) Å, $R_{XB} = 0.904$, $<$(C–Br···N) = 168.85(15)° and d(Br···N) = 2.986(6) Å, $R_{XB} = 0.878$, $<$(C–Br···N) = 175.40(15)°), different from **73·13**$_2$/**83·13**$_2$ but mutually isostructural and consisting of similar 1 : 2 discrete assemblies (Scheme 3.29). Interestingly, for both co-crystal pairs, a lowering of the melting point of *ca.* 70 °C was observed upon changing from diiodo-donor **73** to dibromo-donor **83**. This suggests that the C–I/Br···N halogen bond is the dominant interaction driving the assembly of these co-crystals.

Interestingly, when monotopic donors pentafluoroiodobenzene **48** and bromopentafluorobenzene **127** were used with the same acceptors **13** and **126**, a set of four isostructural co-crystals was obtained (Scheme 3.29).[85] In this case, the structural equivalence of I/Br was combined with that of the N/CH fragments, giving rise to the observed isostructurality.

Next, co-crystallization of donors **73** and **83** with morpholine **38**, piperazine **128**, thiomorpholine **129** and thioxane **130** produced seven isostructural co-crystals—all combinations except that of **83** and **130** successfully yielded co-crystals (Scheme 3.30).[86] In these co-crystals, structural equivalence is established not only between bromine and iodine as the donor atoms, but also between the three different acceptor groups imino NH, ether O and thioether S, in various combinations in the ditopic donors **38, 128–130**. Remarkably, within this set of isostructural co-crystals, the melting points span a range from 55 to 195 °C, as a result of six different types of halogen bonds formed in these systems (C–I···N, C–I···O, C–I···S, C–Br···N, C–Br···O, C–Br···S) and different combinations in which they appear (*e.g.* two C–I···N halogen bonds in **73·128**; one C–I···N and one C–I···S in **73·129**, *etc.*).

Interestingly, two very similar acceptors, 1,4-dioxane **4** and 1,4-dithiane **14** yielded co-crystals with **73**, which were not isostructural with the other ones. In the case of **73·4**, the most obvious difference is that of the halogen bond donor approaching the acceptor group in the equatorial direction, unlike the axial approach in all the other co-crystals. In the co-crystal **73**$_2$**·14**, thioether

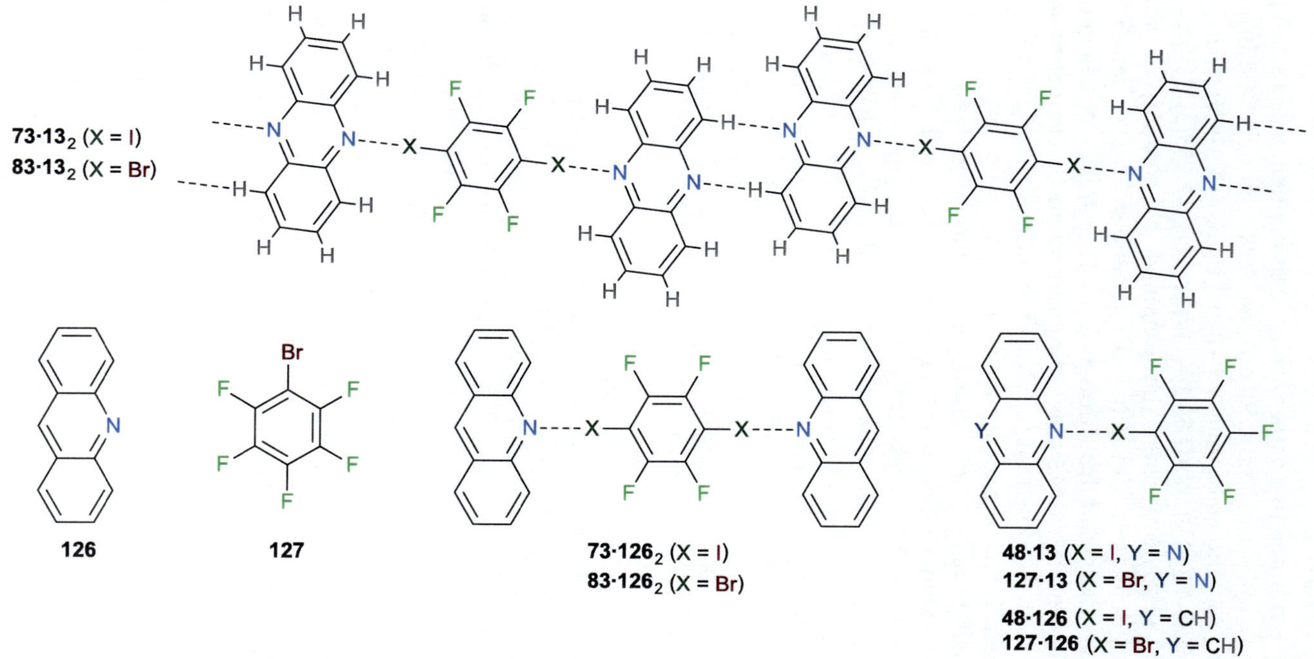

Scheme 3.29 Halogen- and hydrogen-bonded 1-D chains in the co-crystals **73·13$_2$** and **83·13$_2$** and the discrete halogen-bonded assemblies in the co-crystals **73·126$_2$**, **83·126$_2$**, **48·13**, **127·13**, **48·126** and **127·126**.

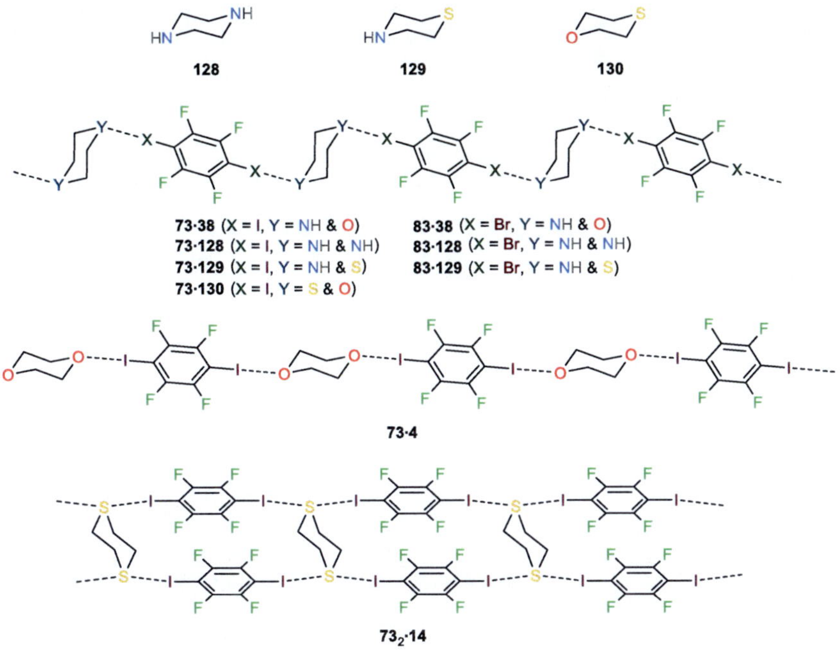

Scheme 3.30 Halogen-bonded 1-D chains in the isostructural co-crystals **73·38**, **73·128**, **73·129**, **73·130**, **83·38**, **83·128** and **83·129** as well as the **73·4** and **73$_2$·14**.

groups act as bifurcated acceptors, forming two halogen bonds with **73** each and giving rise to a completely different 2 : 1 stoichiometry.[87]

3.12 Halogen Bonding Hierarchy

While the isostructurality of co-crystals described above might leave an impression of the absolute structural equivalence of the used donor (Br/I) and acceptor (O, S, NH) moieties, this would not be true, as found by Cinčić *et al.* in their attempts to gain some mechanistic insight into the co-crystallization of **73** and **129** by grinding.[88] While grinding for half an hour yielded the previously described 1 : 1 co-crystal **73·129**, grinding for 4 min yielded a mixture of **73·129** and the 1 : 2 co-crystal **73·129$_2$**, suggesting the latter is an intermediate in the formation of **73·129**. The **73·129$_2$** can be seen as the kinetic product, whereas the **73·129** is the thermodynamic product. Interestingly, the crystal structure of **73·129$_2$** reveals the formation of discrete assemblies through C–I···N (d(I···N) = 2.8595(17) Å, R_{XB} = 0.810, $<$(C–I···N) = 176.34(6)°) halogen bonds only. In the next step, **73·129$_2$** further reacts with an equivalent of **73**, with the discrete assemblies polymerizing through C–I···S halogen bonds to form 1-D halogen-bonded chains of **73·129** (Scheme 3.31; d(I···N) = 2.989(8) Å, R_{XB} = 0.847, $<$(C–I···N) = 175.09(17)° and d(I···S) = 3.171(3) Å, R_{XB} = 0.839, $<$(C–I···S) = 175.49(9)°).

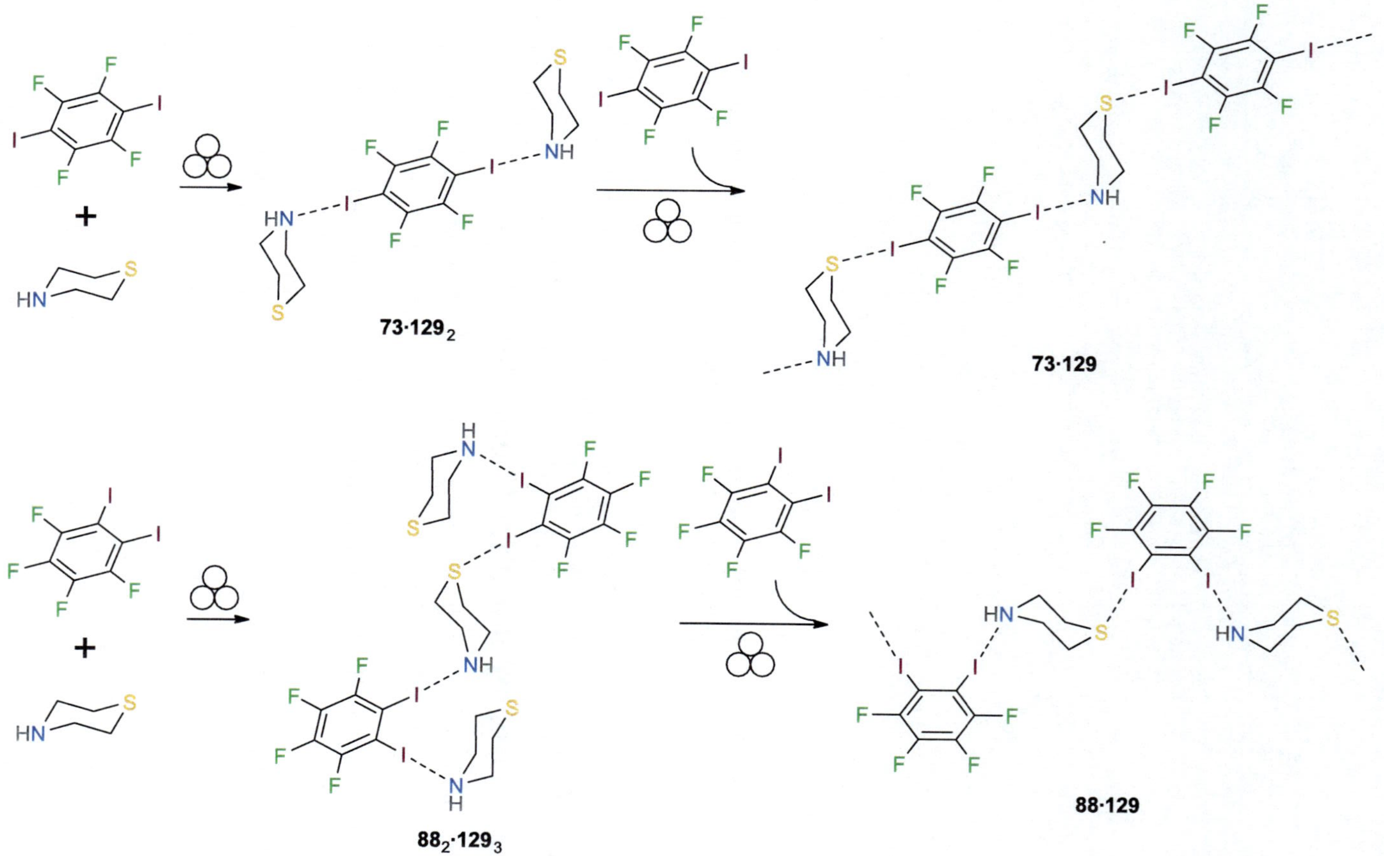

Scheme 3.31 Step-wise formation of halogen-bonded co-crystals **73·129** and **88·129** through respective intermediates **73·129**$_2$ and **88**$_2$·**129**$_3$.

Similarly, the mechanochemical co-crystallization of tetrafluoro-1,2-diiodobenzene **88** and thiomorpholine **129** was also found to proceed through an intermediate, $\mathbf{88_2{\cdot}129_3}$, comprising discrete motifs assembled by three C–I···N and one C–I···S halogen bond each. This intermediate phase then further reacts with another equivalent of **88** to yield **88·129**, which consists of 1-D chains sustained by the same number of C–I···N and C–I···S halogen bonds.

Another example of the competition of different halogen bonding motifs was presented by Russo *et al.*[89] Solution co-crystallization of **73** and **111** in 1 : 2 ratio yielded the co-crystal $\mathbf{73{\cdot}111_2}$ (Scheme 3.32), comprising discrete trimeric assemblies sustained by C–I···N halogen bonds with the pyridine nitrogen (d(I···N) = 2.882(2) Å, R_{XB} = 0.816, <(C–I···N) = 175.03(8)°). On the other hand, changing the ratio of **73** and **111** in the starting solution to 1 : 1 yielded upon evaporation the co-crystal **73·111** (Scheme 3.32). The crystal structure of **73·111** revealed the presence of similar trimeric assemblies sustained by C–I···N halogen bonds (d(I···N) = 2.919(3) Å, R_{XB} = 0.827, <(C–I···N) = 173.81(10)°), which are further linked into 1-D chains by an additional molecule of **73** through C–I···O halogen bonds (d(I···O) = 3.045(3) Å, R_{XB} = 0.870, <(C–I···O) = 165.61(10)°). Thus the presumably stronger C–I···N halogen bonds are preferably formed, while the weaker C–I···O halogen bonds appear only in the presence of a sufficient amount of the donor **73**.

A broad, systematic study of halogen bond selectivity was undertaken by Aakeröy *et al.*[66] Twelve different halogen bond acceptors were selected, each having two distinct types of nitrogen-based acceptor groups (Scheme 3.33). Electrostatic potentials were then calculated to internally rank the acceptor sites within each acceptor and the difference between the electrostatic potentials of the two sites was defined as ΔE. Next, co-crystallization was attempted between these acceptors and nine different iodine-based halogen bond donors **48, 73, 88–90, 93, 98, 99** and **142**, yielding altogether 89 co-crystals out of 108 possible combinations, of which 35 crystal structures were obtained. Another three structures of co-crystals reported by others using the same components were also included in the analysis. All structures were examined for the formation of halogen bonds, specifically whether only the best acceptor site or both acceptor sites were involved in halogen bonding. Based on the crystal structure analysis, the acceptors were then divided into three groups: those with $\Delta E < 35$ kJ mol^{-1}, which showed no selectivity, *i.e.* where no preference could be established for either acceptor site; those with 35 kJ mol^{-1} $< \Delta E <$ 75 kJ mol^{-1}, where no predictions could be made about the presence or absence of selectivity; and those with $\Delta E < 75$ kJ mol^{-1}, where the best acceptor site could be expected to preferably form a halogen bond.

3.13 Hydrogen and Halogen Bonding

Questions about the coexistence of hydrogen and halogen bonding, *i.e.* their cooperativity or competition arose quickly into the renaissance of halogen

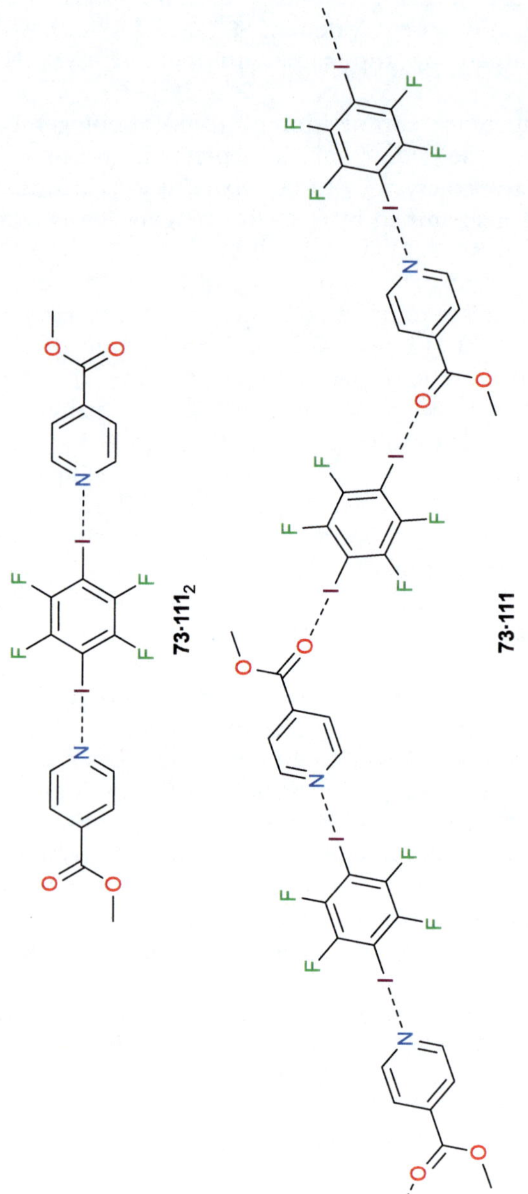

$73{\cdot}111_2$

73·111

Scheme 3.32 Halogen bond hierarchy giving rise to two co-crystals of **73** and **111** with different stoichiometries.

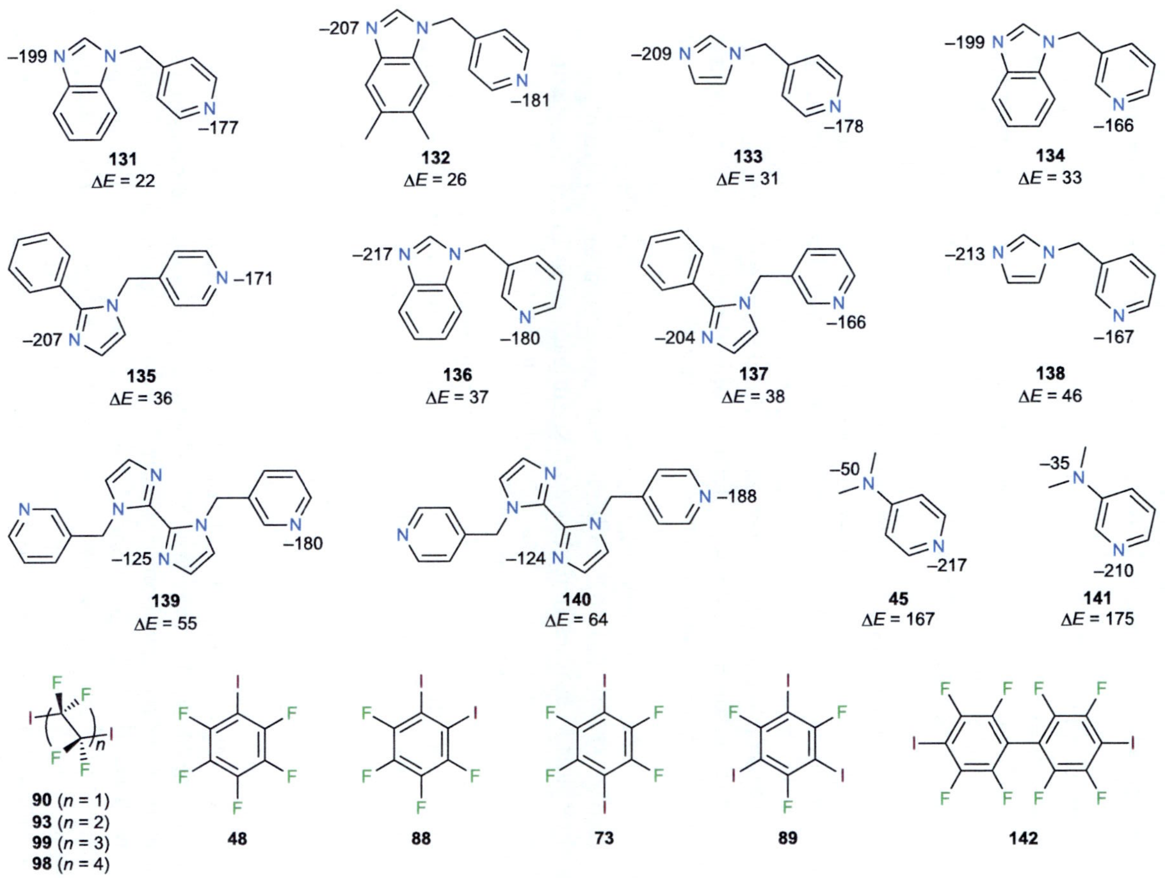

Scheme 3.33 Calculated electrostatic potential values (in kJ mol^{-1}) at the nitrogen acceptor atoms in ditopic halogen bond acceptors **45**, **131–141**, and the difference between the electrostatic potentials of two sites for each acceptor. Halogen bond donors employed in the study: **48**, **73**, **88–90**, **93**, **98**, **99**, **142**.

bonding in the early 2000s. One of the first attempts to clarify the situation was made by Metrangolo, Resnati and co-workers who compared the co-crystallization of 1,2-bis(4-pyridyl)ethane **64** with tetrafluoro-1,4-diiodo-benzene **73** and hydroquinone **143**.[90] Co-crystals were successfully obtained in both cases, with both **73·64** and **143·64** respectively forming halogen or hydrogen-bonded 1-D chains (Scheme 3.34). Interestingly, when the co-crystallization was attempted from an acetone solution containing **64**, **73** and **143** in a 1 : 1 : 1 ratio, the only solid product was **73·64**, with **143** remaining in solution.

Robertson *et al.* expanded on this work by conducting similar experiments with a range of solvents, and with two additional hydrogen bond donors fluorohydroquinone **144** and tetrafluorohydroquinone **145**.[91] Using **73** as the halogen bond donor with **143**, **144** or **145** as the hydrogen bond donor and **64** as the acceptor, the outcome of competitive co-crystallization experiments was shown to depend both on the solvent polarity and the hydrogen bond donor strength, which increases in the series **143** < **144** < **145** due to fluorination. Solvents of higher polarity were shown to preferably yield the halogen-bonded co-crystal **73·64** (Table 3.6). On the other hand, increasing the hydrogen bond donor strength resulted in the crystallization of hydrogen-bonded co-crystals from increasingly polar solvents. Thus, for example the exclusive formation of hydrogen-bonded co-crystal **143·64** was only observed with toluene, whereas **145·64** crystallized exclusively from all solvents except the most polar isopropyl alcohol (Table 3.6). These observations were, to a certain extent, supported by solution studies that showed a significant drop in the association constants between monotopic pyridine and phenol analogues of **64** and **143–145** upon increase in the solvent polarity.

In contrast to the competition for the formation between two co-crystals, a hydrogen and a halogen-bonded one, as described above, the relationship between the two interactions could also be studied within one co-crystal. Thus, Aakeröy *et al.* described two co-crystals of isonicotinamide **146**, with tetrafluoro-1,4-diiodobenzene **73** (**73·146$_2$**) and iodine **10** (**10·146**).[92] The hydrogen-bonded dimers are (expectedly) formed by **146** through the amide–amide homosynthon, with the pyridine group engaged in I···N halogen bonding with either **73** (d(I···N) = 2.858(3)/2.859(3) Å, R_{XB} = 0.810, <(C–I···N) = 177.75(10)/179.04(12)°) or **10** (d(I···N) = 2.442(4) Å, R_{XB} = 0.692, <(C–I···N) = 178.07(9)°). In case of **73·146$_2$**, hydrogen-bonded dimers of **146** are connected by halogen bonding with **73** into 1-D chains. On the other hand, in **10·146** the hydrogen-bonded dimers of **146** are decorated with two molecules of **10**, again halogen-bonded, and the resulting **10$_2$·146$_2$** discrete assemblies engage in short I···I type I interactions (d(I···I) = 3.6314(7) Å, R_{XB} = 0.917, <(I–I···I) = 170.00(2)°) to form 1-D chains (Scheme 3.35). This work showed how the hydrogen and halogen bonds could be rationally combined within the same co-crystal with a predictable outcome and no interference between the hydrogen and halogen bonding synthons.

A possibly even more impressive example of non-interference between the hydrogen and halogen-bonded motifs was also presented by Aakeröy *et al.*

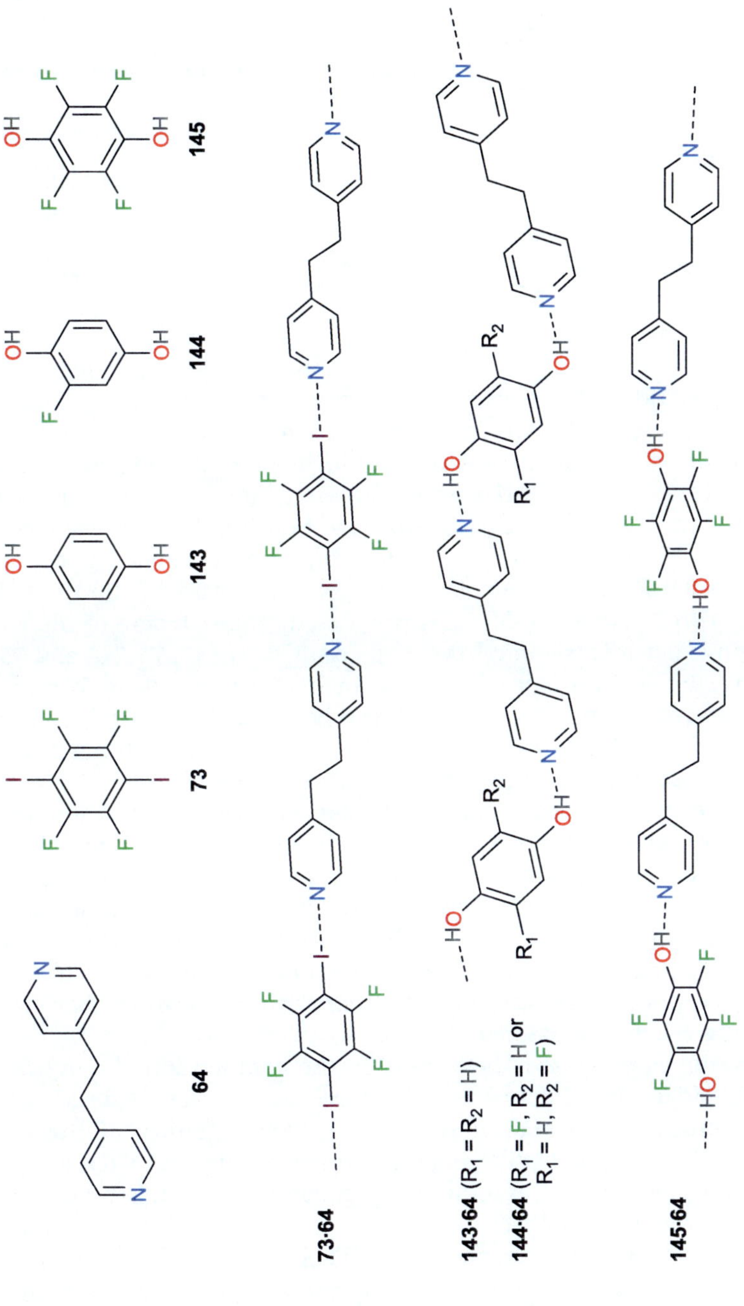

Scheme 3.34 Halogen- or hydrogen-bonded 1-D chains in the co-crystals of **73**, **143**, **144** or **145** with **64**.

Table 3.6 Outcomes of the competitive co-crystallization experiments involving **73**, **143–145** and **64** (HB and XB respectively indicate the formation of hydrogen or halogen-bonded co-crystals).

Solvent	Polarity parameter $E_T(30)$	**73** + **143** + **64**	**73** + **144** + **64**	**73** + **145** + **64**
Toluene	33.9	HB	HB	HB
Chloroform	39.1	HB/XB	HB	HB
Dichloromethane	40.7	HB/XB	HB/XB	HB
Acetone	42.2	XB	XB	HB
Acetonitrile	45.6	XB	XB	HB
Nitromethane	46.3	XB	XB	HB
Isopropyl alcohol	48.4	XB	XB	HB/XB

using **73** as the halogen bond donor and three different acceptors 2-aminopyrazine **147**, 2-amino-5-bromopyrazine **148** and 2-amino-3,5-dibromopyrazine **149**.[93] Here, the acceptor molecules possess two pyridine-type nitrogen atoms, both of which could in principle engage in halogen bonding. However, the *ortho*-amino group enables one of the pyrazine nitrogen atoms to also engage in hydrogen bonding to form a self-complementary hydrogen-bonded dimer like the amide–amide motif in the co-crystals of **146**, leaving the other nitrogen atom available for the halogen bonding with **73**. Remarkably, this was exactly the outcome of the co-crystallization experiments—in all three co-crystals, **73**·**147**, **73**·**148** and **73**·**149**, dimers sustained by a pair of N–H···N hydrogen bonds were formed, connected into 1-D chains by C–I···N halogen bonds (Scheme 3.36). Clearly, the self-complementarity of the 2-aminopyridine moieties in **147–149** directs the halogen bond donor **73** to the other pyridine moiety.

Similar results were also reported by Aakeröy *et al.* with a combination of two heterosynthons, a hydrogen- and a halogen-bonded one, where the carboxylic acid moiety of **150** or **151** formed two hydrogen bonds with the 2-aminopyridine moiety of **147–149**, while the bromine or iodine atoms of **150** or **151** engaged in the halogen bonding with the other pyridine moiety of **147–149** (Scheme 3.36).[94] Interestingly, the observed supramolecular motifs persisted even in **150**·**147** and **151**·**147** where a proton transfer from **150**/**151** to **147** gave rise to salts (Scheme 3.36).

The examples discussed above concern the competition of one hydrogen and one halogen bond donor moiety where the outcome is decided by the complementarity of two discrete binding sites to either of the donors. However, when no such difference between the binding sites exists, and especially in the presence of multiple hydrogen or halogen bond donors, the outcome of the co-crystallization becomes increasingly unpredictable. Such was the case in the work reported by Aakeröy *et al.* where 1-methyl-2-(4-pyridyl)benzimidazole **152** was co-crystallized with pentafluorobenzaldehyde oxime **153**, 4-bromotetrafluorobenzaldehyde oxime **154** and tetrafluoro-4-iodobenzaldehyde oxime **155** to yield three co-crystals **153**·**152**, **154**·**152** and

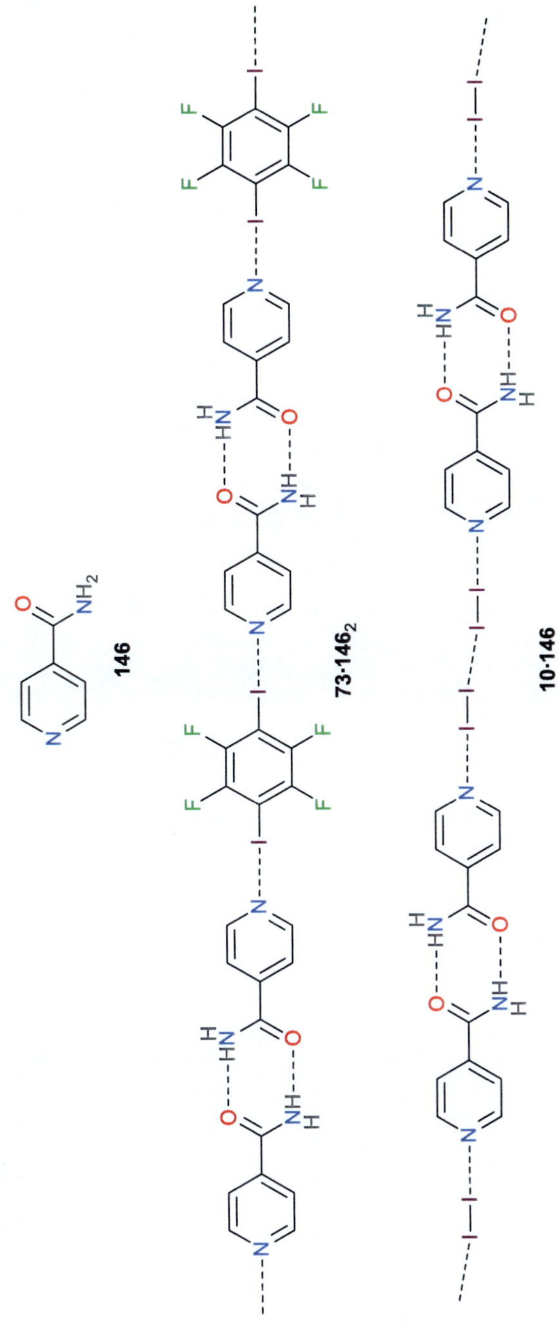

Scheme 3.35 Halogen and hydrogen-bonded 1-D chains in the co-crystals **$73{\cdot}146_2$** and **10·146**.

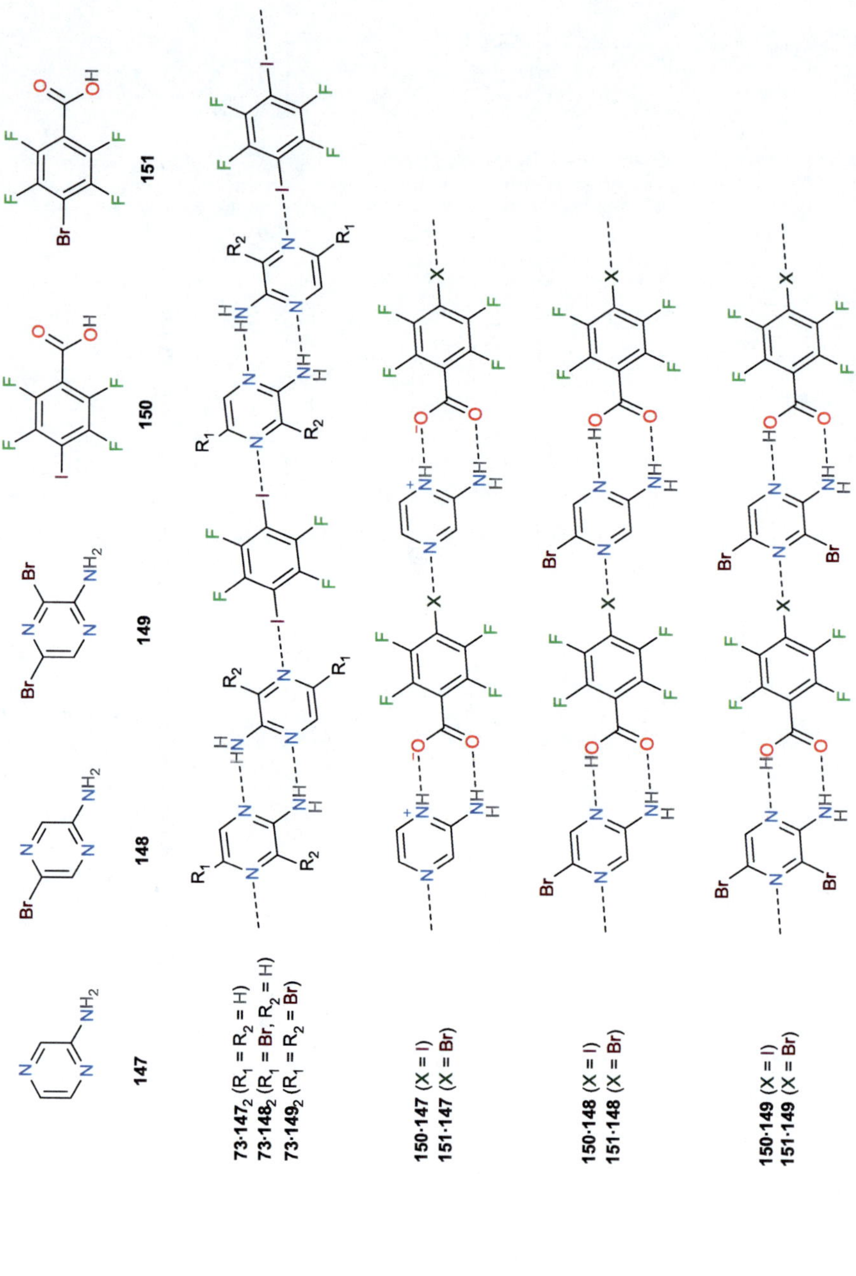

Scheme 3.36 Halogen- and hydrogen-bonded 1-D chains in the co-crystals of **73** with **147–149** and **150** or **151** with **147–149**.

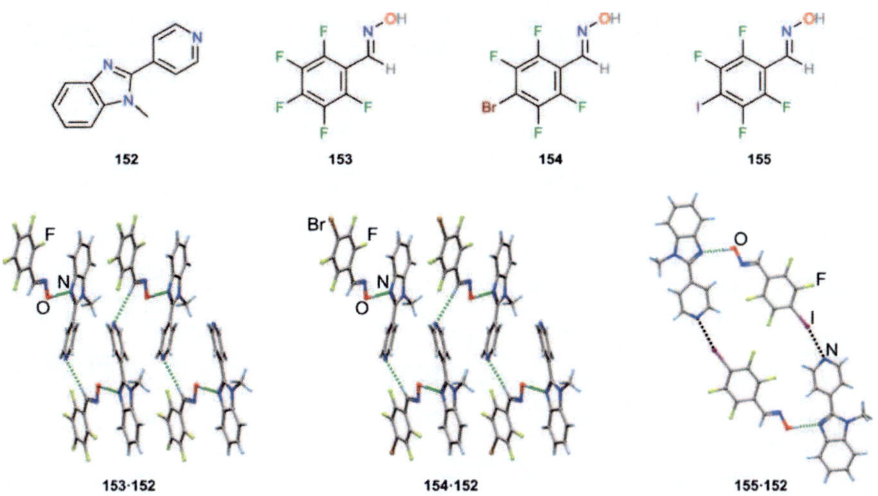

Figure 3.9 Hydrogen-bonded zig-zag chains in the co-crystals **153·152** and **154·152** and the hydrogen- and halogen-bonded discrete assemblies in the co-crystals **155·152**. Halogen and hydrogen bonds are shown as black and green dashed lines, respectively.

155·152 (Figure 3.9).[95] In all three co-crystals, hydrogen bond was formed between the oxime group as the strongest donor and the benzimidazole nitrogen as the strongest acceptor. However, the expected C–X···N halogen bond was only formed in **155·152** (X = I) with pyridine nitrogen atom as the acceptor. In both **153·152** and **154·152** the pyridine nitrogen is involved in C–H···N hydrogen bonds with the aldoxime group instead. While the formation of a C–F···N halogen bond in **153·152** would have been unexpected, the fact that the C–H···N hydrogen bond successfully outcompetes (putative) C–Br···N halogen bonds in **154·152** is somewhat surprising.

In some cases, however, the competition between the hydrogen and halogen bonds can be largely avoided by using motifs, which are essentially orthogonal. This was demonstrated by Pennington and co-workers, who described a number of halogen-bonded co-crystals between various thioureas such as thiourea **156**, methimazole **157**, benzimidazole-2-thione **9** and imidazolidine-2-thione **18** (Figure 3.10).[96,97] All of these co-crystals exhibited various motifs involving N–H···S hydrogen bonding and C–I···S halogen bonding. In **73·156** the molecules of **156** form hydrogen-bonded tapes (d(N–H···S) = 3.427(7)–3.489(7) Å), which are connected by molecules of **73** through C–I···S halogen bonds (Figure 3.10; d(I···S) = 3.2812(19)–3.4036(19) Å, R_{XB} = 0.868–0.900, <(C–I···S) = 171.0(2)–179.6(2)°, <(I···S–C) = 84.6(2)–91.3(2)°). Similarly, in **88·9** the hydrogen-bonded tapes (d(N–H···S) = 3.317(7) and 3.320(8) Å) are decorated by molecules of **88** that form one halogen bond each (Figure 3.10; d(I···S) = 3.309(2) Å, R_{XB} = 0.875, <(C–I···S) = 173.5(2)°, <(I···S–C) = 119.3(3)°). Unlike in **73·156**, in **88·9** the donor **88** only forms one C–I···S halogen bond, with the other iodine not involved in any significant interactions. Moreover, the sulphur

156 157 9 18

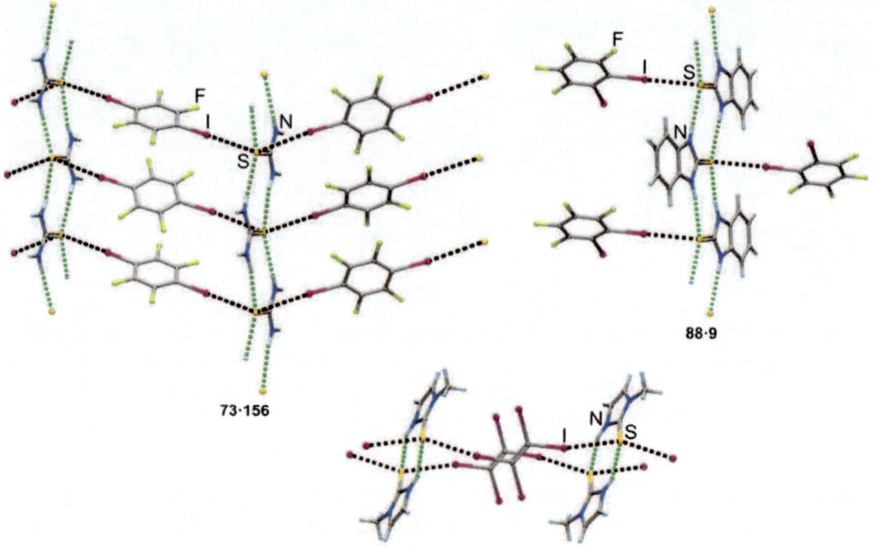

Figure 3.10 Combination of the hydrogen and halogen bonding motifs in the co-crystals **73·156**, **28·157** and **88·9**. Halogen and hydrogen bonds are shown as black and green dashed lines, respectively.

atom of **9** accordingly only forms one C–I···S halogen bond. In the case of **28·157**, hydrogen-bonded dimers of **157** (d(N–H···S) = 3.306(7) Å) are connected by pairs of molecules of **28** through C–I···S halogen bonds (Figure 3.10; d(I···S) = 3.246(3) and 3.302(2) Å, R_{XB} = 0.859 and 0.874, <(C–I···S) = 171.4(2) and 166.9(3)°, <(I···S–C) = 97.0(3) and 99.0(3)°). Similar orthogonality of halogen and hydrogen bonding was described by Vasylyeva *et al.* in the co-crystals of *N*-methylacetamide with different isomers of dibromotetrafluorobenzene and tetrafluorodiiodobenzene.[98]

The orthogonality of halogen and hydrogen bonding observed in these co-crystals was similarly exploited by Topić and Rissanen.[99] A range of halogen bond donors comprising octafluoro-1,4-diiodobutane **93**, tetrafluoro-1,4-diiodobenzene **73**, 1,4-dibromotetrafluorobenzene **83**, pentafluoroiodobenzene **48**, tetrafluoro-1,2-diiodobenzene **88** was combined with thiourea **156** and *N*-methylthiourea **158** as hydrogen bond donors and halogen bond acceptors. However, instead of relying on the self-complementary hydrogen bonding of **156** and **158**, additional hydrogen bond acceptors were introduced, namely 18-crown-6 **159** and benzo-18-crown-6 **160**. The two crown ethers were used for their ability to form a number of hydrogen bonds at the same time, while

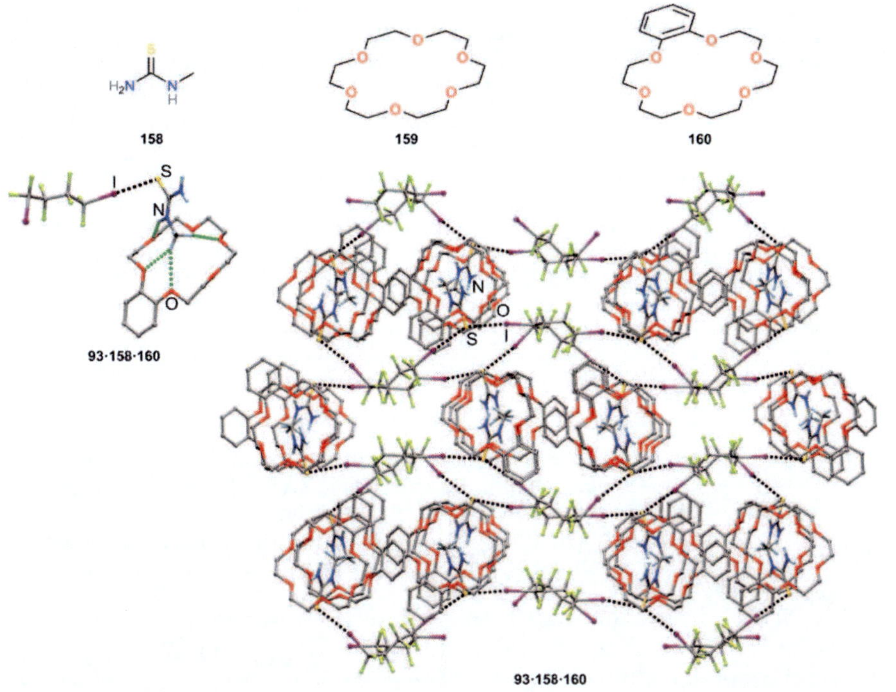

Figure 3.11 Hydrogen and halogen bonding in the co-crystal **93·158·160** (left) and the packing of the same co-crystal (right). Halogen and hydrogen bonds are shown as black and green dashed lines, respectively. Hydrogen atoms of **160** are omitted for clarity. Hydrogen bonds are omitted in the packing of **93·158·160**.

being only weak halogen bond acceptors, unable to compete with thioureas **156** and **158**. This strategy thus describes the use of orthogonal halogen and hydrogen bonding in construction of ternary co-crystals such as **93·158·160** (Figure 3.11).

In some cases, however, the competition between halogen bonding and hydrogen bonding or weaker interactions can emerge unexpectedly and with surprising results. Such was the case in the work of Friščić and co-workers, who studied mechanochemical co-crystallization based on C–I···O halogen bonding.[100] They reported two co-crystals of **73** with methyldiphenylphosphine oxide **161**, exhibiting 1 : 1 (**73·161**) and 1 : 2 (**73·161**$_2$) stoichiometric ratios (Figure 3.12). Contrary to the expectation, in the 1 : 1 co-crystal **73·161**, **161** was found to act as a monodentate acceptor only, forming one C–I···O (d(I···O) = 2.768(2) Å, R_{XB} = 0.791, <(C–I···O) = 171.9(1)°) halogen bond. The other iodine atom of **73** was, on the other hand, found to participate in a less orthodox C–I···π halogen bond[101–104] with the phenyl ring of **161**, with the iodine atom being closest to a C–C bond centroid (3.40 Å, Figure 3.12a). Even more curious was the case of the co-crystal **73·161**$_2$, where the only interaction between **73** and **161** was a C–I···π halogen bond (3.41 Å,

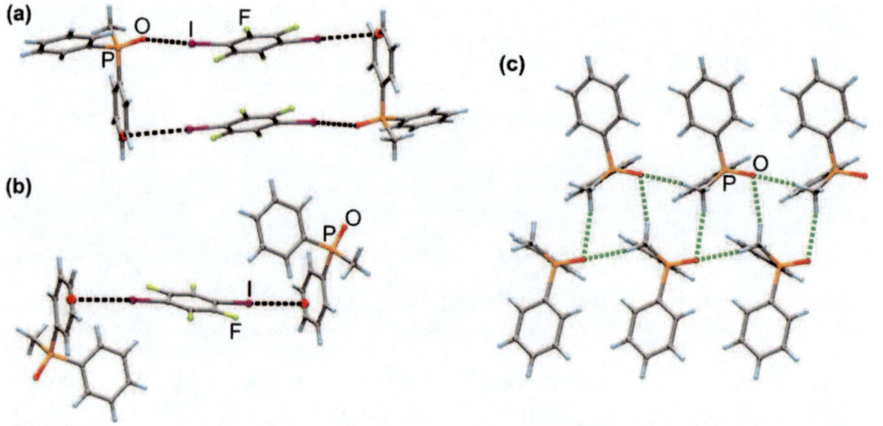

Figure 3.12 (a) Detail from the crystal structure of **73·161**, showing the C–I···O and the C–I···π halogen bonds. (b) Details from the crystal structure of **73·161**$_2$, showing the C–I···π halogen bonds and the (c) C–H···O hydrogen-bonded tapes. Halogen and hydrogen bonds are shown as black and green dashed lines, respectively.

Figure 3.12b), with the structure primarily sustained by the C–H···O hydrogen bonds instead (Figure 3.12c). Interestingly, a bromo-analogue of **73**, 1,4-dibromotetrafluorobenzene **83**, yielded a 1 : 2 co-crystal **83·161**$_2$ isostructural to **73·161**$_2$. On the other hand, no analogue of the 1 : 1 co-crystal could be obtained, suggesting that the putative, weaker C–Br···O interaction cannot overcome the C–H···O hydrogen bonds in **83·161**$_2$. In these co-crystals, the C–I···O halogen bonds are weak enough to be (partially) displaced by non-traditional C–I···π halogen bonds and weak C–H···O hydrogen bonds.

Similarly, using the triphenylphosphine oxide **162** instead of **161** with **73** yielded a 1 : 1 co-crystal **73·162**.[87] The structural analysis revealed an unexpected motif, described by Cinčić *et al.* as a wheel-and-axle motif **73·162**$_2$ held by C–I···O halogen bonding (Scheme 3.37, d(C–I···O) = 2.752(2) Å, R_{XB} = 0.786, <(C–I···O) = 174.6(1)°). While these discrete moieties seem to be linked through C–I···I contacts (Scheme 3.37, d(C–I···I) = 4.0136(4) Å, R_{XB} = 1.014, <(C–I···I) = 156.72(9)°) with another molecule of **73**, these contacts are too long to be classified as halogen bonds. Again, in this case, the C–I···O halogen bonds are overcome by weaker interactions that drive the co-crystal packing. A 2 : 1 co-crystal **73**$_2$**·162**, later obtained and analysed by Bryce and co-workers, further supports this view.[105]

Finally, Kivala, Grimme and co-workers showed that the C–I···O(=P) halogen bond does not have to be a dominant force in the co-crystal formation.[106] Despite being the shortest halogen bond of this type with distance d(C–I···O) = 2.683(5) Å (R_{XB} = 0.767), in the structure of **73·163**$_2$ (Figure 3.13) the halogen bond was only found to, in terms of interaction energy, play second fiddle to the dominant stacking homo- or hetero-interactions between **73** and **163**.

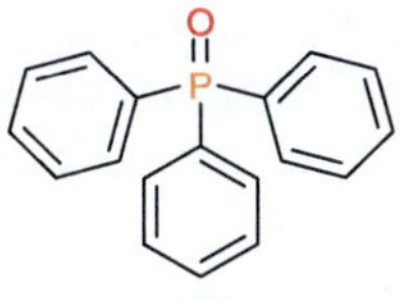

162

73·162

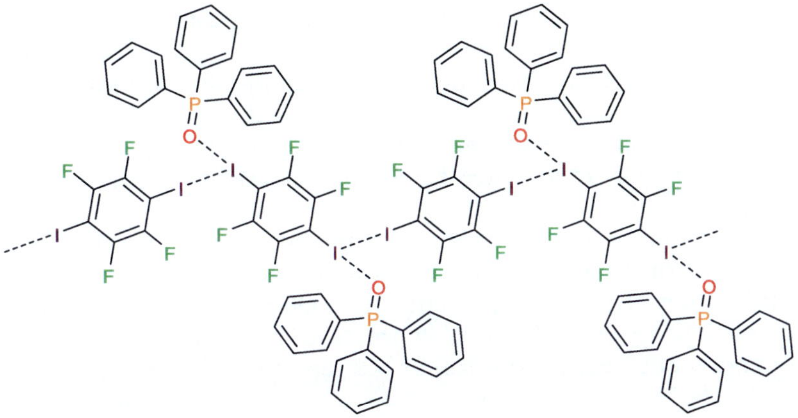

Scheme 3.37 Assembly of the discrete halogen-bonded assemblies in **73·162**.

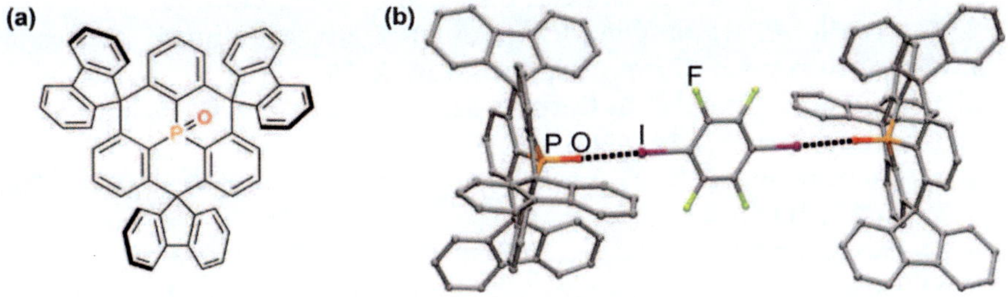

Figure 3.13 (a) Structure of the phosphine oxide halogen bond acceptor **163**. (b) C–I···O halogen-bonded wheel-and-axle assembly in the crystal structure of **73·163**$_2$. Halogen bonds are shown as black dashed lines. Hydrogen atoms are omitted for clarity.

3.14 Conclusions and Outlook

Halogen bonding research has greatly matured in the last two decades, owing to the large amount of work in the field of crystal engineering, *viz.* in the research of co-crystals. Various means of polarizing the halogen bond donor atoms, as covered in this chapter, and the different halogen bond acceptors, have largely debuted as co-crystal formers. This has facilitated not only in-depth probing of their properties through the crystalline solid state, but also high-precision structural studies thanks to X-ray crystallography. In particular, co-crystals have been used to study the unique features of halogen bonding, such as its high directionality, higher than that of the hydrogen bonding, availability of a range of possible donor atoms (iodine, bromine, chlorine, even fluorine) and the possibility of tuning the halogen bonds through the isostructural switching of donor and acceptor atoms. Moreover, the applicability of the halogen bonds, either as a sole interaction or concerted with hydrogen bonds, as well as their very close structural relationship with hydrogen bonds, makes crystal engineering an interesting and very topical research field.

Halogen-bonding interactions have been proved to be designable and to produce well-defined new molecular or solid-state materials. Combining orthogonal interactions such as halogen and hydrogen bonds, enables the construction of 'higher-order' co-crystals, *viz.* ternary, quaternary, *etc.*, allowing much more predictable and straightforward approaches to these intriguing and hard-to-access materials. The ability to master the halogen bonding is a landmark accomplishment that has shed light on the still unexplored nature of halogen bonding. On the other hand, gaining the control over the concerted action of halogen bonding with other weak interactions in the self-assembly processes marks a major future direction for crystal engineering and solid-state supramolecular chemistry.

References

1. L. C. Gilday, S. W. Robinson, T. A. Barendt, M. J. Langton, B. R. Mullaney and P. D. Beer, *Chem. Rev.*, 2015, **115**, 7118.
2. G. Cavallo, P. Metrangolo, R. Milani, T. Pilati, A. Priimagi, G. Resnati and G. Terraneo, *Chem. Rev.*, 2016, **116**, 2478.
3. V. Amico, S. V. Meille, E. Corradi, M. T. Messina and G. Resnati, *J. Am. Chem. Soc.*, 1998, **120**, 8261.
4. P. Metrangolo, H. Neukirch, T. Pilati and G. Resnati, *Acc. Chem. Res.*, 2005, **38**, 386.
5. O. Hassel and J. Hvoslef, *Acta Chem. Scand.*, 1954, **8**, 873.
6. O. Hassel and K. O. Strømme, *Acta Chem. Scand.*, 1958, **12**, 1146.
7. O. Hassel and K. O. Strømme, *Nature*, 1958, **182**, 1155.
8. O. Hassel and K. O. Strømme, *Acta Chem. Scand.*, 1959, **13**, 1775.
9. O. Hassel and H. Hope, *Acta Chem. Scand.*, 1961, **15**, 407.
10. O. Hassel, *Science*, 1970, **170**, 497.
11. G. R. Desiraju, P. S. Ho, L. Kloo, A. C. Legon, R. Marquardt, P. Metrangolo, P. Politzer, G. Resnati and K. Rissanen, *Pure Appl. Chem.*, 2013, **85**, 1711.
12. T. Clark, M. Hennemann, J. S. Murray and P. Politzer, *J. Mol. Model.*, 2007, **13**, 291.
13. P. Politzer, J. S. Murray and T. Clark, *Phys. Chem. Chem. Phys.*, 2010, **12**, 7748.
14. P. Metrangolo, J. S. Murray, T. Pilati, P. Politzer, G. Resnati and G. Terraneo, *CrystEngComm*, 2011, **13**, 6593.
15. P. Metrangolo, J. S. Murray, T. Pilati, P. Politzer, G. Resnati and G. Terraneo, *Cryst. Growth Des.*, 2011, **11**, 4238.

16. P. Politzer, P. Lane, M. C. Concha, Y. Ma and J. S. Murray, *J. Mol. Model.*, 2007, **13**, 305.
17. A. Mukherjee, S. Tothadi and G. R. Desiraju, *Acc. Chem. Res.*, 2014, **47**, 2514.
18. P. Metrangolo and G. Resnati, *IUCrJ*, 2014, **1**, 5.
19. J. P. M. Lommerse, A. J. Stone, R. Taylor and F. H. Allen, *J. Am. Chem. Soc.*, 1996, **118**, 3108.
20. M. Colin, *Ann. Chim.*, 1814, **91**, 252.
21. H. Hartl, H. Bärnighausen and J. Jander, *Z. Anorg. Allg. Chem.*, 1968, **357**, 225.
22. W. T. Pennington, T. W. Hanks and H. D. Arman, in *Halogen Bonding: Fundamentals and Applications*, ed. P. Metrangolo and G. Resnati, Springer Berlin Heidelberg, Berlin, Heidelberg, 2008, 65–104.
23. M. Haukka, P. Hirva and K. Rissanen, in *Non-covalent Interactions in the Synthesis and Design of New Compounds*, ed. A. M. Maharramov, K. T. Mahmudov, M. N. Kopylovich and A. J. L. Pombeiro, John Wiley & Sons, Inc, 2016, 185–197.
24. R. H. Jones, K. S. Knight, W. G. Marshall, S. J. Coles, P. N. Horton and M. B. Pitak, *CrystEngComm*, 2013, **15**, 8572.
25. V. Nemec, K. Lisac, V. Stilinović and D. Cinčić, *J. Mol. Struct.*, 2017, **1128**, 400.
26. L. Koskinen, S. Jääskeläinen, P. Hirva and M. Haukka, *Cryst. Growth Des.*, 2015, **15**, 1160.
27. H. Bock and S. Holl, *Z. Naturforsch., B: J. Chem. Sci.*, 2001, **56**, 111.
28. R. B. Walsh, C. W. Padgett, P. Metrangolo, G. Resnati, T. W. Hanks and W. T. Pennington, *Cryst. Growth Des.*, 2001, **1**, 165.
29. V. Daga, S. Hadjikakou, N. Hadjiliadis, M. Kubicki, J. Z. dos Santos and I. Butler, *Eur. J. Inorg. Chem.*, 2002, **2002**, 1718.
30. G. Y. Chao and J. D. McCullough, *Acta Crystallogr.*, 1960, **13**, 727.
31. G. Y. Chao and J. D. McCullough, *Acta Crystallogr.*, 1961, **14**, 940.
32. S. M. Godfrey, D. G. Kelly, C. A. McAuliffe, A. G. Mackie, R. G. Pritchard and S. M. Watson, *J. Chem. Soc., Chem. Commun.*, 1991, 1163.
33. F. A. Cotton and P. A. Kibala, *J. Am. Chem. Soc.*, 1987, **109**, 3308.
34. Y. Kim, E. J. Mckinley, K. E. Christensen, N. H. Rees and A. L. Thompson, *Cryst. Growth Des.*, 2014, **14**, 6294.
35. L. Koskinen, P. Hirva, E. Kalenius, S. Jaaskelainen, K. Rissanen and M. Haukka, *CrystEngComm*, 2015, **17**, 1231.
36. T. Dahl and O. Hassel, *Acta Chem. Scand.*, 1970, **24**, 377.
37. T. Dahl and O. Hassel, *Acta Chem. Scand.*, 1971, **25**, 2168.
38. D. S. Reddy, D. C. Craig, A. D. Rae and G. R. Desiraju, *J. Chem. Soc., Chem. Commun.*, 1993, 1737.
39. D. S. Reddy, D. C. Craig and G. R. Desiraju, *J. Chem. Soc., Chem. Commun.*, 1994, 1457.
40. S. C. Blackstock, J. P. Lorand and J. K. Kochi, *J. Org. Chem.*, 1987, **52**, 1451.
41. S. C. Blackstock and J. K. Kochi, *J. Am. Chem. Soc.*, 1987, **109**, 2484.

42. T. Dahl and O. Hassel, *Acta Chem. Scand.*, 1968, **22**, 2851.
43. J. A. Webb, P. Liu, O. L. Malkina and N. S. Goroff, *Angew. Chem., Int. Ed.*, 2002, **41**, 3011.
44. P. Liu, L. Li, J. A. Webb, Y. Zhang and N. S. Goroff, *Org. Lett.*, 2004, **6**, 2081.
45. J. Zou, Y. Jiang, M. Guo, G. Hu, B. Zhang, H. Liu and Q. Yu, *Chem. - Eur. J.*, 2005, **11**, 740.
46. C. Perkins, S. Libri, H. Adams and L. Brammer, *CrystEngComm*, 2012, **14**, 3033.
47. A. Sun, J. W. Lauher and N. S. Goroff, *Science*, 2006, **312**, 1030.
48. P. Gagnaux and B. P. Susz, *Helv. Chim. Acta*, 1960, **43**, 948.
49. A. S. Batsanov and J. A. K. Howard, *Acta Crystallogr.*, 2000, **C56**, 252.
50. L. González, N. Gimeno, R. M. Tejedor, V. Polo, M. B. Ros, S. Uriel and J. L. Serrano, *Chem. Mater.*, 2013, **25**, 4503.
51. E. Bosch, *Cryst. Growth Des.*, 2014, **14**, 126.
52. C. B. Aakeröy, T. K. Wijethunga, J. Desper and M. Đaković, *Cryst. Growth Des.*, 2015, **15**, 3853.
53. C. B. Aakeröy, A. M. Beatty and B. A. Helfrich, *J. Am. Chem. Soc.*, 2002, **124**, 14425.
54. C. B. Aakeröy, D. Welideniya and J. Desper, *CrystEngComm*, 2017, **19**, 11.
55. F. H. Allen, B. S. Goud, V. J. Hoy, J. A. K. Howard and G. R. Desiraju, *J. Chem. Soc., Chem. Commun.*, 1994, 2729.
56. S. Tothadi and G. R. Desiraju, *Chem. Commun.*, 2013, **49**, 7791.
57. S. Tothadi, P. Sanphui and G. R. Desiraju, *Cryst. Growth Des.*, 2014, **14**, 5293.
58. V. R. Thalladi, B. S. Goud, V. J. Hoy, F. H. Allen, J. A. K. Howard and G. R. Desiraju, *Chem. Commun.*, 1996, 401.
59. D. Cinčić, T. Friščić and W. Jones, *Chem. Mater.*, 2008, **20**, 6623.
60. J. Syssa-Magalé, K. Boubekeur, J. Leroy, L. Chamoreau, C. Fave and B. Schöllhorn, *CrystEngComm*, 2014, **16**, 10380.
61. C. Präsang, A. C. Whitwood and D. W. Bruce, *Cryst. Growth Des.*, 2009, **9**, 5319.
62. C. B. Aakeröy, M. Baldrighi, J. Desper, P. Metrangolo and G. Resnati, *Chem. - Eur. J.*, 2013, **19**, 16240.
63. C. B. Aakeröy, T. K. Wijethunga and J. Desper, *CrystEngComm*, 2014, **16**, 28.
64. A. C. B. Lucassen, A. Karton, G. Leitus, L. J. W. Shimon, J. M. L. Martin and M. E. van der Boom, *Cryst. Growth Des.*, 2007, **7**, 386.
65. L. C. Roper, C. Präsang, V. N. Kozhevnikov, A. C. Whitwood, P. B. Karadakov and D. W. Bruce, *Cryst. Growth Des.*, 2010, **10**, 3710.
66. C. B. Aakeröy, T. K. Wijethunga, J. Desper and M. Đaković, *Cryst. Growth Des.*, 2016, **16**, 2662.
67. Q. Chu, Z. Wang, Q. Huang, C. Yan and S. Zhu, *J. Am. Chem. Soc.*, 2001, **123**, 11069.
68. F. Fontana, A. Forni, P. Metrangolo, W. Panzeri, T. Pilati and G. Resnati, *Supramol. Chem.*, 2002, **14**, 47.

69. R. Liantonio, P. Metrangolo, T. Pilati, G. Resnati and A. Stevenazzi, *Cryst. Growth Des.*, 2003, **3**, 799.
70. H. Neukirch, E. Guido, R. Liantonio, P. Metrangolo, T. Pilati and G. Resnati, *Chem. Commun.*, 2005, 1534.
71. L. Catalano, P. Metrangolo, T. Pilati, G. Resnati, G. Terraneo and M. Ursini, *J. Fluorine Chem.*, 2017, **196**, 32.
72. C. B. Aakeröy, T. K. Wijethunga, J. Benton and J. Desper, *Chem. Commun.*, 2015, **51**, 2425.
73. K. Raatikainen and K. Rissanen, *CrystEngComm*, 2009, **11**, 750.
74. N. R. Goud, O. Bolton, E. C. Burgess and A. J. Matzger, *Cryst. Growth Des.*, 2016, **16**, 1765.
75. K. B. Landenberger, O. Bolton and A. J. Matzger, *J. Am. Chem. Soc.*, 2015, **137**, 5074.
76. J. C. Bennion, L. Vogt, M. E. Tuckerman and A. J. Matzger, *Cryst. Growth Des.*, 2016, **16**, 4688.
77. E. H. Crowston, A. M. Lobo, S. Parbhakar, H. S. Rzepa and D. J. Williams, *J. Chem. Soc., Chem. Commun.*, 1984, 276.
78. O. Jabay, H. Pritzkow and J. Jander, *Z. Naturforsch., B: Anorg. Chem., Org. Chem.*, 1977, **32**, 1416.
79. K. Raatikainen and K. Rissanen, *CrystEngComm*, 2011, **13**, 6972.
80. V. Stilinović, G. Horvat, T. Hrenar, V. Nemec and D. Cinčić, *Chem. - Eur. J.*, 2017, **23**, 5244.
81. D. Dolenc and B. Modec, *New J. Chem.*, 2009, **33**, 2344.
82. O. Makhotkina, J. Lieffrig, O. Jeannin, M. Fourmigué, E. Aubert and E. Espinosa, *Cryst. Growth Des.*, 2015, **15**, 3464.
83. I. Nicolas, F. Barrière, O. Jeannin and M. Fourmigué, *Cryst. Growth Des.*, 2016, **16**, 2963.
84. F. Bertolotti, G. Cavallo, P. Metrangolo, S. K. Nayak, G. Resnati and G. Terraneo, *Supramol. Chem.*, 2013, **25**, 718.
85. D. Cinčić, T. Friščić and W. Jones, *New J. Chem.*, 2008, **32**, 1776.
86. D. Cinčić, T. Friščić and W. Jones, *Chem. - Eur. J.*, 2008, **14**, 747.
87. D. Cinčić, T. Friščić and W. Jones, *CrystEngComm*, 2011, **13**, 3224.
88. D. Cinčić, T. Friščić and W. Jones, *J. Am. Chem. Soc.*, 2008, **130**, 7524.
89. L. Russo, S. Biella, M. Lahtinen, R. Liantonio, P. Metrangolo, G. Resnati and K. Rissanen, *CrystEngComm*, 2007, **9**, 341.
90. E. Corradi, S. V. Meille, M. T. Messina, P. Metrangolo and G. Resnati, *Angew. Chem., Int. Ed.*, 2000, **39**, 1782.
91. C. C. Robertson, J. S. Wright, E. J. Carrington, R. N. Perutz, C. A. Hunter and L. Brammer, *Chem. Sci.*, 2017, **8**, 5392.
92. C. B. Aakeröy, J. Desper, B. A. Helfrich, P. Metrangolo, T. Pilati, G. Resnati and A. Stevenazzi, *Chem. Commun.*, 2007, 4236.
93. C. B. Aakeröy, P. D. Chopade and J. Desper, *Cryst. Growth Des.*, 2011, **11**, 5333.
94. C. B. Aakeröy, P. D. Chopade, C. Ganser and J. Desper, *Chem. Commun.*, 2011, **47**, 4688.

95. C. B. Aakeröy, M. Fasulo, N. Schultheiss, J. Desper and C. Moore, *J. Am. Chem. Soc.*, 2007, **129**, 13772.
96. J. I. Jay, C. W. Padgett, R. D. B. Walsh, T. W. Hanks and W. T. Pennington, *Cryst. Growth Des.*, 2001, **1**, 501.
97. H. D. Arman, R. L. Gieseking, T. W. Hanks and W. T. Pennington, *Chem. Commun.*, 2010, **46**, 1854.
98. V. Vasylyeva, S. K. Nayak, G. Terraneo, G. Cavallo, P. Metrangolo and G. Resnati, *CrystEngComm*, 2014, **16**, 8102.
99. F. Topić and K. Rissanen, *J. Am. Chem. Soc.*, 2016, **138**, 6610.
100. S. Y. Oh, C. W. Nickels, F. Garcia, W. Jones and T. Friščić, *CrystEngComm*, 2012, **14**, 6110.
101. Q. J. Shen, H. Q. Wei, W. S. Zou, H. L. Sun and W. J. Jin, *CrystEngComm*, 2012, **14**, 1010.
102. H. Y. Gao, X. R. Zhao, H. Wang, X. Pang and W. J. Jin, *Cryst. Growth Des.*, 2012, **12**, 4377.
103. Q. J. Shen, X. Pang, X. R. Zhao, H. Y. Gao, H. Sun and W. J. Jin, *CrystEngComm*, 2012, **14**, 5027.
104. S. d'Agostino, F. Grepioni, D. Braga and B. Ventura, *Cryst. Growth Des.*, 2015, **15**, 2039.
105. Y. Xu, J. Viger-Gravel, I. Korobkov and D. L. Bryce, *J. Phys. Chem. C*, 2015, **119**, 27104.
106. T. A. Schaub, R. Sure, F. Hampel, S. Grimme and M. Kivala, *Chem. – Eur. J.*, 2017, **23**, 5687.

CHAPTER 4

Mechanochemistry in Co-crystal Synthesis

TOMISLAV FRIŠČIĆ

Department of Chemistry, McGill University, 801 Sherbrooke St.W., H3A oB8, Montreal, Canada
Email: tomislav.friscic@mcgill.ca

4.1 Introduction to Mechanochemistry

4.1.1 Relationship Between Mechanochemistry and Supramolecular Synthesis

This chapter is aimed to provide a brief but exhaustive overview of the applications of mechanochemistry, *i.e.* chemical synthesis induced by mechanical force in the form of grinding, milling or shearing,[1] for the assembly of multi-component molecular crystals (co-crystals).[2] Although reports of mechanically induced chemical transformations can be found since ancient times,[3] mechanochemistry is still considered a curiosity in modern chemical synthesis. This view is, however, undergoing a rapid change as the past two decades have witnessed a rapid growth in applications of mechanochemistry across a wide range of chemistry areas, from organic,[4,5] metal–organic,[6,7] main group[8] and organometallic synthesis,[9] to the mechanochemical synthesis of coordination polymers, metal–organic frameworks[10,11] and ultrasmall nanoparticle systems.[12] However, the area where mechanochemistry has been recently applied with extraordinary success is co-crystal formation,[13] wherein two or more molecular species form a new crystalline solid based on a range of supramolecular interactions. Indeed, it may be argued that mechanochemical reaction techniques are particularly compatible with the supramolecular chemistry underlying co-crystal formation,[14] as reactions by ball milling or manual grinding can be conducted in the absence of bulk solvents. This means that mechanochemical reaction conditions provide an excellent opportunity to investigate molecular self-assembly in the absence of solvent competition and concentration effects, solubility differences, and of solvolysis issues that inevitably affect supramolecular processes in solution. Therefore, mechanochemistry may represent an ideal way to study supramolecular chemistry and molecular recognition, as well as to create complex solid-state architectures and materials based on non-covalent interactions. This chapter will try to illustrate and reinforce this view of mechanochemistry as a natural way of conducting supramolecular reactions. Although the scope of this chapter is limited, a number of excellent and authoritative reviews on mechanochemistry have recently been published, addressing topics of organic[15,16] and pharmaceutical synthesis,[17] synthesis of inorganic compounds and materials, coordination compounds, including coordination polymers and metal–organic frameworks, organometallic chemistry, co-crystallization, and more.[4–13]

4.1.2 Definition

Mechanochemistry is defined in the "*Gold Book*" of the International Union for Pure and Applied Chemistry (IUPAC) as a "chemical reaction that is induced by the direct absorption of mechanical energy".[18] However, over the past two decades it has become increasingly obvious that such a definition does not adequately address the entirety of modern mechanochemistry, which now includes not only conventional transformations of organic and inorganic chemistry, but has also expanded to processes of molecular recognition and self-assembly in the solid state. It has, therefore, been proposed to include in the definition of mechanochemistry aspects of supramolecular chemistry and self-assembly that underlie processes of mechanochemical activation (*e.g.* amorphization) or certain types of reactions, such as co-crystal formation or polymorphic transformations.[19] In such a scenario, mechanochemistry would correspond to the use of mechanical force, for example in the form of grinding, milling, shearing, pulling or other types of agitation, as a means to enable, either directly or through an intermediate activation process, the assembly or dissociation of objects ranging in dimensions from nanometers (molecules, molecular complexes) to millimeters (bulk materials).

It is important to note that mechanochemistry is not a uniform area, and can roughly be divided into three directions:[20] (i) the synthesis and transformations of bulk molecules and materials through techniques such as grinding or ball milling; (ii) chemical or materials transformations induced by ultrasonic irradiation[21] and (iii) the cleavage and evaluation of strengths of individual bonds using the tip of an atomic force microscope, as in atomic force spectroscopy.[22] So far, the only area of mechanochemistry relevant to

Scheme 4.1 Illustration of the use of the symbol designating mechanochemical reaction conditions, proposed by the Hanusa group as a means to distinguish mechanochemical reactivity from conventional, typically thermally controlled reactions.[25]

co-crystal formation is bulk mechanochemical transformations of materials by grinding or milling,[23] which are the focus of this chapter. To distinguish mechanochemical reaction conditions, which can sometimes lead to very different outcomes compared to those observed by conventional, thermally-controlled processes,[24] the Hanusa group proposed a symbol which is adopted in this chapter (Scheme 4.1).[25]

4.1.3 Techniques

4.1.3.1 Mechanochemistry by Manual Grinding vs. *Ball Milling*

Mechanochemical transformation of molecular material, including co-crystal formation, salt formation (salification), polymorph transformations, *etc.* are often readily accomplished by manual grinding using a mortar and pestle. Indeed, this was the technique utilized by the Etter group for mechanochemical preparation of a 1 : 1 co-crystal of methyladenine and methylthymine,[26] which was a pioneering demonstration that specific molecular recognition can take place in a reaction between two solids by grinding (Figure 4.1).

Although manual grinding provides a simple entry into mechanochemical transformations, it also has a number of shortcomings. Among these, maybe the most important ones are a lack of control over the reaction environment and also poor reporting of reaction conditions—for example the temperature and atmospheric composition (humidity, CO_2 content) during the course of the mechanochemical experiment, the duration of the grinding process, or the choice of material, size and radius of curvature of the mortar. That these parameters can have a significant effect in the course of a solid-state process is well illustrated in the work of Cinčić *et al.*, who demonstrated that mechanochemical Schiff base formation by grinding together an aromatic aldehyde and an amine is strongly facilitated by atmospheric moisture. The reaction readily proceeded in open air, but no reaction took take place in a dry atmosphere.[27]

An alternative to manual grinding is the use of an electrical ball mill. Whereas there are a number of designs for such automated ball milling,[28] the co-crystallization-related literature so far is vastly dominated by the use of laboratory shaker mills (also known as mixer mills) and planetary mills.

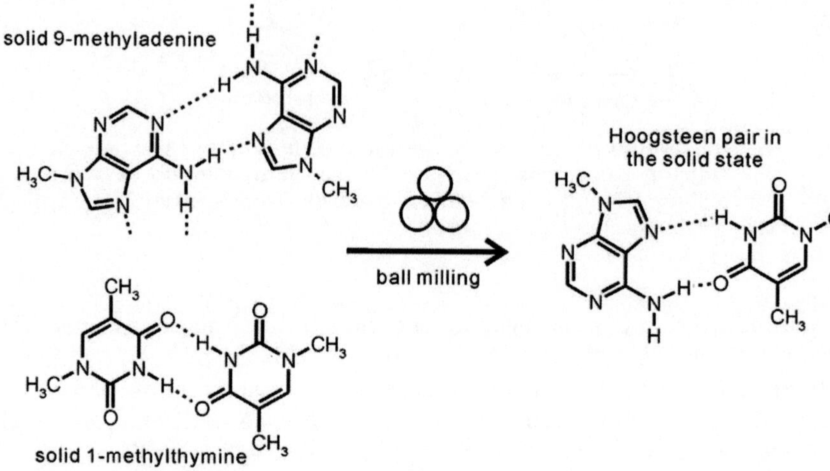

Figure 4.1 Mechanochemical co-crystallization of 9-methyladenine and 1-methylthymine, leading to the formation of a Hoogsteen-type hydrogen-bonded adenine-thymine dimer.[26]

In a laboratory shaker mill design, a sample to be milled is contained within a closed cylindrical vessel, also known as a milling jar, along with milling media in the form of balls made of a particular material. The entire milling assembly is then shaken along an arc, and mechanochemical processes take place as a result of agitation, collisions and shearing due to ball motion. In most laboratory milling equipment, the time of milling and the frequency of shaking are both electronically controlled. An alternative laboratory milling design is the planetary mill, in which the cylindrical milling assembly rotates around its own axis, as well as around a central axis, in that way mimicking the motion of planets around the sun. Such "planetary" motion produces high centrifugal forces on milling media within each milling jar, resulting in mixing, shearing and crushing of the sample. The shaker mill setup is usually applied for sample sizes between 10 mg and 1 gram, while the use of planetary mills is more often reported for larger samples, of sizes between 10 and 100 grams. Consequently, shaker mills are often used in screening for co-crystal formation, whereas planetary mills are more suitable for reaction scaling-up. However, it is important to note that, because of significant differences in mode of operation and instrument design, it is not possible to directly, without any trial-and-error, translate transformations developed within a shaker mill into a planetary milling process. However, once a planetary milling process is developed, further scaling up to industrial scale is expected to be straightforward as the mode of operation of a planetary mill is closely related to that of an industrial roller mill (also known as a gravity mill). In addition to mixer and planetary mills, there is a number of other designs that can be utilized for mechanochemical processing, such as the highly energy-efficient attrition mills or electrical mortars. Descriptions of these can be found in specialized reviews.[28]

The choice of material for the milling assembly can greatly influence the outcome of a mechanochemical experiment. The most popular material is stainless steel, used in a majority of literature reports on ball milling co-crystallization. The more expensive milling assemblies based on softer Teflon® can be utilized in order to avoid harsh impact and amorphization of the milled materials and, conversely, hard and high-density materials such as tungsten carbide (WC, density 16 $g\,cm^{-3}$) can be utilized to achieve harsher milling and higher impact. In the context of synthesizing co-crystals or salts of pharmaceutical ingredients, the use of Teflon or metal-based milling assemblies may lead to concerns of sample contamination. These can be alleviated by the use of zirconia-based milling assemblies, with the high density of zirconia (*ca.* 5.6 $g\,cm^{-3}$) providing a similar reaction environment to stainless steel.

4.1.3.2 Mechanochemistry by Liquid-assisted Grinding (LAG)

The efficiency of a mechanochemical process can be vastly improved through the addition of a small, catalytic amount of a liquid phase. Specifically, the addition of a liquid phase was found to accelerate co-crystal formation, guide the co-crystallization towards the formation of particular types of supramolecular structures or polymorphs, as well as lead to the formation of a highly crystalline material. The ability to accelerate and expand the scope of mechanochemical co-crystallization by addition of a small amount of a liquid into the milling assembly was explored by Shan and co-workers, who reported that mechanochemical co-crystallization of all *cis*-1,3,5-cyclohexanetricarboxylic acids with nitrogen-based hydrogen bond acceptors is strongly accelerated by the addition of small amounts of a liquid phase.[29] For example, while co-crystallization with hexamethylenetetramine took place by neat (dry milling) for 20 min, co-crystal formation with 4,4′-bipyridine led to only partial conversion after 1 h milling, and the use of 4,7-phenanthroline as the hydrogen bond acceptor led to almost no co-crystal formation. In contrast, the addition of a small amount of a liquid phase enabled quantitative co-crystal formation to be accomplished within 20 min in all cases. Importantly, the choice of liquid additive did not significantly affect the outcome of such solvent-drop grinding (SDG) co-crystallization, as the same product was obtained in the presence of water, which readily dissolves both the 1,3,5-cyclohexanetricarboxylic acid and 4,7-phenanthroline, ethyl acetate, which dissolves 4,7-phenanthroline but not the triacid, and acetonitrile, in which both components are only sparingly soluble. The only case in which co-crystallization of 4,7-phenanthroline was not observed was with cyclohexane as the milling additive, which was tentatively explained by complete insolubility of the starting materials.[29]

Later studies have confirmed that the ability of a liquid additive to enable or accelerate mechanochemical co-crystallization is mostly independent of the choice of the liquid phase and, notably, that it is not related to the solubility of the starting materials.[30] This observation, along with the need to establish a more quantitative description of the mechanochemical reaction environment in the presence of a liquid additive, led to the re-naming of the technique from SDG to LAG.[31] A more quantitative description of the LAG reaction environment was accomplished by the introduction of the parameter η, defined as the ratio of the volume of the milling liquid to the weight of reactant substances.[30] Expressed in $\mu l\,mg^{-1}$, the η parameter enables the direct comparison of LAG reactivity to other solvent-mediated phase transformations, such as slurrying or co-crystal formation from solution.[32] In particular, a systematic exploration of co-crystallization of caffeine and theophylline with L-malic and L-tartaric acids as co-crystal formers led to a definition of LAG conditions as *the range of η values over which the outcome of a milling reaction is not affected by reactant solubility*.[30] This study empirically associated LAG with η values between 0 and 2 $\mu l\,mg^{-1}$. Reactions by slurrying were associated with higher content of liquid, with η values approximately between 2 and 12 $\mu l\,mg^{-1}$. Under these conditions the reaction mixtures could be mixed by stirring or sonication, but led to partial conversion or no conversion at all, due to large differences in solubility of starting materials. Finally, higher liquid contents, corresponding to $\eta > 12\ \mu l\,mg^{-1}$, were associated with co-crystal formation from homogeneous solution (Figure 4.2(a)).

The same study also confirmed the beneficial effect of added liquid on the co-crystallization rate, as the conversion to a co-crystal upon 5 min milling of caffeine and L-tartaric acid was greatly enhanced by increasing η (Figure 4.2(b)). Similar observations were also made for milling of theophylline with L-tartaric acid (Figure 4.2(c)).

The above described η-ranges for LAG, slurrying and solution co-crystallization processes are empirical and are likely to depend on the choice of starting materials. Specifically, as the transition from LAG to slurry conditions is associated with the appearance of solubility effects on the co-crystallization outcome, for each reaction system it is possible to define a value η', which is the highest η at which a reaction should provide the co-crystal as the sole product. The η' is dependent on relative solubilities, and for co-crystallization of model compounds **A** and **B** it can be calculated according to the equation:

$$\eta' = \frac{x(\mathbf{A})}{x(\mathbf{A}) + y(\mathbf{B})} \times \frac{1000}{S(\mathbf{A})} \tag{4.1}$$

where x and y are stoichiometric coefficients of components **A** and **B** in a co-crystal, respectively, and $S(\mathbf{A})$ is the equilibrium solubility of the component **A** in the liquid chosen as the additive. If approximate solubilities of co-crystal components in a given liquid phase are known, eqn (4.1) provides a convenient way to design efficient mechanochemical co-crystal screening experiments, and to optimize mechanochemical co-crystallization for complete conversion of reactants.[30]

Besides providing a means to quantify and compare LAG experiments, the η parameter can also be used to affect the polymorphic outcome of a

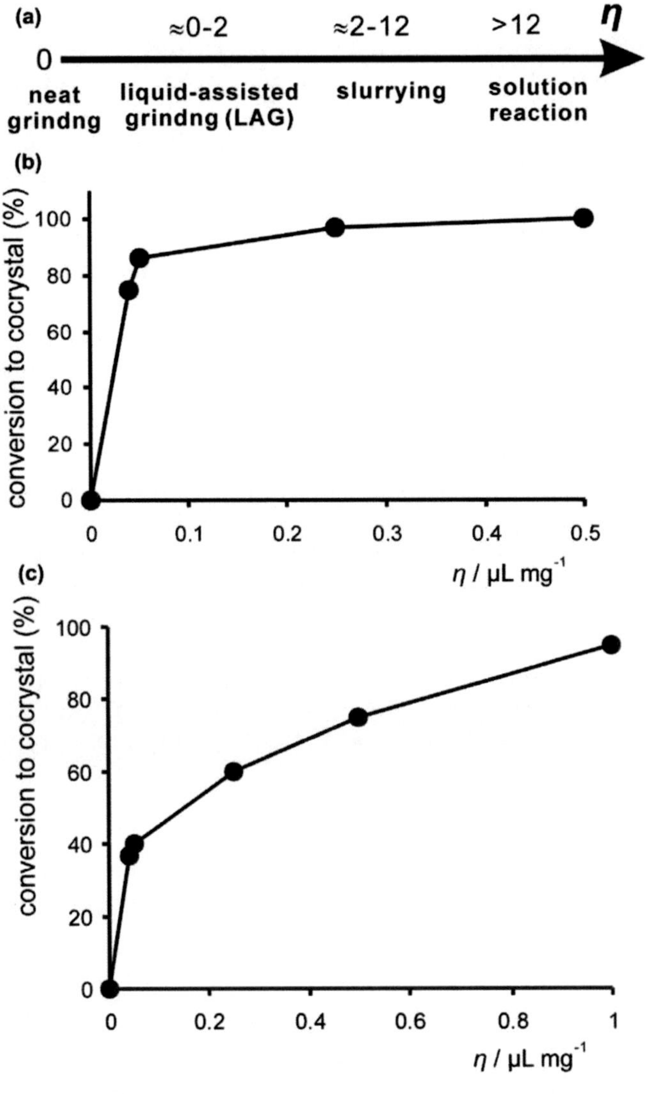

Figure 4.2 (a) Illustration of the η-scale, distinguishing neat grinding, LAG, slurry reactions and co-crystallization from the homogeneous solution. Enhancement in mechanochemical co-crystallization of: (b) caffeine and (c) theophylline with L-tartaric acid upon 5 min milling in the presence of acetonitrile at different η values.
Adapted from ref. 30 with permission from The Royal Society of Chemistry.

mechanochemical co-crystallization, as demonstrated by Hasa and co-workers for the trimorphic co-crystal (caffeine)(anthranilic acid). In one particular example, LAG of anthranilic acid with caffeine in the presence of ethanol as the liquid additive yielded the co-crystal polymorphic form II for η values of

0.05 μl mg^{-1} and 0.1 μl mg^{-1}, a mixture of forms I and II for $\eta = 0.15$ μl mg^{-1}, and form I for η between 0.20 μl mg^{-1} and 0.30 μl mg^{-1}.[33] Very similar results were observed using methanol, *n*-propanol, acetone and 2-butanone as milling liquids. As the polarities of methanol, ethanol, *n*-propanol, acetone and 2-butanone are mutually very different, these results make a strong case for η being the deciding factor controlling polymorphism of (caffeine)(anthranilic) acid co-crystal in these experiments (Figure 4.3).

That LAG conditions enable co-crystallization independent of solubility was first demonstrated by exploring co-crystal formation between caffeine and either L-malic or L-tartaric acids as co-crystal formers in the presence of 15 different liquid additives.[30] The model co-crystallization system was selected because caffeine does not readily form a co-crystal upon dry (neat) milling with either of the two acids. The choice of liquid phases was based on solubilities of caffeine and the two acids, with the intention to explore cases in which: (i) both caffeine and the co-crystal former were readily soluble in the liquid additive (*e.g.* acetonitrile, ethyl acetate); (ii) neither caffeine or the co-crystal former were highly soluble in the added liquid (*e.g.* benzene, cyclohexane, di-isopropylether); (iii) caffeine was highly soluble, but the co-crystal former was not (*e.g.* trifluoroethanol, chloroform, nitromethane), and (iv) the co-crystal former was highly soluble, but the solubility of caffeine in the liquid additive was poor (*e.g.* water, methanol, ethanol). The outcomes of milling experiments, outlined in Tables 4.1 and 4.2, illustrate that co-crystal formation, not observed upon neat milling, readily takes place in the presence of all explored liquid additives with the

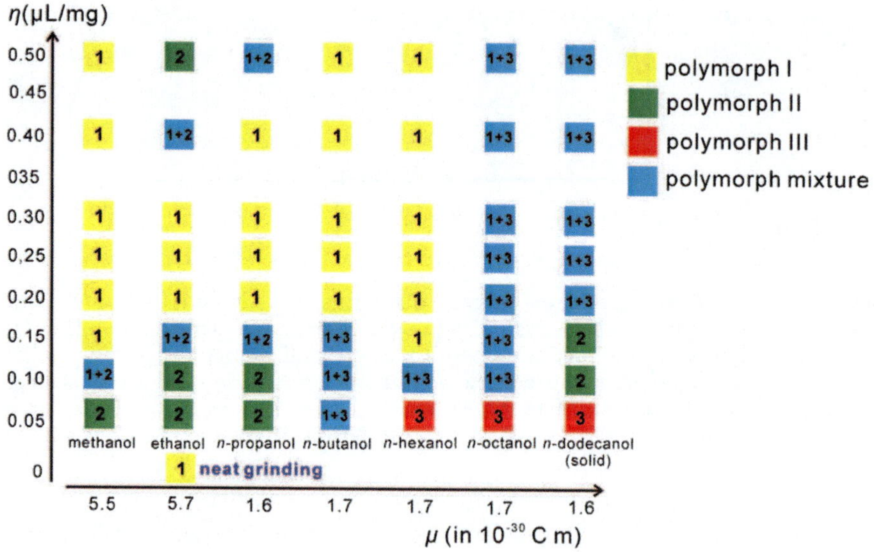

Figure 4.3 Mechanochemical co-crystal screening using different values of η can lead to different polymorphs of the (caffeine)(anthranilic acid) co-crystal, as demonstrated by Hasa and co-workers.[33]

Table 4.1 Solubilities of reactants in the liquid additive and the outcomes of mechanochemical LAG experiments for LAG co-crystallization of caffeine and L-tartaric acid. Co-crystallization experiments were performed by 20 min milling in the presence of each liquid additive ($\eta = 0.25\ \mu l\ mg^{-1}$). Adapted from ref. 30 with permission from The Royal Society of Chemistry.

	Solubility ($mg\ cm^{-3}$)		
Liquid additive	Caffeine	L-Tartaric acid	Product
None (neat grinding)	—	—	No reaction
Water	15	400	**Caffeine** hydrate
Methanol	10	170	Co-crystal
Ethanol	6	72	Co-crystal
iso-Propanol	3	27	Co-crystal
n-Butanol	3	16	Co-crystal
2,2,2-Trifluoroethanol	450	2	Co-crystal + **caffeine**
benzene	<2	<2	Co-crystal + **caffeine**
Fluorobenzene	15	<2	Co-crystal + **caffeine**
Acetonitrile	21	13	Co-crystal
Nitromethane	45	<2	Co-crystal
Chloroform	135	<2	Co-crystal
Ethyl acetate	7	9	Co-crystal
Di-*i*-propyl ether	<2	<2	Co-crystal
Ethyl methyl ketone	8	15	Co-crystal
Cyclohexane	<2	<2	No reaction

Table 4.2 Solubilities of reactants in the liquid additive and the outcomes of mechanochemical LAG experiments for LAG co-crystallization of theophylline and L-malic acid. Co-crystallization experiments were performed by 20 min milling in the presence of each liquid additive ($\eta = 0.25\ \mu l\ mg^{-1}$). Adapted from ref. 30 with permission from The Royal Society of Chemistry.

	Solubility		
Grinding liquid	Theophylline	L-Malic acid	Product
None (neat grinding)	—	—	No reaction
Water	15	800	No reaction
Methanol	10	410	Co-crystal + **theophylline**
Ethanol	6	221	Co-crystal + **theophylline**
iso-Propanol	3	113	Co-crystal
n-Butanol	3	75	Co-crystal + **theophylline**
2,2,2-Trifluoroethanol	450	21	Co-crystal + **theophylline**
Benzene	<2	<2	Co-crystal + **theophylline**
Fluorobenzene	15	<2	Co-crystal + **theophylline**
Acetonitrile	21	54	Co-crystal
Nitromethane	45	<2	Co-crystal
Chloroform	135	<2	Co-crystal
Ethyl acetate	7	31	Co-crystal
Di-*i*-propyl ether	<2	<2	Co-crystal + **theophylline**
Ethyl methyl ketone	8	95	Co-crystal
Cyclohexane	<2	<2	No reaction

exception of cyclohexane, and milling with water led to an alternative product caffeine hydrate.

Similar results were also obtained in several subsequent studies, for example by the Jones group who investigated co-crystallization of caffeine and anthranilic acid in the presence of different liquids,[33] or by Fischer and co-workers who conducted the co-crystallization of theophylline with benzamide in the presence of 18 different liquid additives.[34]

So far, there have been only a very few systematic studies on the role of liquid additives in promoting mechanochemical transformations of molecular solids. Among these, particularly notable is the *in situ* Raman spectroscopy investigation of a mechanochemical reaction of an organic isocyanate with an aniline derivative to form an amide, conducted by Tireli and co-workers. This study is unique as it identifies a property of a liquid additive that can be correlated with reaction enhancement by LAG: a qualitative correlation was established between mechanochemical reactivity and the basicity of the liquid expressed by Gutmann's donor.[35] There was no obvious relationship between reactivity enhancement and some of the other explored physical properties of the liquid, such as polarity (in terms of the molecular dipole moment) or permittivity (Figure 4.4 (top)).

Besides greatly accelerating the course of mechanochemical co-crystallization, the liquid additive also enables the formation of highly crystalline reaction products. This was clearly established by a combined powder X-ray diffraction and terahertz time-domain spectroscopy study of co-crystal formation by milling of mesaconic acid and phenazine by Nguyen and co-workers.[36] Specifically, while X-ray powder diffraction analysis of the reaction mixture after different periods of neat milling revealed the complete disappearance of reactant phases after 60 min, the analysis of thc reaction mixture revealed the presence of only *ca.* 75% crystalline (phenazine)(mesaconic acid) co-crystal. This indicates that the synthesis of the (phenazine)(mesaconic acid) co-crystal by neat grinding leads to at least 25% amorphous material (Figure 4.4 (bottom)). In contrast, X-ray diffraction revealed that LAG with ethanol as the milling additive enables complete conversion of starting materials to the co-crystal within 15 min. Importantly, terahertz analysis was consistent with the product being 100% co-crystal, indicating that the amount of amorphous phase was negligible. The ability to use the milling liquid as a catalyst to obtain products not only quickly, but also with a high degree crystallinity, is important in facilitating crystal structure determination of mechanochemical products directly from X-ray powder diffraction data. This led to the development of a "solvent-free laboratory" concept,[37] applicable not only to co-crystal synthesis, but also to synthesis of organic molecules and other types of materials, wherein the entire synthesis and structural analysis procedure is done without using any bulk solvents at all, relying exclusively on mechanosynthesis and solid-state analysis techniques.

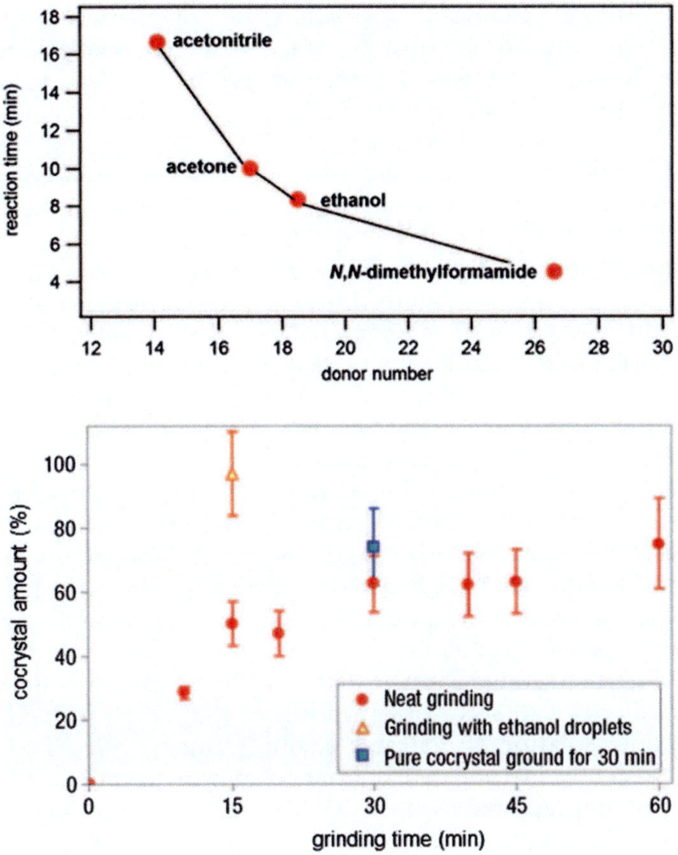

Figure 4.4 (Top) Dependence of reaction time of the Gutmann donor number of the liquid additive in LAG, as established by real-time Raman spectroscopy monitoring of a mechanochemical amide synthesis reaction and (Bottom) the mechanochemical formation of a (phenazine)(mesaconic acid) co-crystal by neat milling and by LAG, monitored *ex situ* by THz-TDS spectroscopy.
(Top) Adapted from ref. 35 with permission from The Royal Society of Chemistry. (Bottom) Adapted with permission from ref. 36.

4.1.3.3 *Further Techniques*

Speed, efficiency and simplicity have rapidly made LAG a highly popular technique for a wide range of applications, from co-crystallization to the synthesis of coordination polymers and active pharmaceutical ingredients. It has also inspired the exploration and development of further techniques relying on the use of catalytic additives to enhance mechanochemical transformations. One of these is ion- and liquid-assisted grinding that utilizes ionic salts,[38] in addition to a liquid additive, to facilitate the activation and reactivity of techniques of inorganic oxides.

More relevant to co-crystal formation is polymer-assisted grinding (POLAG),[39] introduced in 2015 by Hasa and co-workers. The POLAG technique is targeted towards screening for new polymorphs and co-crystals by use of catalytic polymer additives. Its development has been inspired by the observation, exemplified by the work of the Matzger group,[40] that polymer surfaces can influence the nucleation of molecular substances, and in that way enabling the selective formation or the discovery of polymorphs. Consequently, the POLAG technique represents an attempt to merge the efficiency and speed of mechanochemical solid forms screening with control over (co-)crystal nucleation. The first report of the POLAG methodology explored the effect of different poly(ethylene)glycol (PEG) additives on co-crystallization of caffeine and citric acid, caffeine and anthranilic acid and phenazine with mesaconic acid. A particularly remarkable result was the observation that, whereas neat milling of caffeine and citric acid does not yield a co-crystal, the addition of polymers that are either liquids (PEG200, PEG300 and PEG400) or solids at room temperature (PEG3000, PEG6000 and PEG10000) induced the formation of the (caffeine)(citric acid) co-crystal. Moreover, whereas milling of anthranilic acid with caffeine typically produces the thermodynamically most stable form I of the co-crystal, POLAG in the presence of 1% of either liquid PEG200 or solid PEG10000 by weight led to the formation of the metastable form II. Both of these examples illustrate that polymer additives can serve the purposes of inducing mechanochemical reactivity, and isolation of metastable solid forms.

Another recently investigated mechanochemical techniques is seeding-assisted grinding, explored by the Cinčić group,[41] wherein the polymorphic composition of a product of mechanochemical reactions is directed by the addition of pre-synthesized seeds of a particular polymorph of the product. This technique is also applicable for directing the course of mechanochemical co-crystal formation, as demonstrated by the Emmerling group for the (theophylline)(benzamide) co-crystal.[34] Mechanochemical reaction of theophylline and benzamide can produce two different polymorphs of the co-crystal, depending on the milling conditions: neat milling or LAG with non-polar solvents yields the metastable form I, whereas use of polar liquid additives during LAG leads to the thermodynamically preferred form II. However, neat milling reaction can be directed towards the formation of form II with the addition of a small amount of co-crystal seeds. Whereas these examples demonstrate that the course of mechanochemical ball milling reactions can be directed by surface templating effects, all so far reported cases of seeding led to the formation of the thermodynamically more stable form.[34] Over the past 5 years a number of other techniques have begun to emerge, for example milling with variable amounts of a liquid, or milling at different temperatures. An excellent overview of these different mechanochemical techniques, with particular focus on their current or potential applications in pharmaceutical form discovery, has recently been provided by Hasa and Jones.[42]

4.2 Advantages and Challenges of Mechanochemistry in Co-crystal Synthesis

4.2.1 Structural Characterization of Mechanochemical Products

A particularly important advantage of mechanochemistry, compared to other methods for co-crystal screening and synthesis, is the ability to conduct transformations of poorly soluble substances. At the same time, however, products of mechanochemical co-crystal synthesis are powdered materials, which makes their characterization significantly more challenging compared to the single crystal X-ray diffraction analysis that is often applicable to products of co-crystal synthesis in solution. One strategy to circumvent this problem is seeding,[43] *i.e.* the use of mechanochemically prepared materials as seeds for heterogeneous nucleation and growth of single crystals. However, the developments in laboratory instrumentation also offer an opportunity to conduct direct structural characterization of mechanochemical products by powder X-ray diffraction. The efficiency of this strategy was demonstrated in 2007 by Karki and co-workers[44] in the synthesis and characterization of two co-crystals of the natural product theobromine. Co-crystallization of theobromine through solution or thermal (*e.g.* melt) techniques is made very difficult due to its low solubility and high melting point (400 °C). However, milling of equimolar amounts of theobromine with trifluoroacetic acid, or LAG of theobromine with malonic acid readily produced new crystalline materials that were characterized as hydrogen-bonded co-crystals of composition (theobromine)(trifluoroacetic acid) and (theobromine)(malonic acid) through powder X-ray diffraction structure solution and thermal analysis. The high crystallinity and rapid, quantitative formation of LAG products facilitated the structure solution procedure, enabling the entire co-crystal synthesis, and structure solution and refinement protocol to be performed within a day.[44] Similarly, the Emmerling group has utilized mechanochemistry and structure solution from X-ray powder diffraction data to synthesize and characterize a co-crystal of theobromine with oxalic acid. Due to the poor solubility of theobromine, the attempts to synthesize this co-crystal from solution have not been successful.[45]

The above outlined examples demonstrate that the microcrystalline nature of mechanochemical products is not necessary a major hindrance to their efficient characterization. Indeed, as co-crystals are usually based on well-known and often already structurally characterized components, barring the unexpected formation of solvates under LAG conditions, it may be argued that they should be particularly amenable to structure solution from powder diffraction data by using simulated annealing techniques.[46] Structural characterization of mechanochemically prepared materials can also be further facilitated by advanced and increasingly commonplace instrumental techniques, such as solid-state nuclear magnetic resonance spectroscopy,[47] and the emergent theoretical approaches, such as crystal structure prediction (CSP).[48]

In one example,[49] CSP was utilized to predict the most likely crystal structure for a hydrogen-bonded co-crystal of dihydrocarbamazepine and acetic acid. The correctness of the prediction was confirmed by comparing the powder X-ray diffraction pattern that was measured for the mechanochemically prepared material to that simulated for the theoretically generated structure. Overall, these sophisticated but increasingly popular instrumental and theoretical techniques, in combination with quantitative yields and high product crystallinity observed for many LAG reactions, enable the structural characterization of a wide range of mechanochemically prepared materials: not only of co-crystals but also of new organic molecules,[50] coordination polymers,[51] and metal–organic frameworks.[52]

4.2.2 Co-crystal Screening and Stoichiometric Control in Mechanochemistry

The ability to circumvent solubility-related limitations means that, besides expanding the scope of chemical and supramolecular synthesis towards poorly soluble starting materials, mechanochemistry should also provide a highly efficient route for co-crystal synthesis and discovery. Co-crystal screening from solution is often biased towards the formation of the least soluble product, which can be a reactant, solvate or a particular polymorph or stoichiometric composition of a co-crystal. Such effects, however, should not affect co-crystal formation under mechanochemical conditions (including LAG) and, indeed, several comparative studies have demonstrated that the efficiency of mechanochemical techniques in co-crystal screening is on a par with or even superior to that of solution- or melt-based alternatives. For example, a systematic study of a set of 25 co-crystals led the Zaworotko group to establish LAG as a reliable and cost-effective, as well as environmentally friendly, co-crystal screening technique compared to solvent evaporation.[53] Another study, conducted by the Childs and Rodríguez-Hornedo groups, demonstrated that the efficiency of LAG in screening for co-crystals of carbamazepine is comparable to that of reaction co-crystallization, SonicSlurry and evaporative high-throughput screening.[54]

The screening efficiency of mechanochemical co-crystallization is particularly evident in studies of co-crystal stoichiometric variations, *i.e.* co-crystals based on the same set of molecular components, but in different stoichiometric ratios. An early investigation in that context was reported by the Jones group, who described the selective formation of two stoichiometric variations of the co-crystal of caffeine with acetic acid, controlled by the composition of the reaction mixture.[55] Milling of caffeine and acetic acid in a 1 : 1 stoichiometric ratio led to the formation of the co-crystal (caffeine)(acetic acid), in which the two components are held together through a $R_2^2(7)$ hydrogen-bonded synthon involving O–H···N and C–H···O hydrogen bonds. Milling of the two components in a respective 1 : 2 stoichiometric ratio, however, quantitatively yielded a co-crystal of composition (caffeine)(acetic acid)$_2$ in which an additional molecule of acetic acid is associated with a caffeine–acetic acid $R_2^2(7)$ heteromolecular assembly (Figure 4.5(a)).

The efficiency of mechanochemistry in screening for stoichiometric variations of co-crystals was evaluated by Karki and co-workers,[56] who systematically explored co-crystallization of nicotinamide with a family of homologous α,ω-alkanedicarboxylic acids using mechanochemistry (Figure 4.5(b)), and crystallization from the melt and solution growth. Specifically, nicotinamide is expected to form at least two stoichiometrically different types of co-crystals with each dicarboxylic acid: a co-crystal of 1 : 1 stoichiometric composition, in which nicotinamide and the carboxylic acid co-crystal former are associated *via* hydrogen-bonded $R_2^2(7)$ pyridine–carboxylic acid and $R_2^2(8)$ amide–acid heterodimeric synthons, and a co-crystal of 2 : 1 nicotinamide:-dicarboxylic acid, in which the co-crystal components are held together *via* $R_2^2(7)$ pyridine–carboxylic acid heterosynthons and $R_2^2(8)$ amide–amide supramolecular synthons between nicotinamide molecules (Figure 4.5(c)). The results of this study, reproduced in Table 4.3, demonstrate

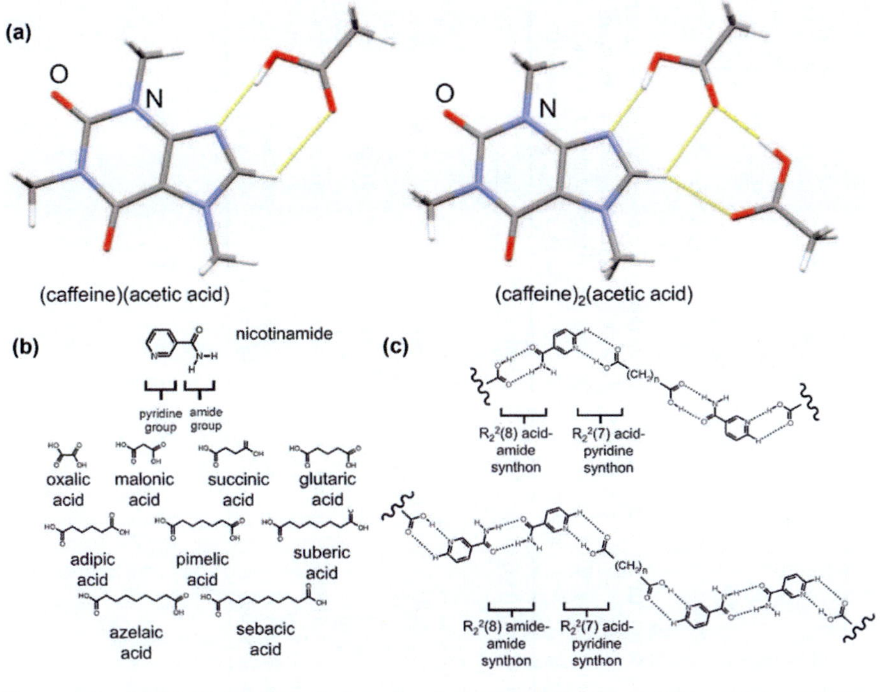

Figure 4.5 (a) Hydrogen-bonded molecular assemblies in stoichiometric variations of the co-crystal of caffeine with acetic acid, selectively prepared by milling caffeine with the acid in a 1 : 1 (left) and 1 : 2 (right) stoichiometric ratio.[55] Schematic representation of: (b) nicotinamide and dicarboxylic acids, as well as (c) corresponding 1 : 1 and 2 : 1 co-crystals investigated by Karki and co-workers.[56]
Parts (b) and (c) were adapted from [56] with permission from The Royal Society of Chemistry.

Table 4.3 Results of screening for stoichiometric variations of co-crystal of nicotinamide (na) with the following dicarboxylic acids: oxalic (oxa), malonic (mal), succinic (suc), glutaric (glu), adipic (adi), pimelic (pim), suberic (sub), azelaic (aze) and sebacic (seb) acid. The bold entries indicate that the screening experiment did not yield a co-crystal with a stoichiometric composition identical to the starting ratio of reactants. Adapted from ref. 56 with permission from The Royal Society of Chemistry.

	Solution co-crystallization		Melt co-crystallization		Neat milling		LAG	
Ratio na : acid	1 : 1	2 : 1	1 : 1	2 : 1	1 : 1	2 : 1	1 : 1	2 : 1
oxa	(na)·(oxa)	**(na)·(oxa)**	—	—	(na)·(oxa)	(na)$_2$·(oxa)	(na)·(oxa)	(na)$_2$·(oxa)
mal	**(na)$_2$·(mal)**	(na)$_2$·(mal)	**(na)$_2$·(mal)**	(na)$_2$·(mal)	**(na)$_2$·(mal)**	(na)$_2$·(mal)	**(na)$_2$·(mal)**	(na)$_2$·(mal)
suc	**(na)$_2$·(suc)**	(na)$_2$·(suc)	**(na)$_2$·(suc)**	(na)$_2$·(suc)	**(na)$_2$·(suc)**	(na)$_2$·(suc)	**(na)$_2$·(suc)**	(na)$_2$·(suc)
glu	(na)·(glu)	**(na)·(glu)**	(na)·(glu)	**(na)·(glu)**	(na)·(glu)	**(na)·(glu)**	(na)·(glu)	**(na)·(glu)**
adi	**(na)$_2$·(adi)**	(na)$_2$·(adi)	**(na)$_2$·(adi)**	(na)$_2$·(adi)	(na)·(adi)	(na)$_2$·(adi)	(na)·(adi)	(na)$_2$·(adi)
pim	(na)·(pim)	**(na)·(pim)**	(na)·(pim)	**(na)·(pim)**	(na)·(pim)	(na)$_2$·(pim)	(na)·(pim)	(na)$_2$·(pim)
sub	(na)·(sub)	**(na)·(sub)**	(na)·(sub)	**(na)·(sub)**	(na)·(sub)	(na)$_2$·(sub)	(na)·(sub)	(na)$_2$·(sub)
aze	(na)·(aze)	**(na)·(aze)**	(na)·(aze)	**(na)·(aze)**	(na)·(aze)	**(na)·(aze)**	(na)·(aze)	**(na)·(aze)**
seb	**(na)$_2$·(seb)**	(na)$_2$·(seb)	**(na)$_2$·(seb)**	(na)$_2$·(seb)	(na)·(seb)	(na)$_2$·(seb)	(na)·(seb)	(na)$_2$·(seb)

mechanochemical LAG and neat grinding as techniques superior to co-crystallization from solution or from the melt: both LAG and neat grinding enabled the synthesis of targeted co-crystal stoichiometric variations in 14 out of 18 (75%) explored cases. In contrast, solution co-crystallization was successful in 9 out of 18 (50%), and co-crystallization from the melt in 8 out of 18 (44%) cases.

Such excellent control over reaction stoichiometry in mechanochemical reactions is not limited to co-crystal formation, but has also been noted in the synthesis of coordination polymers,[51] as well as the assembly of organic molecules.[57]

4.2.3 Polymorphism Control in Liquid-assisted Mechanochemistry

The addition of a small amount of a liquid phase in LAG also provides an attractive opportunity to modify and gain control over the polymorphic composition of the product, including the synthesis of metastable phases. The use of LAG for exploring the polymorphic landscape of a single-component molecular crystal was noted by Trask and co-workers, who reported switching between the three polymorphs of anthranilic acid by using different liquid additives.[58] The applicability of this methodology in the selective synthesis of co-crystal polymorphs was demonstrated using a co-crystal of caffeine and glutaric acid as a model system.[59] There are two known polymorphs of the (caffeine)(glutaric acid) co-crystal, termed form I and form II. Both polymorphs are based on hydrogen-bonded chains of caffeine and the dicarboxylic acid, with the principal difference being in the conformation of the aliphatic chain of glutaric acid in each co-crystal (Figure 4.6(a)). Although co-crystal growth from solution yields both forms as concomitant polymorphs, the selective synthesis of form I is readily achieved by LAG in the presence of a non-polar liquid additive. Conversely, the use of a polar liquid additive leads to selective synthesis of pure form II. The ability to direct the formation of (caffeine)(glutaric acid) polymorphs by switching the polarity of the mechanochemical environment was verified by Hasa and co-workers,[60] who explored co-crystallization of the two components in the presence of different oligo- and poly(ethylene)glycol polymers. In particular, increasing the length of the polymer additive enables control over its polarity: LAG with more polar short chain di- and triethylene glycols selectively yielded co-crystal form II, while POLAG with the less polar PEG200 and PEG10000 led to exclusive formation of form I (Figure 4.6(b)).[60] Another example of the polarity steering co-crystal polymorphs was described by Fischer *et al.* who explored the polymorphism of the co-crystal of theophylline and nicotinamide (Figure 4.6(c)).[34] Specifically, mechanochemical synthesis of (theophylline)(nicotinamide) yielded the form I of the co-crystal upon neat grinding or by LAG in the presence of non-polar solvents pentane, cyclohexane or heptane. Milling with even slightly more polar liquid, such as 1,4-dioxane or *n*-propanol, led to selective formation of form II, which was

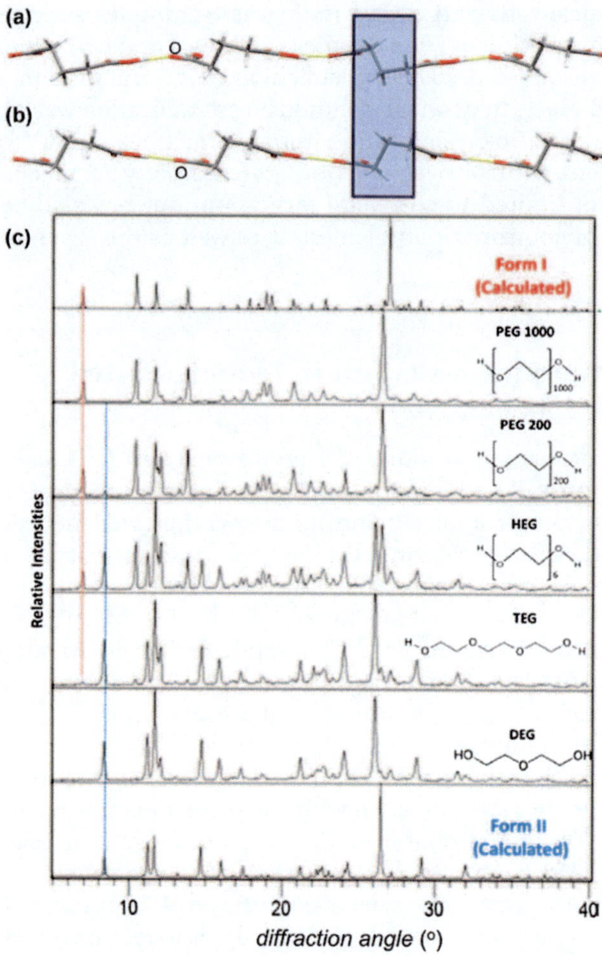

Figure 4.6 Different conformations of the glutaric acid aliphatic chain in polymorphic forms of the (caffeine)(glutaric acid) co-crystal: (a) the monoclinic form I and (b) the triclinic form II.59. (c) Comparison of powder X-ray diffractograms for the product of mechanochemical co-crystallization of caffeine and glutaric acid in the presence of increasingly larger oligo(ethylene)glycol molecules: short-chain, more polar ethyleneglycols lead to the preferred formation of form II, whereas longer, less polar additives lead to co-crystal form I.
Adapted with permission from D. Hasa, E. Carlino and W. Jones, *Cryst. Growth Des*, 2016, **16**, 1772–1779, Copyright 2016 American Chemical Society.[60]

also established to be the thermodynamically more stable material by slurry experiments. Although it is not yet possible to correlate a particular property of the liquid additive to its ability to enhance the mechanochemical reactivity, the above described examples and other studies delineated in

Section 4.3.2 demonstrate that it is possible to identify which properties of the milling liquid (*e.g.* dipole moment or molecular shape and size) are significant in controlling product polymorphism or composition.

4.3 Advances in Mechanistic Studies of Mechanochemical Co-crystallization

4.3.1 Qualitative Description of Mechanochemical Co-crystal Formation

Despite successful applications of mechanochemistry in a number of areas of chemical and materials chemistry, the mechanistic understanding of chemical reactions by ball milling remains poorly developed. A general, qualitative description of mechanochemical transformations has been proposed by Kaupp in the form of a three-step mechanism (Figure 4.7).[61]

The first step of this mechanism involves mechanical grinding and agitation as a means to achieve intimate mixing of reactants and/or activation of their surface. In the next step, the activated phase undergoes a chemical transformation leading to product formation. The final, third step of a mechanochemical process involves product separation and removal, again assisted by mechanical agitation. Mehanochemical reactivity in such a mechanism is mediated by an intermediate phase that may be a solid, a liquid or a gas. Participation of a gas phase is expected to be relevant only in

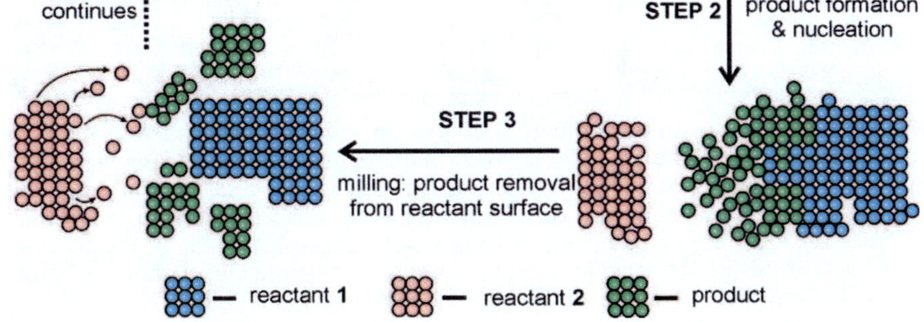

Figure 4.7 Illustration of the qualitative three-step mechanism of mechanochemical reactions put forward by Kaupp.
Adapted with permission from T. Friščić and W. Jones, *Cryst. Growth Des.* 2009, **9**, 1621–1637, Copyright 2009 American Chemical Society.[61]

reactions of materials with relatively high vapour pressures, such as naphthalene, *p*-benzoquinone, *etc.* where the high mobility of molecules greatly facilitates reactant mixing. Notable early studies on this type of mechanochemical reaction have been conducted by Rastogi's group,[62,63] and recent ones have been conducted by the group of Kuroda.[64]

Alternatively, the mixing of reactants in a mechanochemical reaction can take place through the formation of a eutectic phase that subsequently relaxes into the final reaction product.[65] This type of mechanism was explored by the Davey group,[66] who demonstrated that the formation of a co-crystal[67] of diphenylamine and benzophenone is mediated by a eutectic (Figure 4.8).

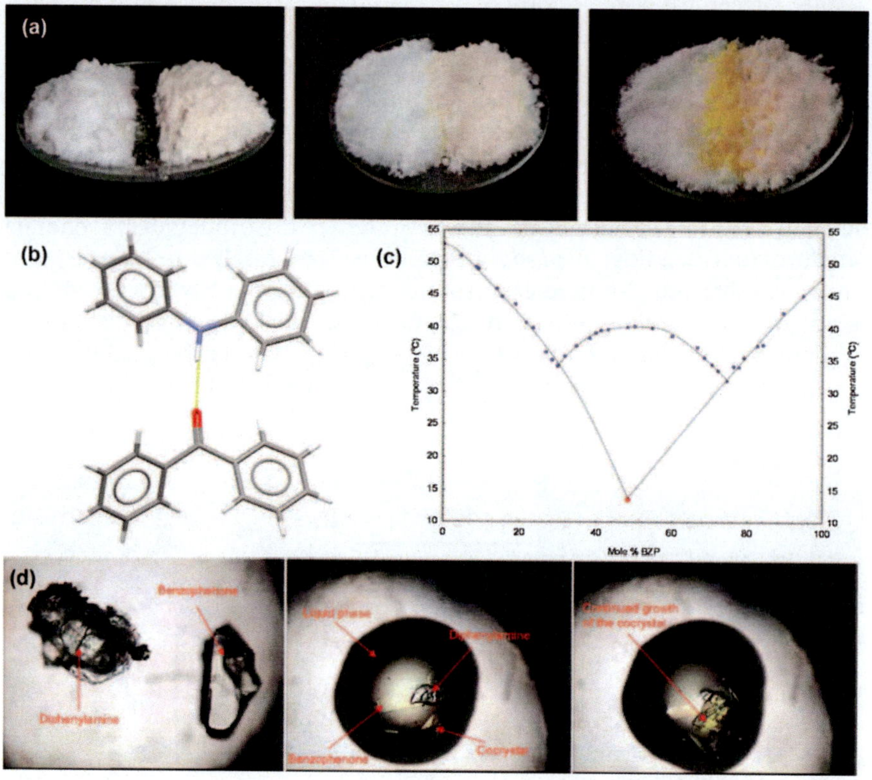

Figure 4.8 (a) Spontaneous formation of a yellow eutectic phase is readily observed when solid diphenylamine and benzophenone come into contact; (b) the hydrogen-bonded assembly of diphenylamine and benzophenone; (c) the phase diagram of mixtures of benzophenone and diphenylamine, revealing an eutectic phase (red) submerged below the co-crystal phase and (d) optical microscopy imaging of mixing solid particles of diphenylamine and benzophenone demonstrating: (left) crystals of individual reactants before contact, (middle) eutectic liquid forming upon contact of the two crystals and (right) growth of the co-crystal phase from the submerged eutectic.
Adapted from ref. 66 with permission from The Royal Society of Chemistry.

In a visually striking process, manual mixing of two colourless solids, benzophenone and diphenylamine, leads to the formation of a yellow liquid phase, which subsequently crystallized to form the hydrogen-bonded co-crystal of composition (diphenylamine)(benzophenone). Analysis of the phase diagram reveals that the liquid phase is a submerged eutectic with a freezing point at 13.3 °C, whose composition is identical to that of the co-crystal that melts at 40.2 °C.

In many cases, however, a mechanochemical reaction can proceed without any observable liquid phase. While such behaviour has sometimes been explained by postulating the existence of an elusive, highly short-lived eutectic phase,[68] it is more likely that such transformations are mediated by an activated, high-energy solid phase. Indeed, the participation of a metastable amorphous material in mechanochemical co-crystallization was indirectly demonstrated by the group of Rodríguez-Hornedo,[69] who utilized cryo-milling (*i.e.* milling at cryogenic temperatures[24,69]) to capture a metastable amorphous intermediate in co-crystallization of carbamazepine with saccharin. At this point, it is worth noting that a mechanochemical co-crystallization may be mediated, but does not necessitate the participation of highly activated amorphous or eutectic phases: simple particle size reduction can be sufficient to ensure sufficient contact and mobility of reactant molecular solids. This was clearly demonstrated by the Blagden group (Figure 4.9),[70] who investigated the rate of co-crystal formation in static mixtures of urea and 2-methoxybenzamide, as well as of caffeine and malonic acid. The mixtures were prepared by mixing together powdered reactants with controlled particle sizes, revealing that reduction in particle size leads to significant rates of spontaneous co-crystal formation with particle sizes of roughly 100 μm or less.

4.3.2 Real-time and *In Situ* Studies of Mechanochemical Co-crystallization

An important instrumental and conceptual advance in studies of mechanochemical reaction mechanisms was the introduction of first techniques for direct, real-time monitoring of ball milling processes.[71] The first report of this methodology was published in 2013 and demonstrated the ability to use high energy, short wavelength (*ca.* 0.1 Å) synchrotron X-ray powder diffraction to observe the transformations of crystalline phases in the synthesis of metal–organic frameworks.[72] The highly penetrating nature of such radiation enabled the observation of mechanochemical reactivity in vessels made of steel, aluminum or poly(methyl)methacrylate plastic (PMMA) (Figure 4.10(a)). The latter material also enabled X-ray diffraction monitoring of model pharmaceutical co-crystals.[73] Specifically, the use of *in situ* X-ray diffraction enabled the direct observation of the amorphization of the reaction mixture of saccharin and carbamazepine, confirming the appearance of an amorphous phase indirectly detected by the group of Rodríguez-Hornedo. In contrast,

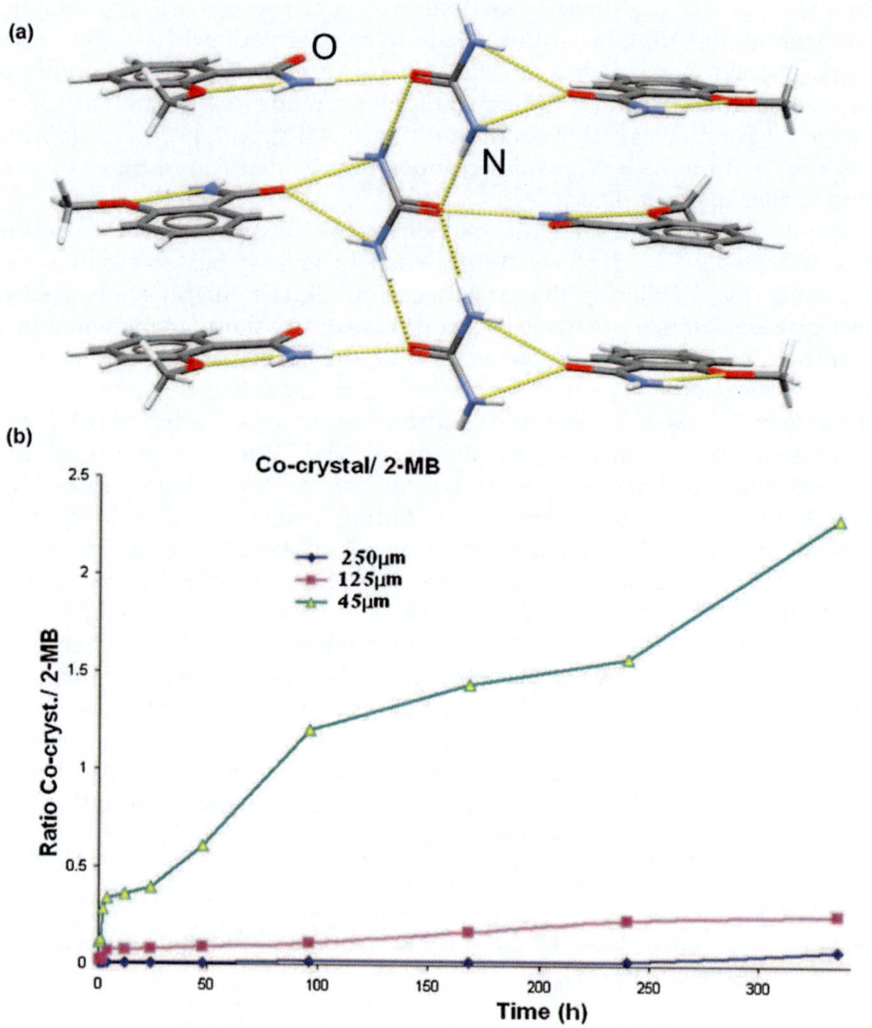

Figure 4.9 (a) Fragment of the crystal structure of the co-crystal of urea with 2-methoxybenzamide (2-MB), investigated by the Blagden group and (b) spontaneous formation of the co-crystal from non-stirred mixtures of solid reactants with controlled particle sizes, evaluated through measuring the time-dependent change in the amount of co-crystal relative to 2-MB reactant.
Part (b) adapted from ref. 70 with permission from The Royal Society of Chemistry.

real-time monitoring of an analogous LAG process revealed rapid co-crystal formation, demonstrating the quantitative assembly of form I of (carbamazepine)(saccharin) co-crystal in less than 5 min (Figure 4.10(b)–(d)).

The application of *in situ* X-ray diffraction monitoring to mechanochemical co-crystallization of nicotinamide and suberic acid revealed that

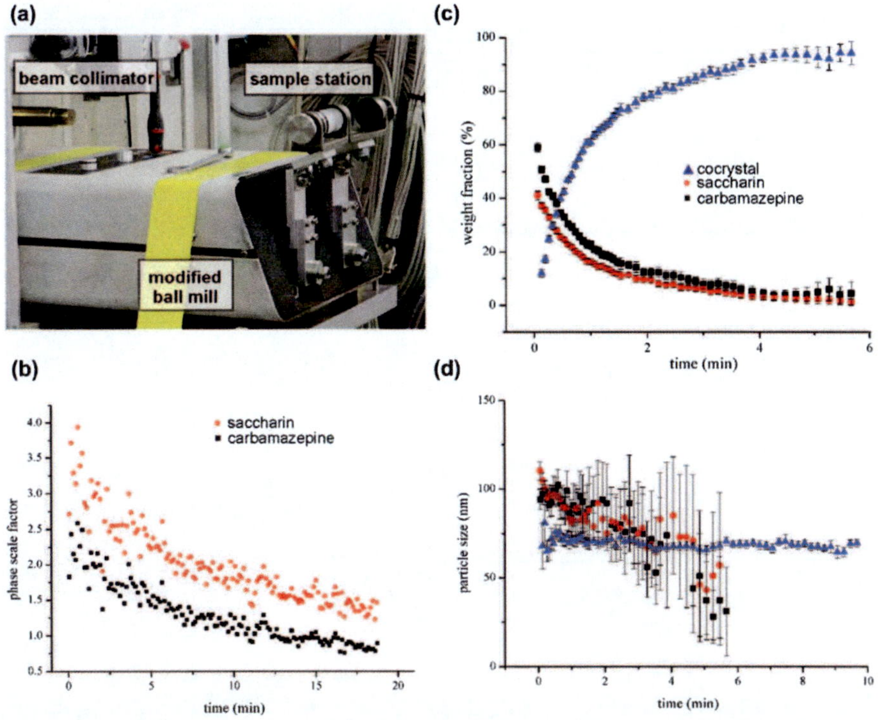

Figure 4.10 (a) View of a setup for *in situ* monitoring of mechanochemical reactions by synchrotron X-ray powder diffraction. Outcomes of real time X-ray diffraction monitoring of: (b) neat milling of saccharin and carbamazepine; (c) LAG of saccharin and carbamazepine and (d) particle size development during the LAG reaction described in (b).
Adapted from ref. 73 with permission from John Wiley and Sons, Copyright © 2013 WILEY-VCH Verlag GmbH & Co. KGaA, Weinheim.

milling of the two components in a 2 : 1 stoichiometric ratio leads to the intermediate formation of a co-crystal with composition (nicotinamide)(suberic acid) (Figure 4.11). The intermediate transforms within five min of further milling into the final reaction product (nicotinamide)$_2$(suberic acid). A similar but significantly faster stepwise process was observed by monitoring the analogous LAG reaction, in which the (nicotinamide)(suberic acid) intermediate disappeared within 3 min. However, monitoring of the LAG transformation also revealed the appearance of a previously not observed crystalline material, with a lifetime of around 3 min.[73]

A stepwise structural transformation was also observed by the Emmerling group in real-time synchrotron X-ray diffraction monitoring of milling co-crystallization of pyrazinamide and malonic acid, using radiation with a longer wavelength of 1.0 Å.[74] The *in situ* diffraction revealed that milling of the two co-crystal components first led to the formation of a metastable form II of the (pyrazinamide)(malonic acid) co-crystal, followed by the co-crystal form I as the final product. An identical, but greatly accelerated

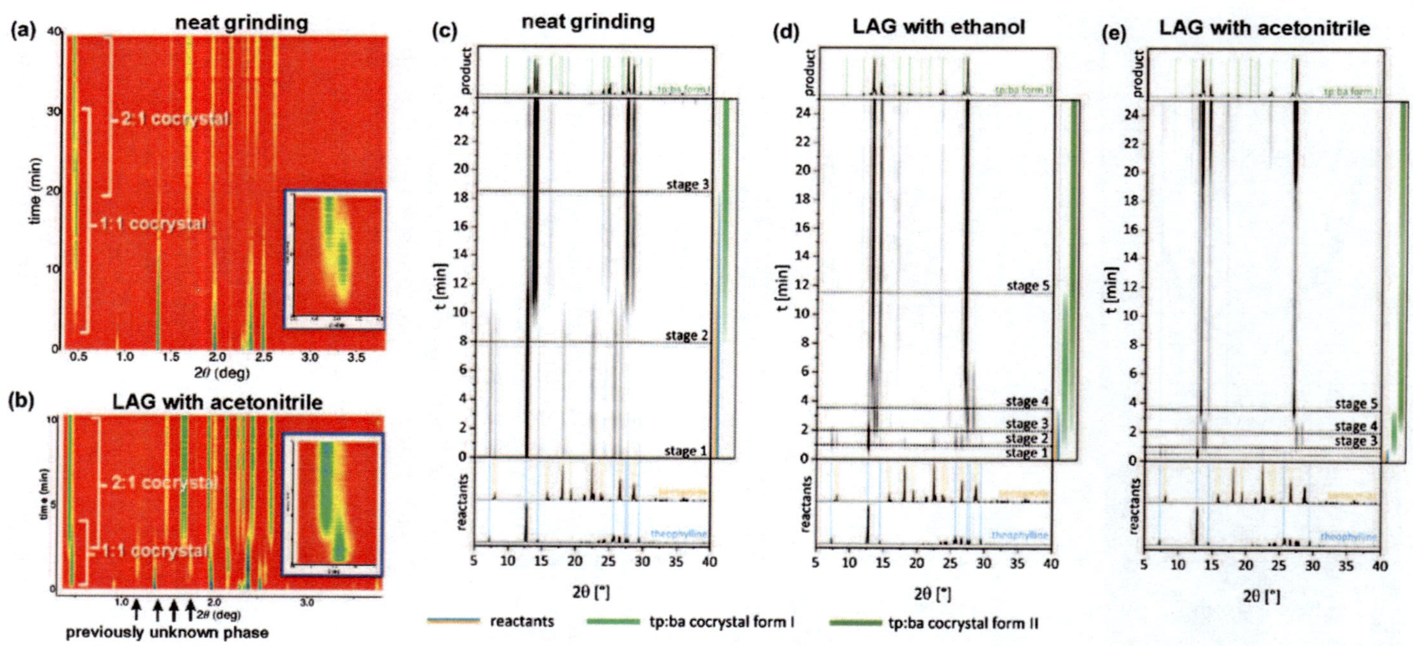

Figure 4.11 Examples of time-resolved powder X-ray diffractograms for mechanochemical co-crystallization reactions: (a) co-crystallization of nicotinamide and suberic acid in a 2 : 1 respective stoichiometric ratio proceeds through the formation of the (nicotinamide)(suberic acid) co-crystal as an intermediate; (b) the analogous LAG process is significantly faster and also involves an additional reaction intermediate; (c) co-crystallization of theophylline and nicotinamide by neat grinding proceeds without detectable intermediates to yield the form I polymorphs of the co-crystal, while LAG with (d) ethanol and (e) acetonitrile yield co-crystal form II, with form I as an intermediate phase.
Parts (a) and (b) adapted from ref. 73 with permission from John Wiley and Sons, Copyright © 2013 WILEY-VCH Verlag GmbH & Co. KGaA, Weinheim. Part (c) adapted with permission from F. Fischer, A. Heidrich, S. Greiser, S. Benemann, K. Rademann and F. Emmerling *Cryst. Growth Des.* 2016, **16**, 1701–1707, Copyright 2016 American Chemical Society.[34]

transformation was observed for the LAG process: the form II intermediate persists in the reaction mixture for over 10 min upon neat milling, but in LAG the conversion to form I is complete within *ca.* 5 min.[74] Besides changes in reaction kinetics, switching from neat milling to LAG can also bring about significant modifications to product formation and the appearance of intermediate phases, as observed for mechanochemical co-crystallization of theophylline and benzamide (Figure 4.11).[34] Neat milling of the two compounds in the 1:1 stoichiometric ratio leads to the appearance of the (theophylline)(benzamide) polymorph form I within 10 min, and complete conversion of reactants within *ca.* 19 min. In contrast, milling in the presence of a catalytic amount of ethanol leads to the complete disappearance of the diffraction signals of reactant solids within 3 min of milling, accompanied by the almost simultaneous appearance of co-crystal polymorph forms I and II. Within *ca.* 10 min of milling, however, the product is exclusively co-crystal form II. Replacing ethanol with less polar, non-protic acetonitrile as the milling liquid induces a noticeable change in the reaction course and kinetics, as the reactant solid phases selectively convert to co-crystal form I within 1 min and form II appears only subsequently, after 2 min milling (Figure 4.11). Consequently, it appears that changes in choice of milling liquid can switch from the concomitant to sequential appearance of co-crystal polymorphs during milling.[34]

Tracking the progress of mechanochemical co-crystallization is hindered by the inability of powder X-ray diffraction to directly detect amorphous phases, and quantitative analysis of reaction mixtures is often not possible without prior knowledge of all phases participating in the reaction. These limitations can be overcome by the application of Raman spectroscopy for monitoring mechanochemical reactions, which was introduced in 2014 by Halasz *et al.*, using as model systems the co-crystallization of nicotinamide with suberic acid and formation of coordination polymers based on cyanoguanidine.[75] This technique was coupled with simultaneous X-ray powder diffraction monitoring by Batzdorf and co-workers, enabling the direct and real-time observation of mechanochemical reactions of synthesizing inorganic and metal–organic materials, as well as a (theophylline)(benzoic acid) as a model co-crystal.[76] The usefulness of this tandem monitoring methodology in analyzing complex mechanochemical co-crystallization systems was illustrated by Lukin and co-workers,[77] who explored the formation of co-crystals involving nicotinamide and benzoic acid. Despite the simplicity of these components, LAG synthesis of the corresponding co-crystal revealed very different mechanisms depending on the liquid additive, involving a total of four crystalline phases. As only two of these phases have been structurally characterized *via* powder X-ray diffraction, and identified as co-crystal forms I and II, the reaction kinetics could not be quantified by Rietveld analysis. Quantification of the reaction progress was, however, possible through simultaneous *in situ* Raman measurements, that enabled the description of the reaction pathways, which involved structurally not yet characterized crystal forms III and IV (Figure 4.12).[77]

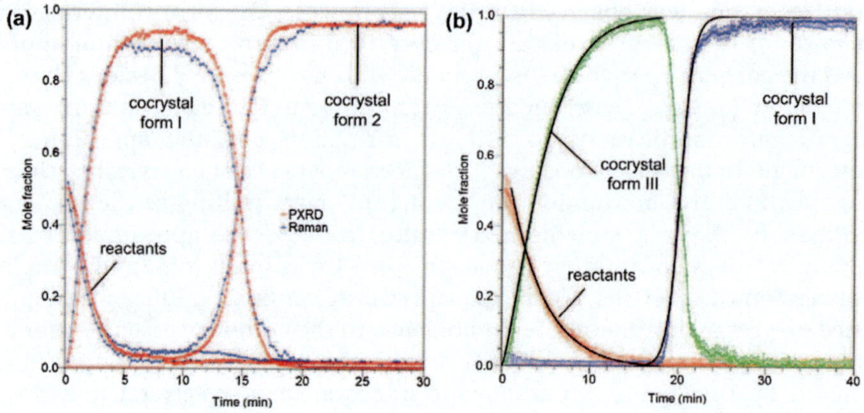

Figure 4.12 Examples of using Raman spectroscopy for real-time quantitative monitoring of the kinetics and mechanisms of mechanochemical co-crystallization reactions involving intermediate phases with unknown structures: (a) matching of X-ray powder diffraction and Raman spectroscopy data for the formation and interconversion of forms I and II of the (nicotinamide)(benzoic acid) co-crystal upon LAG with water and (b) Raman spectroscopy evaluation of the formation and interconversion of (nicotinamide)(benzoic acid) co-crystal forms I and III upon neat grinding.
Adapted from ref. 77 with permission from John Wiley and Sons, © 2017 Wiley-VCH Verlag GmbH &Co. KGaA, Weinheim.

4.4 Mechanochemical Synthesis of Complex Molecular Solids

4.4.1 Mechanochemical Synthesis of Three-component Co-crystals

By avoiding the limitations of solution-based synthesis, mechanochemistry can circumvent one of the most persistent challenges of solid-state supramolecular chemistry: the synthesis of co-crystals containing more than two different constituents. Although supramolecular designs for the construction of three-component (ternary) co-crystals have been devised,[78–80] the synthesis of such materials from solution is often hampered in practice by differences in the solubility of the different components or by the formation of alternative co-crystal or solvate phases, illustrating to the crystal engineer that *the best laid plans of mice and men often go awry.*[81] Consequently, attempts to construct a multi-component co-crystal are in many cases likely to produce simpler solid phases that happen to be less soluble in a given solvent, for example crystals of pure components, corresponding solvates, or binary co-crystals (Figure 4.13).

In principle, mechanochemistry should provide an excellent environment to explore designs for the synthesis of three-component or even more

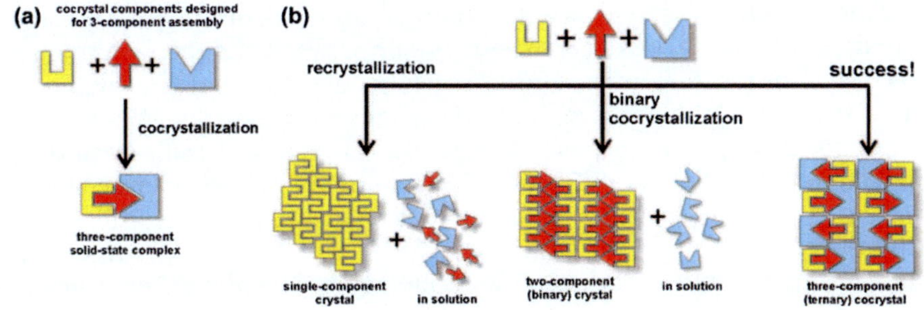

Figure 4.13 The design of three-component co-crystals in solution, illustrated in (a), is often hampered by solubility of simpler substances illustrated in (b), such as solid starting materials or alternative binary co-crystals or solvates. Adapted from ref. 14 with permission from The Royal Society of Chemistry.

complex co-crystal phases, by avoiding limitations related to solubility of individual co-crystal components. Pioneering studies in that context have been conducted by the group of Kuroda, who utilized manual grinding to explore the mechanochemical formation of three-component co-crystals involving racemic and enantiomerically pure derivatives of bis-β-naphthol (BINOL) as building blocks.[82,83] The targeted co-crystals were all held together by a combination of strong O–H···O hydrogen bonds and charge-transfer π···π stacking interactions, and were generally composed of *p*-benzoquinone as an electron-deficient component, a BINOL derivative as an electron-rich component, and an additional arene. In most cases, the formation of co-crystals was readily recognized by a striking change in colour of the reaction mixture, as the resulting co-crystals were all highly coloured due to extensive charge–transfer interactions. For example, manual grinding of *p*-benzoquinone with racemic BINOL and naphthalene rapidly led to the formation of the blue three-component co-crystal of composition $(rac\text{-BINOL})_2\cdot(p\text{-benzoquinone})\cdot(\text{naphthalene})_2$. The co-crystal was also readily obtained from solution, which enabled structural characterization by single crystal X-ray diffraction. Crystal structure analysis revealed that the co-crystal is based on a three-component π-stacked supramolecular motif consisting of a benzoquinone molecule "sandwiched" between two electron rich bis-β-naphthol moieties (Figure 4.14(a)). The neighboring three-component motifs, which were identified as the chromophoric component of the co-crystal, are further associated into chains through O–H···O hydrogen bonds. The stacking of such hydrogen-bonded chains produces an open host structure in which the guest naphthalene molecules are held *via* C–H···π contacts. In contrast, replacing the *rac*-BINOL in the mechanochemical reaction with the optically pure *R*-isomer leads to the formation of a red three-component co-crystal. The composition of the co-crystal was established to be $(R\text{-BINOL})_2\cdot(p\text{-benzoquinone})_2\cdot(\text{naphthalene})_3$, and the same material was also readily obtained through co-crystallization from solution. Following crystal structure determination, the difference in colour between the three-component co-crystals of racemic and enantiomerically pure BINOL was explained by the distortion of the chromophoric three-membered stacks, which, in the chiral co-crystal, assemble into helical chains (Figure 4.14a). Overall, the appearance of strikingly different colours upon manual mechanochemical formation of the three-component co-crystals of *p*-benzoquinone and naphthalene with racemic (blue) and chiral (red) BINOL illustrate a very simple, but effective, solid-state methodology to visually differentiate between a chiral and a racemic form of a compound.[82]

Grinding of anthracene with *p*-benzoquinone and *rac*-BINOL leads to the formation of a red three-component solid of composition (*rac*-BINOL)$_2$·(*p*-benzoquinone)$_2$·(anthracene). However, attempts to obtain the same material from solution were not successful, leading to a different, blue-coloured three-component co-crystal, having the composition (*rac*-BINOL)$_2$·(*p*-benzoquinone)···(anthracene)$_2$. Single crystal X-ray structure analysis revealed that this co-crystal consists of similar π-stacked and hydrogen-bonded chains that were observed in the above described blue ternary co-crystal involving naphthalene as a guest arene (Figure 4.14b). However, since the red (*rac*-BINOL)$_2$·(*p*-benzoquinone)$_2$·(anthracene) could only be obtained by a solid–solid reaction, its structural analysis *via* single crystal X-ray diffraction was impossible. Structure determination of the three-component co-crystal was successfully accomplished by Kuroda, Harris and co-workers in a pioneering report[83] on structure determination using X-ray powder diffraction data. Following structural characterization, the red colour of the co-crystal was explained by the absence of the blue chromophore BINOL–benzoquinone–BINOL stacks. Instead, the structure was found to contain four-component BINOL–benzoquinone–benzoquinone–BINOL stacks that are connected into chains by O–H···O hydrogen bonds (Figure 4.14(b)). The anthracene component is placed in cavities between the chains, and is held in the structure by extensive C–H···π bonding.

The systematic investigation of the relative efficiencies of neat grinding, LAG and solution growth for the synthesis of ternary complexes explored the inclusion compounds of a hydrogen-bonded host of caffeine and succinic acid.[31,84] Although caffeine and succinic acid do not form a binary hydrogen-bonded co-crystal, crystallization from dioxane readily yields a ternary co-crystal of composition (caffeine)·(succinic acid)·0.66(dioxane) in which dioxane guests are included in channels of a self-assembled hydrogen-bonded host (caffeine)·(succinic acid). A systematic exploration of 25 different guest molecules, including dioxane, revealed the formation of a solid-state ternary complex with four potential guests when using solution-based methods. In contrast, neat grinding led to the formation of the ternary complex in 15 cases and LAG enabled the synthesis of 18 ternary solids, clearly demonstrating the advantages of mechanochemical grinding in the assembly of complex hydrogen-bonded solids.[31]

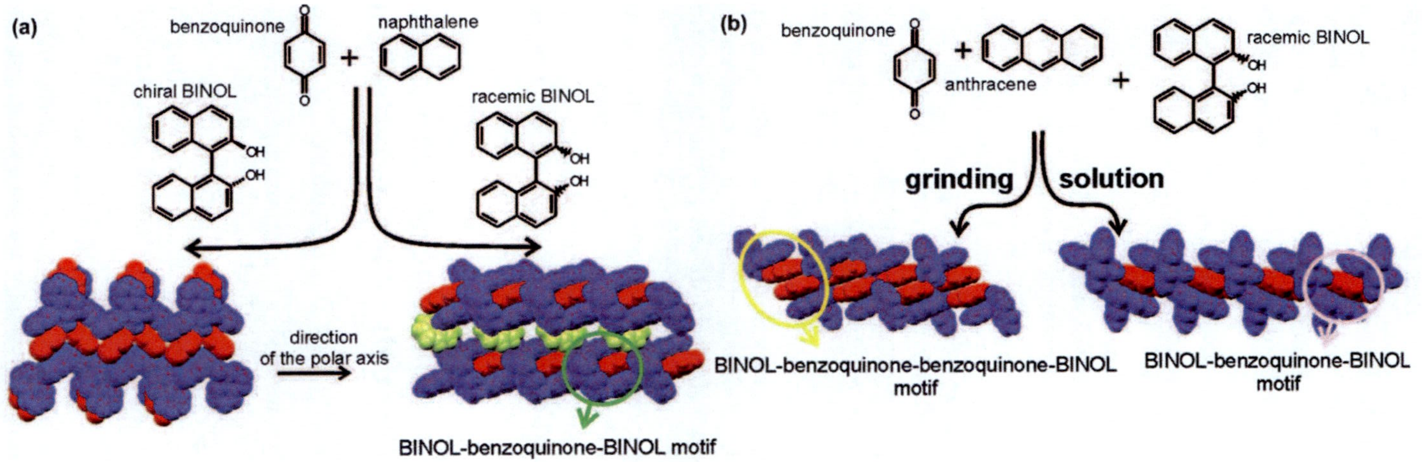

Figure 4.14 (a) Reactivity of a mixture of benzoquinone (red) and naphthalene (yellow) towards enantiomerically pure (left, with omitted naphthalene molecules) and racemic (right) BINOL (blue) in solution or in solid state and (b) difference in reactivity of a mixture of benzoquinone (red), anthracene and racemic BINOL (blue) upon mechanochemical grinding (left) and in solution (right). For clarity, anthracene molecules have been omitted and chromophoric supramolecular motifs are indicated by circles.[82,83]
Adapted from ref. 14 with permission from The Royal Society of Chemistry.

4.4.2 Combining Different Types of Molecular Self-assembly

The ability to conduct multiple, orthogonal self-assembly processes provides an attractive route to synthesize increasingly complex structures from simple building blocks. Solid-state mechanochemical reactions are readily applicable for such synthetic strategies, as demonstrated by Friščić *et al.*[85] who screened for lattice inclusion compounds of self-assembled metal–organic hosts by ball milling of mononuclear nickel(II) or cobalt(II) dibenzoylmethanate complexes with isonicotinamide and a potential liquid guest. The milling led to the formation of new inclusion compounds through a combination of three different types of self-assembly processes, notably ligand exchange on the metal centre, hydrogen bond-driven self-assembly of the resulting coordination complexes, and the formation of an inclusion compound with the liquid guest (Figure 4.15).

Combinations of other types of self-assembly processes under mechanochemical conditions have been explored by the James group, who combined the reversible aldimine condensation reaction with coordination driven self-assembly to produce a zinc complex of the popular "salen" ligand by a one-pot reaction of zinc oxide with ethylenediamine and salicylaldehyde.[86] A similar process was also investigated by Cinčić and co-workers who demonstrated the one-pot four-component synthesis of halogen-bonded co-crystals of a metal–organic complex by milling together an amine, an aldehyde, a metal salt, and 1,4-diiodotetrafluorobenzene as a halogen bond acceptor.[87] The combination of reactants led to the formation of a Schiff base chelating ligand by aldimine condensation, its coordination to a metal salt and, finally, the halogen bond-driven co-crystallization of the resulting metal–organic complex. A five-component mechanochemical process was described by Chow and co-workers, by milling of magnesium oxide, salicylic acid, 15-crown-5, acetonitrile and water, leading to a four-component crystal.[88]

4.5 Understanding Molecular and Biomolecular Recognition Through LAG Co-crystallization

4.5.1 Screening for Molecular Recognition

Because it is efficient and requires short reaction times (typically between 5 and 20 min), LAG is an excellent method to systematically explore factors underlying molecular recognition and templating in molecular solids. This potential of LAG was first systematically utilized in exploring the structure-templating properties of guests in the formation of hydrogen-bonded host of caffeine and succinic acid.[31,84] Notably, these two components were observed to form two types of open hydrogen-bonded frameworks upon milling with different liquid additives: a 1 : 1 framework with composition (caffeine)·(succinic acid),[31] that was also used as a model system for evaluating the relative efficiencies of LAG, neat grinding and solution co-crystallization in

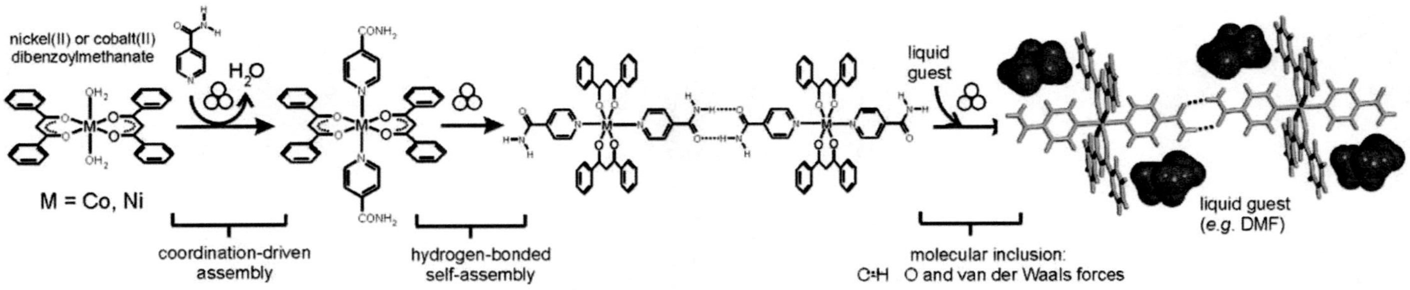

Figure 4.15 Schematic representation of a mechanochemical process that combines different types of self-assembly in one pot: formation and cleavage of coordination bonds, formation of hydrogen bonds, and the assembly of a host–guest inclusion compound held together by van der Waals forces.
Adapted from ref. 85 with permission from John Wiley and Sons, Copyright © 2009 WILEY-VCH Verlag GmbH & Co. KGaA, Weinheim.

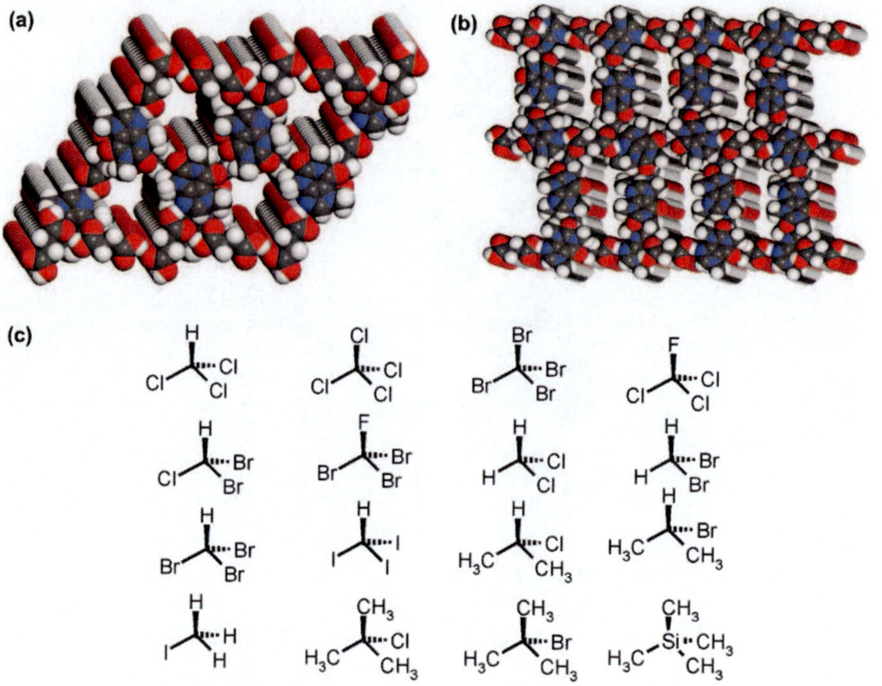

Figure 4.16 Fragments of supramolecular hydrogen-bonded host frameworks based on caffeine and succinic acid, with guest molecules omitted for clarity: (a) (caffeine)·(succinic acid) and (b) (caffeine)$_4$·(succinic acid). (c) The library of potential guest molecules used in LAG exploration of molecular parameters responsible for directing the formation of different host architectures.[31,84]
Adapted with permission from T. Friščić and W. Jones, *Cryst. Growth Des.* 2009, **9**, 1621–1637, Copyright (2009) American Chemical Society.[61]

screening for three-component crystals, and an alternative 4:1 host framework of composition (caffeine)$_4$·(succinic acid) (Figure 4.16(a) and (b)).[84]

Systematic LAG screening for co-crystallization of caffeine and succinic acid in the presence of potential guests (shown in Figure 4.16(c)), which were carefully chosen to address differences in molecular size, shape or functionality, revealed that the formation of the (caffeine)·(succinic acid) host was the most likely event for guests with molecular size between 3.5 Å and 3.8 Å. The exception was haloform-type guests whose presence induced the formation of (caffeine)$_4$·(succinic acid) (Figure 4.17(a)).[84] The selective formation of (caffeine)$_4$·(succinic acid) host upon milling with haloform molecules was subsequently rationalized by crystallographic analysis of the inclusion compound containing bromoform. Specifically, structural analysis the ability of bromoform molecules to template host framework formation through short and directional C–H···O hydrogen bonds and Br···N halogen bonds (Figure 4.17(b)). That both of these interactions are needed for the

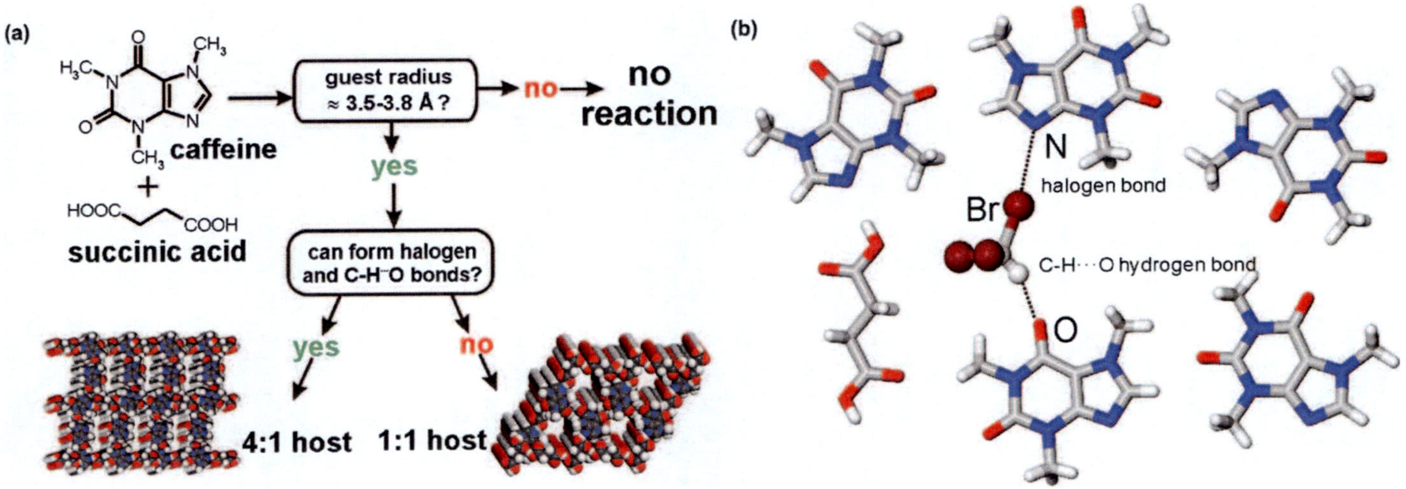

Figure 4.17 (a) The mechanochemically established relationships between the guest structure and the preference for the formation of different hydrogen-bonded host frameworks based on caffeine and succinic acid and (b) the hydrogen C–H···O and halogen Br···N bonds in the bromoform inclusion compound of the $(caffeine)_4$·(succinic acid) host framework.[31,84] Adapted from ref. 14 with permission from The Royal Society of Chemistry.

formation of the $(caffeine)_4$·(succinic acid) host was confirmed by mechanochemical screening experiments involving guests in which either the haloform hydrogen atom was replaced with a similarly sized fluorine, or the halogen atom was replaced with a methyl group of similar size.[31,84]

4.5.2 Screening for Recognition Motifs of Steroids

The mechanochemical screening approach described in Section 4.5.1 was subsequently applied to investigate the molecular recognition properties of steroids, leading to the unexpected discovery of a previously unknown self-assembly motif of steroid sex hormones. This study was inspired by the pioneering investigations of steroid co-crystallization by Eger and Norton,[89] which indicated the existence of co-crystals of androgenic steroids and naphthalene, despite the absence of obvious sites for molecular recognition between the molecules. This inspired a systematic investigation of the propensity of structurally similar steroids progesterone, pregnenolone, β-estradiol and estrone towards co-crystal formation using substituted and non-substituted aromatic molecules (Figure 4.18(a)).[90] Screening for co-crystal formation by mechanochemical LAG of each of the steroids with a library of 24 aromatic countermolecules (Figure 4.18(b)) revealed that, unlike the other three steroids in the test, progesterone exhibited an outstanding affinity to form co-crystals with arenes. Indeed, the formation of co-crystals with progesterone, which was detected by X-ray powder diffraction immediately after each milling experiment, was observed to take place independent of the presence of any strong hydrogen bonding functionalities on the arene. From all arenes in the test library, only perfluoronaphthalene and the xanthines did not yield a co-crystal with progesterone (Figure 4.18(b)). Such mechanochemically established propensity of progesterone for co-crystallization with arenes was rationalized through a combined structural and computational study involving single crystal and powder X-ray diffraction structure determination, molecular shape analysis and crystal structure prediction. Structural analysis and CSP computational studies revealed the persistent formation of three-component supramolecular assemblies in which the steroid fragments form a "sandwich" around the arene (Figure 4.18(c)).[90] This type of molecular assembly revealed a new motif of steroid recognition designated the $\alpha\cdots\pi$ interaction. The proposed $\alpha\cdots\pi$ interaction[90,91] occurs between the α-surface of the steroid molecule and the π-system of the arene as a result of highly complementary electrostatic molecular surface potentials. In contrast to the persistent $\alpha\cdots\pi$ stacking motif that was observed in co-crystals of progesterone, no such structural regularity was observed for the several prepared co-crystals involving steroids pregnenolone, β-estradiol and estrone. Specifically, mechanochemical screening for co-crystals of pregnenolone yielded a co-crystal held together by hydrogen bonds, whereas the exploration of β-estradiol co-crystallization revealed a binary co-crystal with pyrene based on an open hydrogen-bonded framework of β-estradiol with molecules of pyrene included as guests.

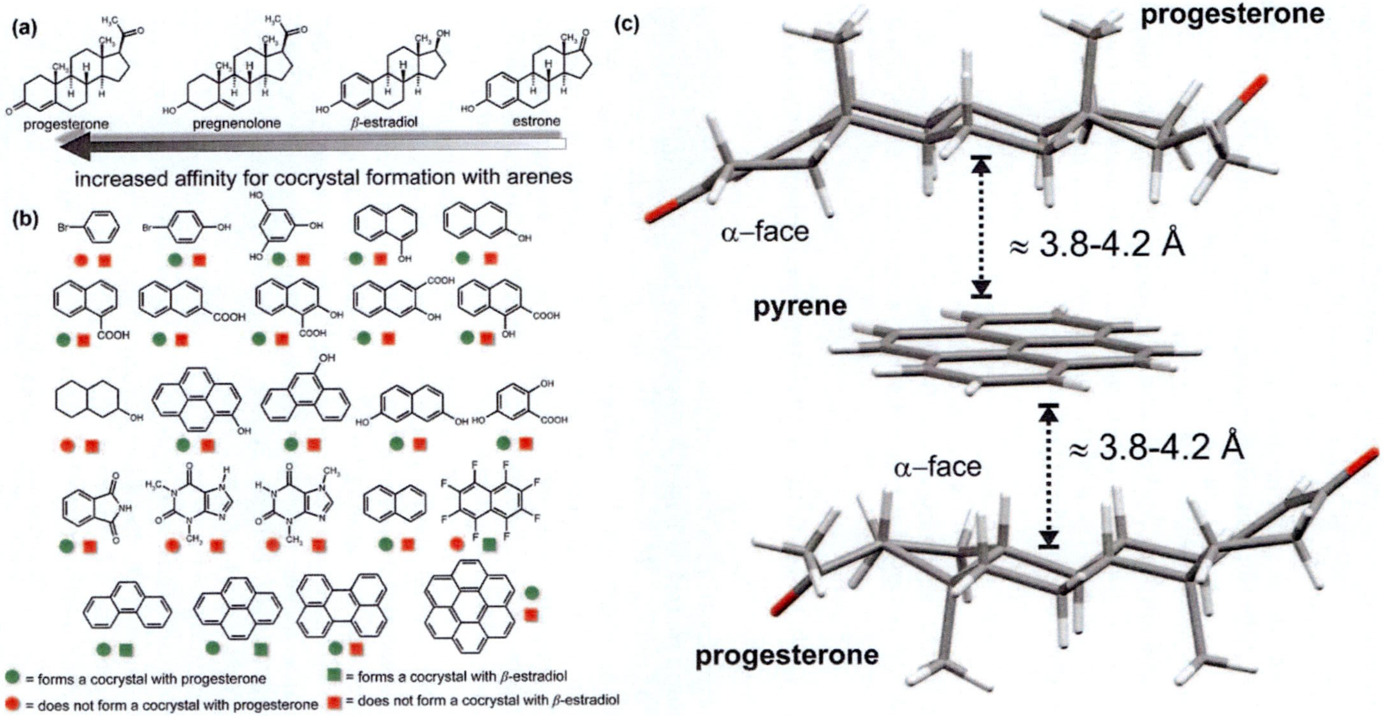

Figure 4.18 (a) Explored steroids and (b) arene co-crystallization partners with indicated ability to form co-crystals with progesterone and β-estradiol. (c) A progesterone–arene "sandwich" complex involving in the co-crystal with pyrene.[90] Adapted from ref. 14 with permission from The Royal Society of Chemistry.

A subsequent report by Ardila-Fierro and co-workers[92] described a systematic mechanochemical screen for co-crystallization of another steroid, β-estradiol, with carefully selected aromatic molecules. Similar to mechanochemical exploration of progesterone co-crystallization and formation of inclusion compounds of caffeine and succinic acid hosts, this screening also revealed very clear relationships between the structure of the aromatic molecules and co-crystal formation. Notably the formation of co-crystals took place only for arene structures larger than just two naphthalene rings, specifically: phenanthrene, pyrene and different isometric phenantridines. Subsequent crystal structure analysis provided an explanation of that trend, by revealing that the co-crystals are, in fact, host–guest lattice compounds based on a three-dimensional (3-D) lattice host of β-estradiol. The host framework is held by O–H···O hydrogen bonds between steroid molecules, and exhibits channels with a square cross-section, occupied by arene molecules as guests. That the dimensions of the lattice host channels are largely determined by the geometry of the β-estradiol molecule, provided an explanation for the observed preferences in co-crystallization with arenes.[92]

4.6 Mechanochemical Synthesis of Halogen-bonded Co-crystals

Mechanochemistry has found extensive application in the synthesis of molecular or ionic materials based on hydrogen bonds, including co-crystals, salts, salt co-crystals, *etc.* However, in the context of halogen-bonded co-crystal systems, synthesis by milling or grinding has been much less exploited. One of the first reports of mechanochemical formation of halogen-bonded structures was provided by Cinčić and co-workers,[93] who utilized 1,4-dibromo- or 1,4-diiodoperfluorobenzene as halogen bond donors, in combination with six-membered oxygen-, nitrogen-, or sulfur-containing heterocycles as acceptors. Ball milling of different combinations of donors or acceptors in a 1 : 1 stoichiometric ratio led to rapid, quantitative assembly of a set of eight isostructural halogen-bonded co-crystals. As the co-crystals were isostructural, the evaluation of their thermal stability (*i.e.* melting points) provided a simple entry to evaluate the relative strengths of different types of halogen bonding interactions. This approach revealed the following ordering of halogen bonds, in order of decreasing strength: I···N > Br···N > I···O ≈ I···S > Br···O.

Subsequent study of the reaction of 1,4-diiodotetrafluorobenzene or 1,2-dioodotetrafluorobenzene with thiomorpholine revealed a stepwise reaction mechanism, dictated by the kinetic formation of stronger N···I halogen bonds.[94] Further sets of isostructural co-crystals based on 1,4-diiodotetrafluorobenzene or 1,4-diiodotetrafluorobenzene were also obtained by milling with either acridine or phenazine as the halogen bond acceptor.[95] In all cases, the differences in melting points of isostructural co-crystals indicated that I···N halogen bonding arrangements are more stable than corresponding

Br···N ones. The reliable isostructurality of co-crystals based on I···N and Br···N interactions led to a conclusion that the halogen-bonded iodine and bromine groups are structurally equivalent. This concept was subsequently combined with the established structural equivalency of aromatic nitrogen and C–H groups to deliberately design a family of four isostructural co-crystals, by co-crystallization of iodo- or bromopentaflurobenzene as halogen bond donors with phenazine or acridine as halogen bond acceptors.[96] The co-crystals were readily obtained by milling the halogen bond acceptor and the halogen bond donor in a 1:1 stoichiometric ratio, and their iso-structurality was confirmed by X-ray powder diffraction. However, attempts to obtain the co-crystals from solution were successful only for combinations of iodo- and bromopentafluorobenzene with acridine, as well as for the combination of iodopentafluorobenzene with phenazine. The co-crystal of bromopentafluorobenzene and phenazine could not be obtained from solution, which was rationalized due to low solubility or phenazine combined with the weakness of Br···N halogen bonds (Figure 4.19).

Mechanochemical milling has also been demonstrated as a versatile methodology to synthesize halogen-bonded co-crystals involving metal–organic complexes as acceptors. The first report on this topic was by Lapadula and co-workers, who utilized a specifically substituted metal beta-diketonate complex as a halogen bond acceptor.[97] Recently, the Cinčić group demonstrated the use of a carbonyl group as halogen bond acceptor functionality, enabling the mechanochemical synthesis of diverse halogen-bonded co-crystals involving complexes of copper, zinc, cobalt and nickel.[98]

Recent studies have systematically explored mechanochemistry as a route to synthesize halogen-bonded co-crystals with the known brominated reagent N-bromosuccinimide (NBS) as a halogen bond donor. Whereas solution co-crystallization is often found to lead to side-reactions involving bromination of the halogen bond acceptors based on pyridine, mechanochemistry enables circumventing such reactivity, leading to selective formation of co-crystals.[99,100]

4.7 Co-crystal-catalyzed Photo-mechanochemical Reactions

The MacGillivray group demonstrated how mechanochemistry can be combined with template-directed photochemical [2 + 2] photodimerization[101] into an entirely solvent-free and organocatalytic process for regiospecific solid-state photodimerization. Specifically, Atkinson and co-workers have utilized[102] neat grinding to assemble pyridine-substituted olefin reactants and ditopic molecular templates into co-crystals composed of four-membered hydrogen-bonded assemblies in which pairs of olefin reactant molecules are aligned for topochemical [2 + 2] photodimerization (Figure 4.20(a)). Exposing the mechanochemically prepared co-crystals to ultraviolet (UV) radiation led to anticipated regioselective photochemical

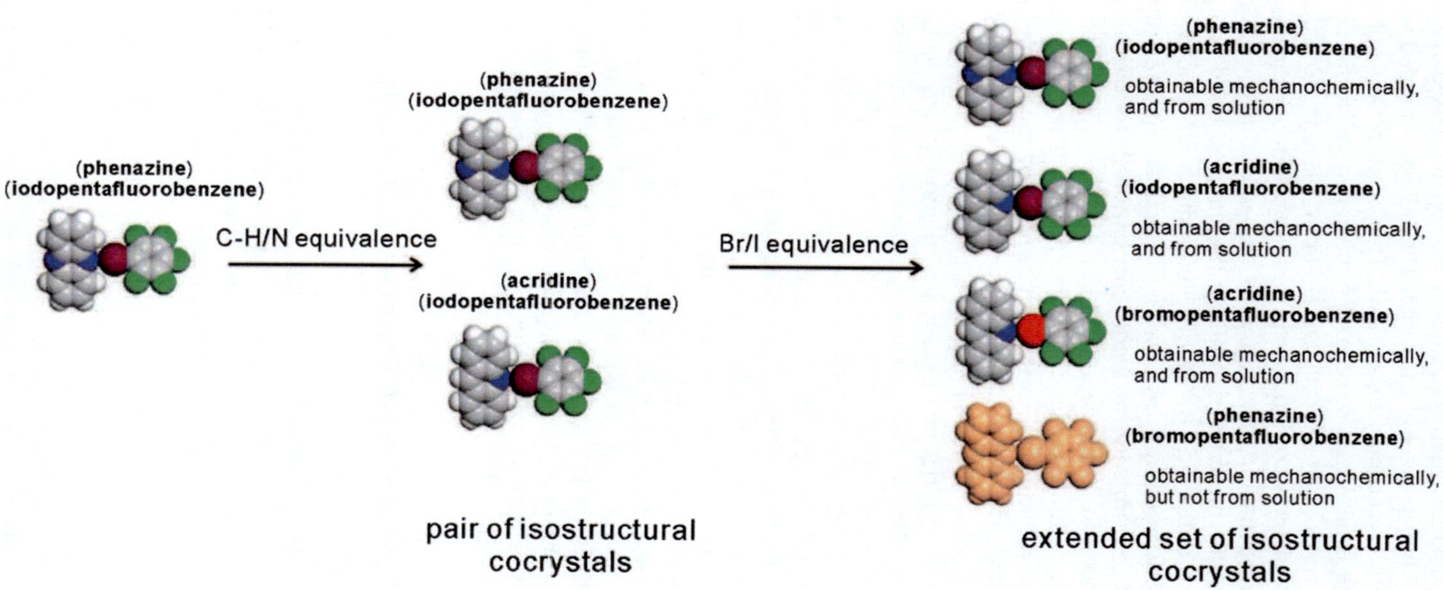

Figure 4.19 Schematic representation of the design for the synthesis of a family of isostructural co-crystals demonstrated by Cinčić and co-workers through structural equivalence of aromatic C–H and N groups, as well as halogen-bonded Br and I donor atoms. The synthesis of the entire isostructural set required the use of mechanochemistry, as attempts to obtain the co-crystal of phenazine and bromopentafluorobenzene from solution were unsuccessful.[96]

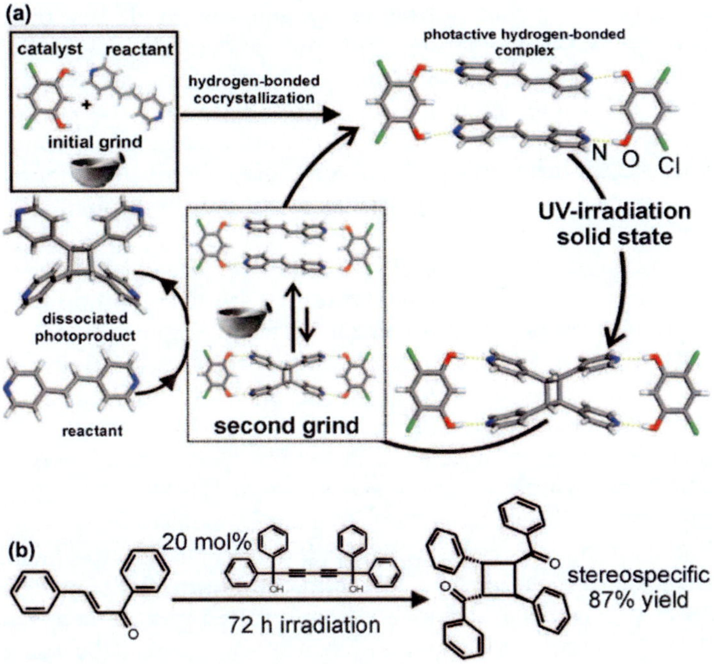

Figure 4.20 (a) The catalytic mechano- and photochemical dimerization process developed by Sokolov *et al.*[103] and (c) the catalytic mechanochemical photodimerization of a chalcone reported by Toda.[107]
Adapted from ref. 14 with permission from The Royal Society of Chemistry.

synthesis of *rctt*-1,2,3,4-tetrasubstituted cyclobutanes. Subsequently, it was also shown that such mechanochemical co-crystallization can be utilized to achieve catalytic photochemical [2 + 2] photodimerization in the solid-state (Figure 4.20(a)).[103] Specifically, grinding together the reactant pyridine-substituted olefin with a sub-stoichiometric amount (*ca.* 10–50 mol%) of a resorcinol as the hydrogen-bonding template, followed by irradiation of the resulting solid mixture to UV light resulted in the formation of photoactive assemblies and template-controlled photocyclodimerization. Upon repeated grinding, however, the resorcinol template re-assembled with unreacted olefin, allowing the entire process to be repeated. In that way, through repeated consecutive mechanochemical and photochemical steps, it was possible to regiospecifically obtain 99% yield of the *rctt*-cyclobutane product, using 10 mol% of the catalyst over 80 h. The catalytic behaviour was explained through theoretical calculations, which indicated that the supramolecular hydrogen-bonded assemblies of the resorcinol template with the reactant olefins should be *ca.* 4.8 kJ mol^{-1} more stable than the analogous assemblies of the cyclobutane photoreaction product. The calculated energy difference provided an explanation of the driving force for the mechanochemical regeneration of the photoactive assemblies upon grinding of the irradiated mixture. It was subsequently demonstrated that the mechanochemical co-crystal formation and the photochemical [2 + 2] photodimerization in the co-crystal could be conducted simultaneously, by vortexing of solid-state mixtures of the template and the olefin reactant in the presence of milling media.[104] This approach, first applied for the synthesis of 1,2,3,4-tetrakis(4-pyridyl)cyclobutane, was subsequently adapted for the photo-mechanochemical synthesis of a *para*-cyclophane.[105]

In 2016 Zhu and co-workers[106] designed a "green" process for [2 + 2] photodimerization of vitamin K_3, based on the observation that it forms a photo-reactive co-crystal by grinding with halogen bond donors 1,4-diiodo- or 1,4-dibromotetrafluorobenzene. The alignment of olefin groups of vitamin K_3 in such a co-crystal ensures stereospecific and quantitative formation of the vitamin K_3 *syn*-dimer upon UV-irradiation. The halogen bond donor can then be readily removed from the resulting material by sublimation, allowing it to be re-used for the next cycle of photochemical dimerization by grinding with fresh reactant.[106] The described use of mechanochemical co-crystallization to enable catalytic solid-state [2 + 2] photodimerization was, in a way, anticipated by Toda's group[107] who noticed that mixing and UV-irradiation of an olefin-containing chalcone derivative in the presence of a wheel-and-axle diol host resulted in a stereospecific head-to-tail [2 + 2] photodimerization, which was rationalized by the transient formation of a photoactive host–guest inclusion compound (Figure 4.20(b)).

4.8 Mechanochemical Reactions of Co-crystals

4.8.1 Supramolecular Metathesis and Co-crystal–Co-crystal Reactions

The mechanochemical environment provides a unique opportunity to conduct chemical and supramolecular transformations of pre-assembled complex solid-state structures, specifically co-crystals, that would in solution normally dissociate into constituents. An early exploration of mechanochemical reactivity of molecular co-crystals was reported by Caira's group, who observed that grinding of anthranilic acid with a co-crystal of the active pharmaceutical ingredient sulfadimidine and salicylic acid led to a metathesis reaction, yielding a new co-crystal of sulfadimidine with anthranilic acid, as well as solid salicylic acid, and a small amount of a co-crystal of anthranilic and salicylic acid.[108] The formation of the co-crystal of sulfadimidine with anthranilic acid, which was also observed in competitive co-crystallization experiments conducted by grinding three-component mixtures of solid sulfadimidine, anthranilic acid and either salicylic, acetylsalicylic or benzoic acid, was explained by higher stability of the (sulfadimidine)(anthranilic acid) co-crystal. The Braga group[109] explored such co-crystal metathesis reactions by kneading of pyrazine co-crystals with chiral (D or L), racemic (DL) or diastereomeric (*meso*) tartaric acid. Co-crystals of pyrazine with chiral or racemic tartaric acid were found to undergo acid exchange by kneading with *meso*-tartaric acid. However, no reaction was observed upon grinding the co-crystal of pyrazine and *meso*-tartaric acid with either chiral or racemic forms of tartaric acid. The outcomes of these metathesis reactions indicated greater stability for the co-crystal of *meso*-tartaric acid which was indeed confirmed by slurry experiments.[109]

Another type of co-crystal reactivity involves milling together of pre-assembled co-crystals, as demonstrated by solid-state reactions of enantiomeric co-crystals of the model APIs caffeine or theophylline with either L- or D-forms of tartaric acid.[110] Two outcomes of such reactivity have been observed: the formation of a centrosymmetric co-crystal based on racemic tartaric acid as the co-crystal former, and the dismantling of both co-crystals with the formation of pure racemic tartaric acid in the crystalline state (Figure 4.21(a)). In particular, grinding equimolar amounts of the $(\text{theophylline})_2\cdot$(D-tartaric acid) and $(\text{theophylline})_2\cdot$(L-tartaric acid) co-crystals produced the centrosymmetric co-crystal $(\text{theophylline})_2\cdot$(DL-tartaric acid) (Figure 4.21(b)). This co-crystal may formally be considered a three-component co-crystal, with a composition $(\text{theophylline})_2\cdot(\text{D-tartaric acid})_{0.5}\cdot(\text{L-tartaric acid})_{0.5}$. In contrast, grinding of the equimolar mixture of (caffeine)·(D-tartaric acid) and (caffeine)·(L-tartaric acid) co-crystals led to the formation of physical mixtures of crystalline caffeine and DL-tartaric acid. A subsequent study[111] addressed the co-crystal–co-crystal reactivity of enantiomeric co-crystals based on theophylline and either L- or D-malic acid. In this case, milling of enantiomeric co-crystals led to the assembly of the racemic co-crystal. The conversion was, however, difficult to detect by X-ray powder diffraction, as the enantiomerically pure and racemic forms of the co-crystals were isostructural. A different pattern of reactivity, involving both the formation of a centrosymmetric co-crystal and partial dismantling of the

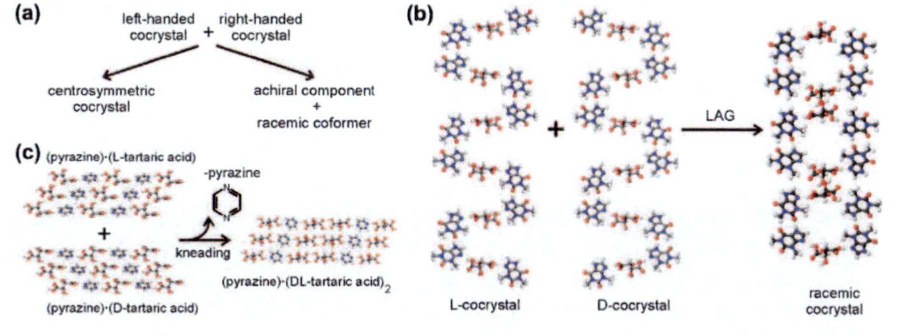

Figure 4.21 (a) Two types of mechanochemical reactions observed between enantiomerically-related co-crystals; (b) the reaction between co-crystals of theophylline with D- and L-forms of tartaric acid, leading to a racemic co-crystal[110,111] and (c) the combination of mechanochemical co-crystal dismantling and merging observed by Braga's group for reactions of enantiomeric co-crystals involving pyrazine and tartaric acid.[109] Adapted from ref. 14 with permission from The Royal Society of Chemistry.

co-crystal was observed by Braga's group, who explored kneading reactions of enantiomeric co-crystals of pyrazine with D- and L-tartaric acids (Figure 4.21(c)).[109] Kneading of equimolar amounts of chiral co-crystals (pyrazine)·(L-tartaric acid) and (pyrazine)·(D-tartaric acid) led to the formation of the centrosymmetric (pyrazine)·(DL-tartaric acid)$_2$ co-crystal, accompanied with the release of volatile solid pyrazine. The course of this reaction was strongly affected by the rapid evaporation of the generated pyrazine: grinding in the presence of a large excess of pyrazine did not lead to co-crystal degradation; instead, the reaction yielded a stoichiometrically different centrosymmetric co-crystal with composition (pyrazine)·(DL-tartaric acid).[109]

4.9 Conclusions

This chapter provided a very brief overview of different aspects of mechanochemistry in co-crystal synthesis, outlining the basic concepts, while highlighting advantages in co-crystal screening efficiency, diversity of co-crystallization systems accessible to mechanochemical techniques in terms of component selection and complexity, and recent advances in mechanistic understanding of mechanochemical co-crystal formation. Admittedly, there remains a significant number of very recent topics that have been omitted from this succinct overview, including approaches to scale-up and high-throughput screening through conventional, as well as more innovative experimental approaches. These include twin screw extrusion,[112] a methodology suitable for scale-up and continuous manufacture of co-crystals, as well as organic molecules[113] and metal–organic phases,[114] as well as resonant acoustic mixing (RAM), an emergent technique for co-crystal synthesis on a small and large scale.[115,116] All these developments show that mechanochemical synthesis is an active, continuously evolving and dynamic area of co-crystal research. I hope that this brief overview will provide an inspiration the further development of this unique area of "using crystals to make crystals".[117]

References

1. S. L. James, C. J. Adams, C. Bolm, D. Braga, P. Collier, T. Friščić, F. Grepioni, K. D. M. Harris, G. Hyett, W. Jones, A. Krebs, J. Mack, L. Maini, A. G. Orpen, I. P. Parkin, W. C. Shearouse, J. W. Steed and D. C. Waddell, *Chem. Soc. Rev.*, 2012, **41**, 413–447.
2. S. Aitipamula, R. Banerjee, A. K. Bansal, K. Biradha, M. L. Cheney, A. R. Choudhury, G. R. Desiraju, A. G. Dikundwar, R. Dubey, N. Duggirala, P. P. Ghogale, S. Ghosh, P. K. Goswami, N. R. Goud, R. R. K. R. Jetti, P. Karpinski, P. Kaushik, D. Kumar, V. Kumar, B. Moulton, A. Mukherjee, G. Mukherjee, A. S. Myerson, V. Puri, A. Ramanan, T. Rajamannar, C. M. Reddy, N. Rodríguez-Hornedo, R. D. Rogers, T. N. Guru Row, P. Sanphui, N. Shan, G. Shete, A. Singh,

C. C. Sun, J. A. Swift, R. Thaimattam, T. S. Thakur, R. K. Thaper, S. P. Thomas, S. Tothadi, V. R. Vangala, N. Variankaval, P. Vishweshwar, D. R. Weyna and M. J. Zaworotko, *Cryst. Growth Des.*, 2012, **12**, 2147–2152.

3. L. Takacs, *Chem. Soc. Rev.*, 2013, **42**, 7649–7659.
4. A. Bruckmann, A. Krebs and C. Bolm, *Green Chem.*, 2008, **10**, 1131–1141.
5. G.-W. Wang, *Chem. Soc. Rev.*, 2013, **42**, 7668–7700.
6. A. L. Garay, A. Pichon and S. L. James, *Chem. Soc. Rev.*, 2007, **36**, 846–855.
7. C. Mottillo and T. Friščić, *Molecules*, 2017, **22**, 144.
8. A. A. Gečiauskaitė and F. Garcia, *Beilstein J. Org. Chem.*, 2017, **13**, 2068–2077.
9. N. R. Rightmire, T. P. Hanusa and A. L. Rheingold, *Organometallics*, 2014, **33**, 5952–5955.
10. T. Friščić, *J. Mater. Chem.*, 2010, **20**, 7599–7605.
11. P. A. Julien, C. Mottillo and T. Friščić, *Green Chem.*, 2017, **19**, 2729–2747.
12. P. Baláž, M. Achimovičová, M. Baláž, P. Billik, Z. Cherkezova-Zheleva, J. M. Criado, F. Delogu, E. Dutková, E. Gaffet, F. J. Gotor, R. Kumar, I. Mitov, T. Rojac, M. Senna, A. Streletskii and K. Wieczorek-Ciurowa, *Chem. Soc. Rev.*, 2013, **42**, 7571–7637.
13. D. Braga, L. Maini and F. Grepioni, *Chem. Soc. Rev.*, 2013, **42**, 7638–7648.
14. T. Friščić, *Chem. Soc. Rev.*, 2012, **41**, 3493–3510.
15. V. Štrukil, *Beilstein J. Org. Chem.*, 2017, **13**, 1828–1849.
16. J. G. Hernández and C. Bolm, *J. Org. Chem.*, 2017, **82**, 4007–4019.
17. D. Tan, L. Loots and T. Friščić, *Chem. Commun.*, 2016, **52**, 7760–7781.
18. *IUPAC Compendium of Chemical Terminology*, 2nd edn, (the "Gold Book"). Compiled by A. D. McNaught, A. Wilkinson, Blackwell, Scientific Publications, Oxford, 1997.
19. T. Friščić, S. L. James, E. V. Boldyreva, C. Bolm, W. Jones, J. Mack, J. W. Steed and K. S. Suslick, *Chem. Commun.*, 2015, **51**, 6248–6256.
20. K. S. Suslick, *Faraday Discuss.*, 2014, **170**, 411–422.
21. S. You, M.-W. Chen, D. D. Dlott and K. S. Suslick, *Nat. Commun.*, 2015, **6**, 6581.
22. M. K. Beyer and H. Clausen-Schaumann, *Chem. Rev.*, 2005, **105**, 2921–2948.
23. J.-L. Do and T. Friščić, *Synlett*, 2017, **28**, 2066–2092.
24. T. N. Drebuschak, A. A. Ogienko and E. V. Boldyreva, *CrystEngComm*, 2011, **13**, 4405–4410.
25. N. R. Rightmire and T. P. Hanusa, *Dalton Trans.*, 2016, **45**, 2352–2362.
26. M. C. Etter, S. M. Reutzel and C. G. Choo, *J. Am. Chem. Soc.*, 1993, **115**, 4411–4412.
27. D. Cinčić, I. Brekalo and B. Kaitner, *Chem. Commun.*, 2012, **48**, 11683–11685.

28. C. F. Burmeister and A. Kwade, *Chem. Soc. Rev.*, 2013, **42**, 7660–7667.
29. N. Shan, F. Toda and W. Jones, *Chem. Commun.*, 2002, 2372–2373.
30. T. Friščić, S. L. Childs, S. A. A. Rizvi and W. Jones, *CrystEngComm*, 2009, **11**, 418–426.
31. T. Friščić, A. W. Trask, W. Jones and W. D. S. Motherwell, *Angew. Chem., Int. Ed.*, 2006, **45**, 7546–7550.
32. G. G. Z. Zhang, R. F. Henry, T. B. Borchardt and X. Lou, *J. Pharm. Sci.*, 2007, **96**, 990–995.
33. D. Hasa, E. Miniussi and W. Jones, *Cryst. Growth Des.*, 2016, **16**, 4582–4588.
34. F. Fischer, A. Heidrich, S. Greiser, S. Benemann, K. Rademann and F. Emmerling, *Cryst. Growth Des.*, 2016, **16**, 1701–1707.
35. M. Tireli, M. Juribašić Kulcsár, N. Cindro, D. Gracin, N. Biliškov, M. Borovina, M. Ćurić, I. Halasz and K. Užarević, *Chem. Commun.*, 2015, **51**, 8058–8061.
36. K. L. Nguyen, T. Friščić, G. M. Day, L. F. Gladden and W. Jones, *Nat. Mater.*, 2007, **6**, 206–209.
37. V. Štrukil, M. D. Igrc, L. Fábián, M. Eckert-Maksić, S. L. Childs, D. G. Reid, M. J. Duer, I. Halasz, C. Mottillo and T. Friščić, *Green Chem.*, 2012, **14**, 2462–2473.
38. T. Friščić, D. G. Reid, I. Halasz, R. S. Stein, R. E. Dinnebier and M. J. Duer, *Angew. Chem., Int. Ed.*, 2010, **49**, 712–715.
39. D. Hasa, G. Schneider Rauber, D. Voinovich and W. Jones, *Angew. Chem., Int. Ed.*, 2015, **54**, 7371–7375.
40. A. D. Grzesiak, F. J. Uribe, N. W. Ockwig, O. M. Yaghi and A. J. Matzger, *Angew. Chem., Int. Ed.*, 2006, **45**, 2553–2556.
41. D. Cinčić, I. Brekalo and B. Kaitner, *Cryst. Growth Des.*, 2012, **12**, 44–48.
42. D. Hasa and W. Jones, *Adv. Drug. Delivery Rev.*, 2017, **117**, 147–161.
43. D. Braga, S. L. Giaffreda, F. Grepioni, A. Pettersen, L. Maini, M. Curzi and M. Polito, *Dalton Trans.*, 2006, 1249–1263.
44. S. Karki, L. Fábián, T. Friščić and W. Jones, *Org. Lett.*, 2007, **9**, 3133–3136.
45. F. Fischer, G. Scholz, L. Batzdorf, M. Wilke and F. Emmerling, *CrystEngComm*, 2015, **17**, 824–829.
46. M. Tremayne, B. M. Kariuki and K. D. M. Harris, *Angew. Chem., Int. Ed.*, 1997, **36**, 770–772.
47. R. W. Schurko, *Acc. Chem. Res.*, 2013, **46**, 1985–1995.
48. G. M. Day, *Crystallogr. Rev.*, 2011, **17**, 3–52.
49. A. J. Cruz-Cabeza, G. M. Day, W. D. S. Motherwell and W. Jones, *J. Am. Chem. Soc.*, 2006, **128**, 14466–14467.
50. V. Štrukil, D. Gracin, O. V. Magdysyuk, R. E. Dinnebier and T. Friščić, *Angew. Chem., Int. Ed.*, 2015, **54**, 8440–8443.
51. V. Štrukil, L. Fábián, D. G. Reid, M. J. Duer, G. J. Jackson, M. Eckert-Maksić and T. Friščić, *Chem. Commun.*, 2010, **46**, 9191–9193.
52. K. Fujii, A. L. Garay, J. Hill, E. Sbircea, Z. Pan, M. Xu, D. C. Apperley, S. L. James and K. D. M. Harris, *Chem. Commun.*, 2010, **46**, 7572–7574.

53. D. R. Weyna, T. Shattock, P. Vishweshwar and M. J. Zaworotko, *Cryst. Growth Des.*, 2009, **9**, 1106–1123.
54. S. L. Childs, N. Rodríguez-Hornedo, L. S. Reddy, A. Jayasankar, C. Maheshwari, L. McCausland, R. Shipplett and B. C. Stahly, *CrystEngComm*, 2008, **10**, 856–864.
55. A. V. Trask, J. van de Streek, W. D. S. Motherwell and W. Jones, *Cryst. Growth Des.*, 2005, **5**, 2233–2241.
56. S. Karki, T. Friščić and W. Jones, *CrystEngComm*, 2009, **11**, 470–481.
57. V. Štrukil, M. D. Igrc, M. Eckert-Maksić and T. Friščić, *Chem. Commun.*, 2012, **48**, 9705–9707.
58. A. V. Trask, N. Shan, W. D. S. Motherwell, W. Jones, S. Feng, R. B. H. Tan and K. J. Carpenter, *Chem. Commun.*, 2005, 880–882.
59. A. V. Trask, W. D. S. Motherwell and W. Jones, *Chem. Commun.*, 2004, 890–891.
60. D. Hasa, E. Carlino and W. Jones, *Cryst. Growth Des.*, 2016, **16**, 1772–1779.
61. T. Friščić and W. Jones, *Cryst. Growth Des.*, 2009, **9**, 1621–1637.
62. R. P. Rastogi and N. B. Singh, *J. Phys. Chem.*, 1968, **72**, 4446–4449.
63. R. P. Rastogi and N. B. Singh, *J. Phys. Chem.*, 1966, **70**, 3315–3324.
64. R. Kuroda, K. Higashiguchi, S. Hasebe and Y. Imai, *CrystEngComm*, 2004, **6**, 464–468.
65. G. Rothenberg, A. P. Downie, C. L. Raston and J. L. Scott, *J. Am. Chem. Soc.*, 2001, **123**, 8701–8708.
66. K. Chadwick, R. Davey and W. Cross, *CrystEngComm*, 2007, **9**, 732–734.
67. H. H. Lee and J. C. Warner, *J. Am. Chem. Soc.*, 1933, **55**, 209–214.
68. O. Dolotko, J. W. Wiench, K. W. Dennis, V. K. Pecharsky and V. P. Balema, *New J. Chem.*, 2010, **34**, 25–28.
69. A. Jayasankar, A. Somwangthanaroj, Z. J. Shao and N. Rodríguez-Hornedo, *Pharm. Res.*, 2006, **23**, 2381–2392.
70. A. Y. Ibrahim, R. T. Forbes and N. Blagden, *CrystEngComm*, 2011, **13**, 1141–1152.
71. K. Užarević, I. Halasz and T. Friščić, *J. Phys. Chem. Lett.*, 2015, **6**, 4129–4140.
72. T. Friščić, I. Halasz, P. J. Beldon, A. M. Belenguer, F. Adams, S. A. J. Kimber, V. Honkimäki and R. E. Dinnebier, *Nat. Chem.*, 2013, **5**, 66–73.
73. I. Halasz, A. Puškarić, S. A. J. Kimber, P. J. Beldon, A. M. Belenguer, F. Adams, V. Honkimäki, R. E. Dinnebier, B. Patel, W. Jones, V. Štrukil and T. Friščić, *Angew. Chem., Int. Ed.*, 2013, **52**, 11538–11541.
74. H. Kulla, S. Greiser, S. Benemann, K. Rademann and F. Emmerling, *Cryst. Growth Des.*, 2017, **17**, 1190–1196.
75. D. Gracin, V. Štrukil, T. Friščić, I. Halasz and K. Užarević, *Angew. Chem., Int. Ed.*, 2014, **53**, 6193–6197.
76. L. Batzdorf, F. Fischer, M. Wilke, K.-J. Wenzel and F. Emmerling, *Angew. Chem., Int. Ed.*, 2015, **54**, 1799–1802.

77. S. Lukin, T. Stolar, M. Tireli, M. V. Blanco, D. Babić, T. Friščić, K. Užarević and I. Halasz, *Chem. - Eur. J.*, 2017, **23**, 13941–13949.
78. C. B. Aakeröy, A. M. Beatty and B. A. Helfrich, *Angew. Chem., Int. Ed.*, 2001, **40**, 3240–3242.
79. C. B. Aakeröy, J. Desper and M. Smith, *Chem. Commun.*, 2007, **38**, 3936–3938.
80. F. Topić and K. Rissanen, *J. Am. Chem. Soc.*, 2016, **138**, 6610–6616.
81. R. Burns, "To a Mouse, on Turning Her Up in Her Nest With the Plough," 1785.
82. Y. Imai, N. Tajima, T. Sato and R. Kuroda, *Org. Lett.*, 2006, **8**, 2941–2944.
83. E. Y. Cheung, S. J. Kitchin, K. D. M. Harris, Y. Imai, N. Tajima and R. Kuroda, *J. Am. Chem. Soc.*, 2003, **125**, 14658–14659.
84. T. Friščić, A. V. Trask, W. Jones and W. D. S. Motherwell, *Cryst. Growth Des.*, 2008, **8**, 1605–1609.
85. T. Friščić, E. Meštrović, D. Škalec-Šamec, B. Kaitner and L. Fábián, *Chem. - Eur. J.*, 2009, **15**, 12644–12652.
86. M. Ferguson, N. Giri, X. Huang, D. Apperley and S. L. James, *Green Chem.*, 2014, **16**, 1374–1382.
87. D. Cinčić and T. Friščić, *CrystEngComm*, 2014, **16**, 10169–10172.
88. E. H. H. Chow, F. C. Strobridge and T. Friščić, *Chem. Commun.*, 2010, **46**, 6368–6370.
89. C. Eger and D. A. Norton, *Nature*, 1965, **208**, 997–999.
90. T. Friščić, R. W. Lancaster, L. Fábián and P. G. Karamertzanis, *Proc. Natl. Acad. Sci. U. S. A.*, 2010, **107**, 13216–13221.
91. Z. Luo, T. Friščić and R. Z. Khaliullin, *Phys. Chem. Chem. Phys.*, 2018, **20**, 898–904.
92. K. J. Ardila-Fierro, V. André, D. Tan, M. T. Duarte, R. W. Lancaster, P. G. Karamertzanis and T. Friščić, *Cryst. Growth Des.*, 2015, **15**, 1492–1501.
93. D. Cinčić, T. Friščić and W. Jones, *Chem. - Eur. J.*, 2008, **14**, 747–753.
94. D. Cinčić, T. Friščić and W. Jones, *J. Am. Chem. Soc.*, 2008, **130**, 7524–7525.
95. D. Cinčić, T. Friščić and W. Jones, *Chem. Mater.*, 2008, **20**, 6623–6626.
96. D. Cinčić, T. Friščić and W. Jones, *New J. Chem.*, 2008, **32**, 1776–1781.
97. G. Lapadula, N. Judaš, T. Friščić and W. Jones, *Chem. - Eur. J.*, 2010, **16**, 7400–7403.
98. V. Nemec, L. Fotović, T. Friščić and D. Cinčić, *Cryst. Growth Des.*, 2017, **17**, 6169–6173.
99. J. Mavračić, D. Cinčić and B. Kaitner, *CrystEngComm*, 2016, **18**, 3343–3346.
100. M. Eraković, V. Nemec, T. Lež, I. Porupski, V. Stilinović and D. Cinčić, *Cryst. Growth Des.*, 2018, **18**, DOI: 10.1021/acs.cgd.7b01651.
101. L. R. MacGillivray, G. S. Papaefstathiou, T. Friščić, T. D. Hamilton, D.-K. Bučar, Q. Chu, D. B. Varshney and I. G. Georgiev, *Acc. Chem. Res.*, 2008, **41**, 280–291.

102. M. B. J. Atkinson, D. K. Bučar, A. N. Sokolov, T. Friščić, C. Robinson, M. Bilal, N. G. Sinada, A. Chevannes and L. R. MacGillivray, *Chem. Commun.*, 2008, 5713–5715.
103. A. N. Sokolov, D.-K. Bučar, J. Baltrusaitis, S. X. Gu and L. R. MacGillivray, *Angew. Chem., Int. Ed.*, 2010, **49**, 4273–4277.
104. J. Stojaković, B. S. Farris and L. R. MacGillivray, *Chem. Commun.*, 2012, **48**, 7958–7960.
105. J. Stojaković, B. S. Farris and L. R. MacGillivray, *Faraday Discuss.*, 2014, **170**, 35–40.
106. B. Zhu, J.-R. Wang, Q. Zhang and X. Mei, *CrystEngComm*, 2016, **18**, 6327–6330.
107. F. Toda, K. Tanaka and A. Sekikawa, *Chem. Commun.*, 1987, 279–280.
108. R. Caira, L. R. Nassimbeni and A. F. Wildervanck, *J. Chem. Soc., Perkin Trans. 2*, 1995, 2213–2216.
109. D. Braga, F. Grepioni and G. I. Lampronti, *CrystEngComm*, 2011, **13**, 3122–3124.
110. T. Friščić, L. Fábián, J. C. Burley, W. Jones and W. D. S. Motherwell, *Chem. Commun.*, 2006, 5009–5011.
111. T. Friščić, L. Fábián, J. C. Burley, D. G. Reid, M. J. Duer and W. Jones, *Chem. Commun.*, 2008, 1644–1646.
112. D. Daurio, K. Nagapudi, L. Li, P. Quan and F.-A. Nunez, *Faraday Discuss.*, 2014, **170**, 235–249.
113. D. E. Crawford, C. K. Miskimmin, J. Cahir and S. L. James, *Chem. Commun.*, 2017, **53**, 13067–13070.
114. D. Crawford, J. Casaban, R. Haydon, N. Giri, T. McNally and S. L. James, *Chem. Sci.*, 2015, **6**, 1645–1649.
115. K. Nagapudi, E. Y. Umanzor and C. Masui, *Int. J. Pharm.*, 2017, **521**, 337–345.
116. D. Tan and T. Friščić, *Eur. J. Org. Chem.*, 2018, 18–33.
117. D. Braga and F. Grepioni, *Chem. Commun.*, 2005, 3635–3645.

CHAPTER 5

Pharmaceutical Co-crystals—Molecular Design and Process Development

SRINIVASULU AITIPAMULA,[a] YU ZAIQUN,[a] PUI SHAN CHOW[a] AND REGINALD B. H. TAN*[a,b]

[a] Crystallization and Formulation Science, Institute of Chemical and Engineering Sciences, A*STAR (Agency for Science, Technology and Research), 1 Pesek Road, Jurong Island, Singapore 627833; [b] Department of Chemical & Biomolecular Engineering, National University of Singapore, 4, Engineering Drive 4, Singapore 117576
*Email: reginald@nus.edu.sg

5.1 Introduction

Identification of a suitable solid form of an active pharmaceutical ingredient (API) that possesses desirable physicochemical properties will have a profound impact on the overall drug development process.[1] Hence, preformulation stages of drug development focus on identification of as many solid forms as possible and evaluation of their physicochemical properties. An API can exist in a variety of solid forms and drug developers often call upon amorphous solids, polymorphs, hydrates, solvates, salts, and co-crystals (Figure 5.1).[2] The solid state nature of each of these solid forms determines their properties and plays a significant role in drug development.

Amorphous solids typically do not exhibit long range order at a molecular level and this often results in a faster dissolution rate and higher

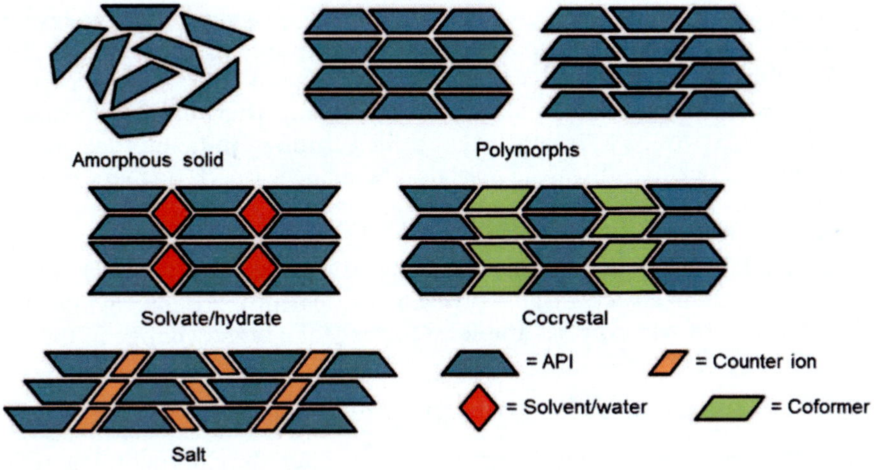

Figure 5.1 Most common solid forms of pharmaceuticals. Multicomponent crystals (salts, solvates/hydrates, and co-crystals) can also exist in polymorphic forms.[4]

bioavailability than their crystalline form. However, amorphous solids are often not the preferable solids for formulation development due to their poor physical stability and tendency to convert to the crystalline state. Hence, the majority of marketed drug formulations consist of APIs in crystalline form. Owing to their stability and ease of handling, crystalline forms are the preferred solids for formulation development. A crystalline form can exist in multiple crystalline phases (called polymorphs), which have different arrangements of molecules in the crystal lattice.[3,4] The properties of polymorphic forms rely on their crystal structure and confer unique characteristics, such as stability, solubility, dissolution rate, and bioavailability. A major concern with polymorphic forms is that the stability of a particular polymorph is controlled by external factors such as temperature and pressure, and unstable (metastable) polymorphs tend to convert to the stable form upon external stimuli.[5] Inclusion of water or solvent molecules in the crystal lattice leading to the formation of hydrates or solvates is a common phenomenon in organic crystals and APIs. Although some of these solids have been used in marketed drugs, the volatility (hence poor stability) and toxicity of the solvents limit their usage as precursors for the development of novel drug formulations. Pharmaceutical salts, derivatives of proton transfer from an acid to the base, are far more common than the neutral solid forms in the development of drug formulations.[6] It has been estimated that 50% of marketed drugs contain the active ingredients in their salt form. The ionic nature of salts confers them unique properties such as improved solubility and dissolution rates, which can have a direct impact on the bioavailability. Even though salt formation is the most preferred route for development of drug products, it is only applicable when the target API has an ionizable functional group. Over the past two decades, interest in exploring co-crystals for pharmaceutical use has grown significantly. Pharmaceutical co-crystals consist of an API and a pharmaceutically acceptable co-former that is a solid under ambient conditions.[7] A major advantage with co-crystals is that the technique can be applied to all APIs regardless of their ionizable capacity and the properties of APIs can be controlled through judicious selection of co-formers. Pharmaceutical co-crystals are made with co-formers that are chosen from the list of Generally Recognized As Safe (GRAS) chemicals. These constitute a wide range of chemicals that include carboxylic acids, amides, aldehydes, alcohols, *etc.* Therefore, the diversity of GRAS chemicals in terms of their structure and physicochemical properties provides additional means of selecting a suitable co-former for a target change in the API.

5.1.1 Role of Co-crystals in Drug Development

Approximately 80% of the drugs currently under development belong to Class II or IV of the biopharmaceutical classification system (BCS).[8] Poor solubility is the characteristic nature of the drugs that fall into these classes. Drug formulators often rely on techniques such as micronization, solid dispersion, encapsulation, salt formation, amorphous forms, *etc.* to address solubility issues.[9] However, these techniques have inherent drawbacks with respect to manufacturing and potential stability problems. Co-crystals have been proved effective in addressing solubility issues of BCS Class II and IV drugs. In addition, many solid-state problems such as solubility, physical and chemical stability, hygroscopicity, melting point, mechanical and flow properties could be modified by selecting an appropriate co-former (Figure 5.2).[10,11] From an intellectual property perspective, the emergence of co-crystals as novel solid forms provided new avenues for extending patent life of parent drugs. Co-crystals that show improved physicochemical properties satisfy the three criteria required for a patent: novelty, non-obviousness, and utility. In this regard, a review by Trask is noteworthy, which highlights the patentability aspects of pharmaceutical co-crystals.[12] With the interest in pharmaceutical co-crystals growing, there is also a plethora of patent applications on co-crystals and some have been granted. For innovator companies, pharmaceutical co-crystals provide an additional means of extending patent life of their APIs through identification of novel co-crystals and thus results in greater economic benefit.

A greater impetus for development of pharmaceutical co-crystals as drug products has been received from the regulatory bodies such as the US Food and Drug Administration (FDA, 2011)[13] and the European Medicines Agency (EMA, 2014),[14] which have recognized the importance of pharmaceutical co-crystals and released draft guidance on their regulatory classification. While the FDA classifies pharmaceutical co-crystals as drug product intermediates and the co-formers used to make the co-crystals are defined as excipients, the EMA defines co-crystals as being solid state variants of the APIs, aligning them with salts, polymorphs, hydrates or solvates. The views of the regulatory bodies, in particular the FDA, are contrary to what the current understanding

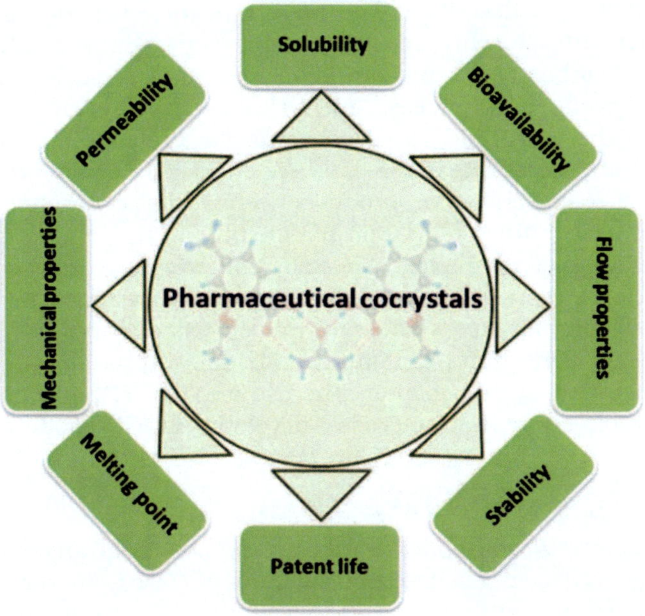

Figure 5.2 Impact of pharmaceutical co-crystals on drug properties and development.

of the concept of pharmaceutical co-crystals is: that the second component in a co-crystal is a pharmaceutically acceptable co-former. Therefore, to address the industry concerns and also to ease the regulatory burden, the FDA has most recently reclassified co-crystals as a special class of solvates in which the second component is non-volatile.[15] Furthermore, the FDA made sure that the regulatory classification of co-crystals is similar to that of a polymorph of the API, meaning that co-crystals do not require regulatory approval, as is required by a new API or a drug intermediate.

5.1.2 Process Development and Scale-up of Co-crystallization

Various methods have been reported for co-crystals preparation on the bench scale, including milling, melt extrusion, solution mediated transformation, evaporation, solution co-crystallization, *etc.* When it comes to commercial production, solution crystallization has advantages over other methods in terms of purity control and scalability. It can be integrated with current manufacturing processes seamlessly by utilizing existing crystallization facilities.

Understanding the mechanisms and techniques involved is a prerequisite for successful crystallization development. In this chapter, a very brief introduction to the major concepts and mechanisms in crystallization is given as the foundation to rationalize development efforts. Then a general crystallization development procedure adopting the principles of quality by design (QbD) is described. The procedure includes solvent screening, measurement of solubility and meta-stable zone width (MSZW), process optimization, robustness study, and considerations in scale-up. An understanding of product and process, and risk assessment is emphasized.

However, there are additional concerns for the development of a co-crystallization process due to the presence of a second component. For one, the purity of co-crystallization products needs to be considered. Under certain operating conditions, crystals of constituent single components may precipitate out to contaminate co-crystals. This concern can be addressed by determination of a ternary solid–liquid phase diagram. The traditional definition of solubility for a single component system does not apply to co-crystals. A new definition of solubility and supersaturation is required and will be illustrated in this chapter. The use of ATR-FTIR and process modelling for the co-crystallization process development will also be highlighted.

5.2 Co-crystal Design

Crystal structure is the representation of the spatial arrangement of molecules interconnected by intermolecular interactions. In the case of co-crystals, hydrogen bonds play a great role in linking different molecules together to form a co-crystal. The inherent directionality of hydrogen bonds makes them reliable design elements in the co-crystal screening process. Hence, knowledge of intermolecular interactions between the molecular components is a prerequisite for successful co-crystal design. Co-crystals are defined as multi-component crystals composed of two or more solid components in stoichiometric ratio.[7] Design of co-crystals for a given molecule starts with analysing the functional groups available on the target molecule and finding complementary functional groups that would likely form predictable intermolecular interactions with the functional groups of the target molecule (Figure 5.3).[16] Etter formulated a set of hydrogen bonding rules that provide valuable insight into molecular recognition events between

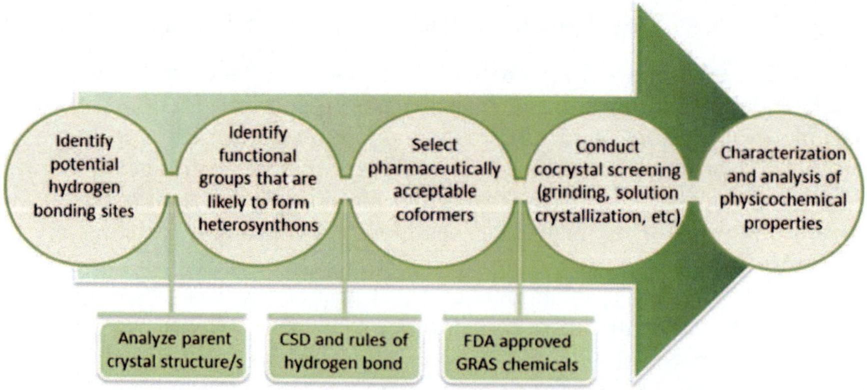

Figure 5.3 Flow chart for the design and synthesis of pharmaceutical co-crystals.

hydrogen bond donors and acceptors.[17] According to these rules, association of two or more molecules in the crystal lattice would be determined by the hydrogen bond donor and acceptor capability of functional groups. Co-crystal design necessitates the knowledge of hierarchy of hydrogen bonds, and hence Etter's hydrogen bond rules provide guiding principles for the co-crystal design. Crystal engineering, which is defined by Desiraju, as "*the understanding of intermolecular interactions in the context of crystal packing and in the utilization of such understanding in the design of new solids with desired physical and chemical properties*" finds potential relevance in the design of co-crystals.[18] Knowledge of intermolecular interactions forms the basis for strategizing co-crystal formation between two or more molecules and, thus, crystal engineering strategies facilitate design of co-crystals through wise selection of potential co-formers. In this regard, structural store houses, such as the Cambridge Structural Database (CSD),[19] serve as libraries of structural information, and analysis of relevant crystal structures deposited in the CSD is used in the selection of co-formers.

Many strategies have been devised to understand and strategize the molecular recognition between two or more different molecules and thereby predict the likelihood of co-crystal formation. Some of these are described with representative examples in the following sections.

5.2.1 Synthon-based Design Strategies

Recurring intermolecular interactions in the crystal structures are recognized by Desiraju as supramolecular synthons.[20] Synthons that are formed between the same functional groups are classified as homosynthons and the ones that involve different but complementary functional groups are called heterosynthons. Synthon-based crystal engineering strategies have been proved effective in the design of co-crystals. These strategies generally rely on the hydrogen bond capability of functional groups present in the components of co-crystals and often demand strong and directional hydrogen bonds for successful co-crystal formation. The existing vast literature on co-crystals established the role of strong intermolecular interactions in the successful co-crystal design. For example, an anticonvulsant drug, carbamazepine, forms more than 50 co-crystals with a diverse set of co-formers, which could be attributed to its ability to form strong hydrogen bonds with the functional groups present in the co-formers.[21] The study has also emphasized the role of molecular shape packing interactions in the design of novel co-crystals of carbamazepine.

A synthon-based co-crystal design strategy for a compound of pharmaceutical interest is demonstrated in Figure 5.4. Triflusal (TFA) is a platelet aggregation inhibitor that poses serious challenges in formulation development because it is practically insoluble in water.[22] In addition, it has a short half-life of 0.5 h and rapidly converts to its deacetylated metabolite. The rapid degradation of TFA also makes it incompatible with all the common excipients. Therefore, the marketed formulations do not contain any

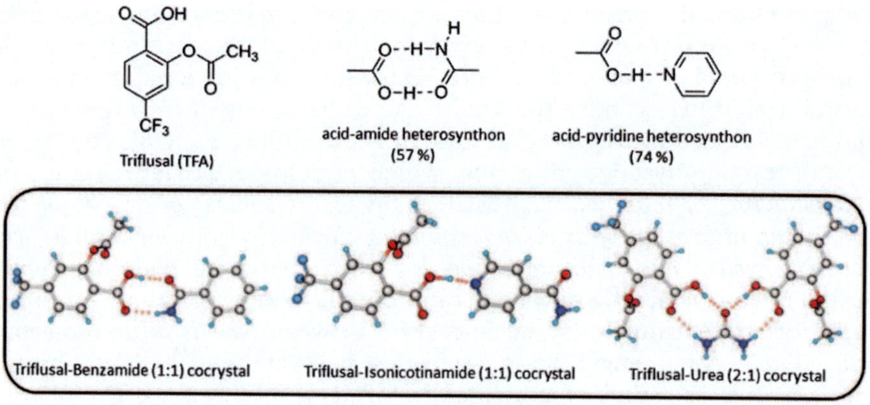

Figure 5.4 A crystal engineering strategy for making co-crystals of triflusal with molecules that contain amide and pyridine functional groups. Adapted from ref. 23 with permission from The Royal Society of Chemistry.

excipient. Co-crystallization could be a possible means to overcome these problems with TFA. The molecular diagram of TFA suggests that it has a carboxylic acid group that forms an acid–acid dimer synthon in its parent crystal structure. Therefore, if one were to make co-crystals of TFA, the first step would be to identify complementary functional groups that would likely form intermolecular interactions with the acid group. Analysis of the crystal structures deposited in the CSD revealed that the carboxylic acid group preferentially forms heterosynthons such as acid–amide and acid–pyridine heterosynthons. Therefore, co-crystal screening of TFA was conducted with amide or pyridine functionalized compounds selected from the list of GRAS chemicals.[23] Crystal structure analysis of the co-crystals revealed that all the co-crystals feature the expected heterosynthons. Some of the co-crystals have been subsequently proved stable at different experimental conditions and claimed as promising candidates for developing stable TFA formulations.

5.2.2 Co-crystal Design for Molecules That Are Devoid of Hydrogen Bonding Sites: Trial and Error Methods

Notwithstanding the wide applications of a synthon-based strategy, it fails when the target molecule is devoid of potential hydrogen bonding sites. Such molecules with no donors or minimal acceptors pose serious challenges in finding reliable synthons for co-crystal design. Crystal structures of these molecules predominately consist of close packing of molecules *via* van der Waals, C–H···π, and π···π interactions. Unlike in the case of molecules with potential hydrogen bonding sites, co-crystal screening for these molecules has only resulted in limited success. For example, griseofulvin (GF) is a poorly soluble antifungal drug that only possesses hydrogen bond acceptors, barring potential hydrogen bonding donors.

A search for crystal structures that contain GF resulted in a few solvated structures. In all the structures, the acceptor, carbonyl, groups partake in weak C–H···O interactions, which further reiterates the difficult nature of the design of co-crystals for GF compared to molecules with hydrogen bond donors. We have conducted a co-crystal screening by choosing a variety of co-formers that consist of carboxylic acids, pyridines, amides, and phenols.[24] A total of 40 co-formers consisting of functional groups that can interact with carbonyl groups were chosen and the co-crystal screening was conducted using solution crystallization and solid-state grinding. Interestingly, most of these experiments either resulted in a physical mixture or the GF and co-former crystallizing separately, with the exception of GF and an artificial sweetener, acesulfame (AceH), which gave a co-crystal in 2 : 1 stoichiometric ratio.[24] The co-crystal was characterized by thermal and X-ray diffraction techniques and found to be in its monohydrate form. As shown in Figure 5.5, water molecules play an important role in the co-crystal formation; they form strong O–H···O hydrogen bonds with all other components of the co-crystal. The co-crystal was found to improve the

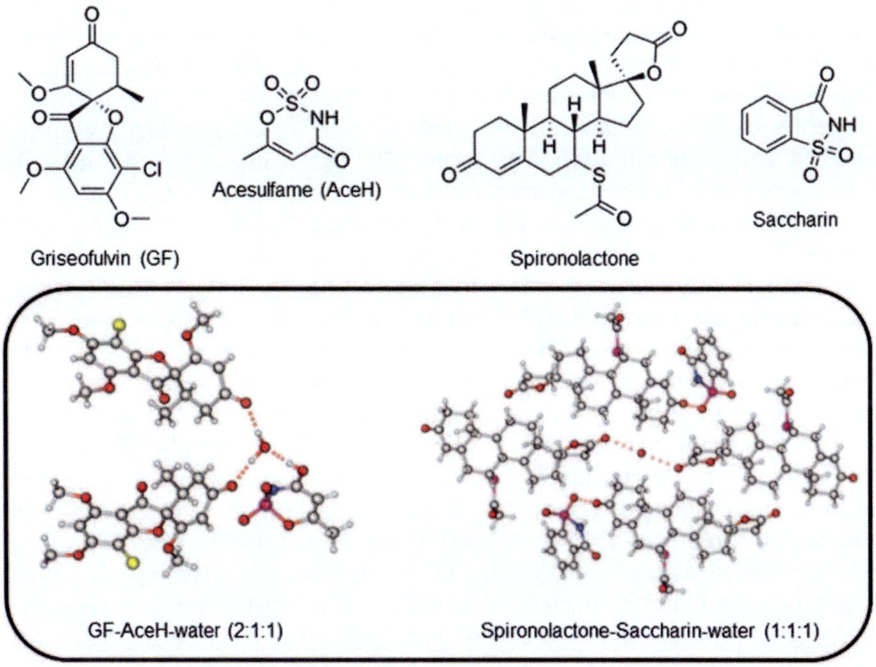

Figure 5.5 Examples of co-crystal for the molecules with limited hydrogen bonding sites. (left) (GF)2-Ace-H-H_2O, (right) Spironolactone-Saccharin hydrate. Adapted with permission from S. Aitipamula, V. R. Vangala, P. S. Chow and R. B. H. Tan, *Cryst. Growth Des.*, 2012, **12**, 5858–5863, Copyright 2012 American Chemical Society, and from N. Takata, R. Takano, H. Uekusa, Y. Hayashi and K. Terada, *Cryst. Growth Des.*, 2010, **10**, 2116–2122, Copyright 2010 American Chemical Society.

solubility and dissolution rate of GF. Similarly, an exhaustive co-crystal screening on spironolactone, a steroidal aldosterone agonist, resulted in a co-crystal hydrate with saccharin as a co-former (Figure 5.5).[25] Crystal structure analysis of the co-crystal hydrate revealed that the API and the co-former form linear channels in which the water molecules are located. It has also been found that the co-crystal hydrate undergoes single-crystal to single-crystal dehydration, which facilitated the determination of the crystal structure of the dehydrated co-crystal.

For both the molecules in these case studies, the co-crystals were found primarily by trial and error, and, hence, the co-crystal screening demands extensive experimental work, which results in huge development costs pertaining to characterization and research resources. The lack of design scope for this type of molecules necessitates knowledge of additional factors such as the method of co-crystallization, solubility of the components, and molecular size complementarity that would likely influence the success or failure of co-crystal screening.

5.2.3 Molecular Descriptor Based Strategy

The aforementioned examples underscore the importance of knowledge of intermolecular interactions in the design of co-crystals. Therefore, the prediction of co-crystal formation between two molecules prior to experimental screening is valuable for reducing the number of trials and development costs. Computational tools to predict the association of two molecules, based on the structural information that can be found in CSD, have been developed recently. One such predictive methodology has been proposed by Fábián.[26] The prediction methodology utilizes knowledge of molecular complementarity and relies on molecular descriptors that are usually used in quantitative structure–activity relationship studies. In this method, each molecule is defined by a set of descriptors based on size, shape, polarity, hydrogen bond donor and acceptor counts, *etc.* It was proposed that if a descriptor was similar enough for the two molecules, a co-crystal would likely form. Some of the salient findings of this study include: (a) molecules preferably form co-crystals with partners of similar polarity, (b) matching of molecular shapes is more important than the matching of absolute molecular dimensions for co-crystal formation, (c) counting donors and acceptors in constituent molecules is insufficient to describe their complementarity, and the formation of synthons is governed by the strength of hydrogen bonds between co-crystal formers rather than by the number of available donors and acceptors.[26] The proposed statistical approach has been applied to an antimalarial drug, artemisinin—a molecule that contains only acceptor atoms—to assess the likelihood of co-crystal formation (Figure 5.6).[27] The method predicted co-crystal formation between artemisinin and 41 of the 75 chosen co-formers. However, despite exhaustive screening using solid-state grinding methods, only two co-crystals with co-formers, resorcinol and orcinol, were experimentally produced. The

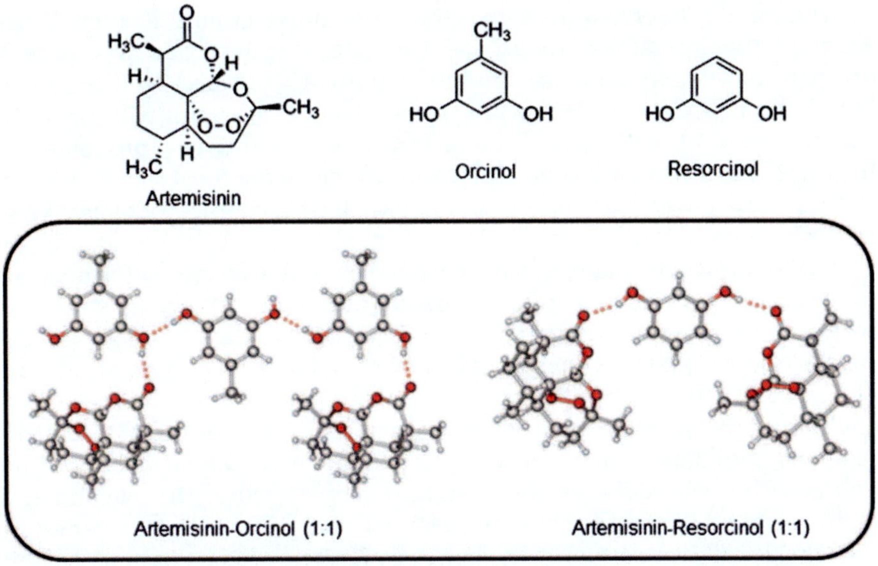

Figure 5.6 Co-crystals of artemisinin as observed from experimental methods. Adapted from ref. 27 with permission from the Royal Society of Chemistry.

co-crystals were structurally characterized and found that the hydroxyl groups of the co-formers form O–H···O hydrogen bonds with the carbonyl of the artemisinin. Although the method represents a rational approach for streamlining the selection of co-formers for co-crystallization, the experimental results suggest that the method is, as yet, not best suited to molecules with few or limited hydrogen bonding functional groups.

5.2.4 Knowledge-based Strategy

With the wealth of knowledge encompassed within the crystal structures deposited in the CSD along with the various software components provided by the Cambridge Crystallographic Data Centre (CCDC), it is possible to examine how a particular functional group interacts with other hydrogen bond donors and acceptors. In this regard, an alternative approach—a knowledge-based methodology—to those previously described has been developed to predict hydrogen bond formation between acceptor and donor atoms and has been applied for the selection of co-formers for co-crystal formation.[16] The method comprises interaction and fragment searching in IsoStar[28] and Conquest,[29] as well as full interaction mapping and contact searching with specific interaction motifs in Mercury.[30] Many of these software programs allow the environment to be easily varied and manipulated, with the inclusion and modification of restraints, to best describe the chemical space of interest. Each functional group is treated individually for hydrogen bond propensity analysis and then the likelihood of formation of intermolecular interactions with other functional groups is determined. Once all the interaction maps are available, the next step would be to compare the likelihood of hetero-interactions with respect to those of the homo-interactions in order to judge how likely a co-crystal will form. The combinations of functional groups that have the highest probability of hydrogen bond formation are chosen for experimental trials.

To illustrate applications of the knowledge-based methodology, Wood *et al.* chose paracetamol as a model system and applied the methodology to predict co-crystal formation between paracetamol and the co-formers that were used in previous co-crystal screening attempts.[16] Previous studies revealed that paracetamol forms co-crystals with 13 of the 35 co-formers chosen for the co-crystal screening. As shown in Figure 5.7, paracetamol has phenol and secondary amide groups as potential hydrogen bonding sites. An analysis of the three known polymorphs of paracetamol revealed that the phenol group forms O–H···O hydrogen bond with the secondary amide group. Using the software tools in CSD and Mercury, the likelihood of hydrogen bond formation between the functional groups of paracetamol and the functional groups found in the 35 co-formers was evaluated. A multi-component (MC) score was calculated based on the probability of the hetero- and homo-interactions between the paracetamol and the chosen co-formers. Figure 5.8 shows the results of the MC analysis. Experimentally observed co-formers are shown as green diamonds whereas the unsuccessful co-formers are shown as red triangles. Some of the successful co-formers that formed co-crystals with paracetamol are 4,4-bipyridine, 1,2-bis(4-pyridyl)ethane, phenazine, citric acid, piperazine, and morpholine, while the unsuccessful co-formers include, caffeine, pyrazine, malic acid, malonic acid, and benzoic acid. Based on this study, Wood *et al.* further concluded that the synthon-based or hydrogen bond propensity methods are more successful when the hydrogen bonding is more dominant, while a combination of methods is valuable when performing knowledge-based co-former screening. Various steps that are encompassed in knowledge-based methodology are detailed with a representative example in the next section.

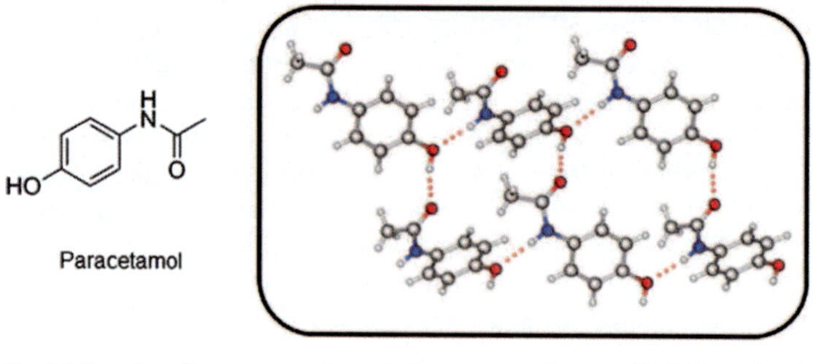

Figure 5.7 Molecular diagram and crystal structure (monoclinic form) of paracetamol.
Adapted from ref. 16 with permission from The Royal Society of Chemistry.

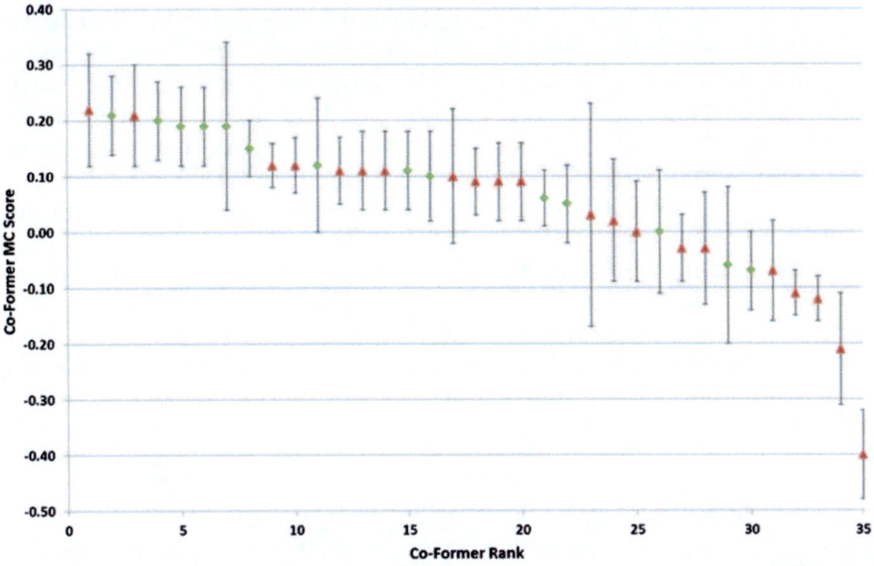

Figure 5.8 Results of paracetamol hydrogen bond propensity analysis as predicted by the knowledge-based methodology; experimentally observed co-formers are shown as green diamonds whereas the unsuccessful co-formers are shown as red triangles.
Reproduced from ref. 16 with permission from The Royal Society of Chemistry.

5.2.5 Case Study—Design of Propyphenazone Co-crystals

As discussed in the previous sections, prediction of co-crystal formers for molecules with strong hydrogen bond donors and acceptors is a relatively easy exercise using the methodologies that have evolved over time. However, these methodologies are only effective for molecules containing strong hydrogen bond donors and acceptors. Molecules that are dominated by hydrophobic groups generally form crystals that feature weaker intermolecular interactions such as C–H···O, C–H···π or π–π interactions. As prediction of these interactions is highly challenging and requires knowledge of various other factors that influence molecular recognition, application of a combination of multiple prediction methodologies is necessary for a knowledge-based co-crystal screen for molecules with no potential hydrogen bonding sites. We have recently conducted a co-crystal screen for propyphenazone (Propy), an antipyretic and analgesic agent, which falls into the aforementioned category of molecules lacking hydrogen bond donor groups and containing only one acceptor atom.[31] Knowledge of intermolecular interactions through a combination of crystal structures deposited in the CSD and additional knowledge that was gained from structurally related molecules were used as guiding principles for the selection of suitable co-formers.

Propy is known to exist in three polymorphic forms and crystal structures of two of them have been obtained from powder X-ray diffraction data. In both the polymorphs, methyl substituent of the Propy forms a C–H···O interaction with a carbonyl group. Crystal structure of a drug–drug co-crystal involving Propy and pyrithyldione has also been reported and the crystal structure of the co-crystal features a $N–H_{pyrithyldione}\cdots O_{carbonyl}$ hydrogen bond.[32] A number of co-crystals for two of structurally similar (antipyrine and edaravone) molecules have been reported (Figure 5.9).[33,34] Interestingly, most of the co-formers that formed co-crystals with these two targets feature a carboxylic acid group. An analysis of intermolecular interactions in these systems revealed that the main interactions of the carbonyl group involve carboxylic acid and amine donor groups. In addition to this preliminary structural information, an in-depth knowledge of potential interactions that can occur for Propy has been deduced using the software tools incorporated in CSD and Mercury.

5.2.5.1 *Interaction Searching Using IsoStar*

Crystal structures that contain a carbonyl functional group were retrieved from the CSD and investigated for their intermolecular interactions. Different models of the carbonyl group were tested to achieve a balance of correct description of the chemical environment and neighbouring atoms of the carbonyl in Propy whilst ensuring that sufficient hits were returned to draw reliable conclusions. The IsoStar search tool (IsoStar v2.2.3),[28] which was incorporated in Mercury was used to analyse the probability of intermolecular interactions between the pairs of functional groups selected for the search. Figure 5.10 depicts the donor functional groups utilized in the IsoStar search. These contain several differing descriptions of the CO_2H, OH, NH_2 and NH groups, which are general donors and encompass a number of more specific functionalities (*e.g.* amides, imides, and various ring systems). The search returned the largest number of structures that contained the interactions involving an aromatic amine, aliphatic hydroxyl, or carboxylic acid as the donor group. This provided a starting point for co-former

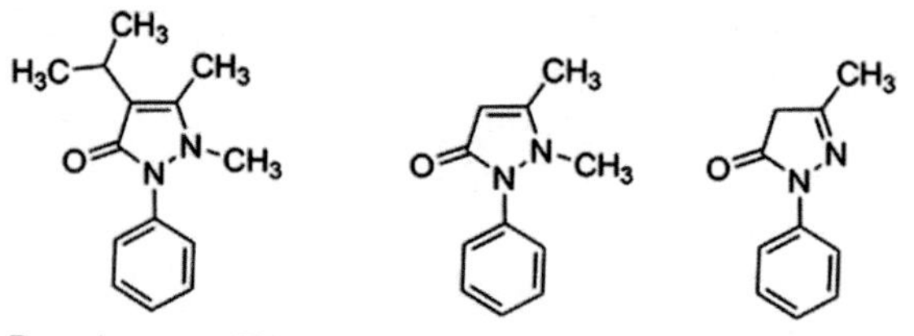

Figure 5.9 Molecular structures of Propy and the structurally related antipyrine and edaravone.
Reprinted with permission from L. K. Mapp, S. J. Coles and S. Aitipamula, *Cryst. Growth Des.*, 2017, **17**, 163–174, Copyright 2017 American Chemical Society.

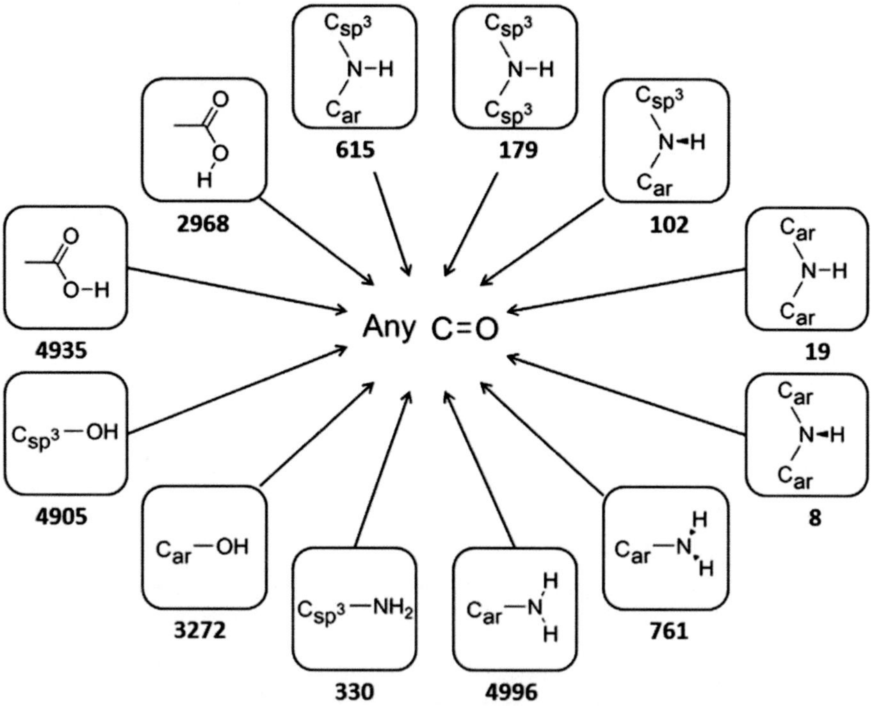

Figure 5.10 Schematic of the donor functional groups implemented in the Isostar search with the number of structures returned which contain an interaction to a carbonyl group.
Reprinted with permission from L. K. Mapp, S. J. Coles and S. Aitipamula, *Cryst. Growth Des.*, 2017, **17**, 163–174, Copyright 2017 American Chemical Society.

selection and an idea of the most ideal functional groups that would be present in the co-former molecules for co-crystal formation.

5.2.5.2 Motif Contact Search

Interaction searching (CSD-Materials Suite and Mercury) provides a more detailed and specific approach and allows greater flexibility in the description of the chemical environment of the functional groups. In this study, a number of different models with differing chemical environments for the carbonyl group were utilized (Table 5.1) as it was found that there are very few examples of structures that contain a carbonyl in a similar fragment as that found in Propy. Therefore, the environment around the carbonyl group was modified by relaxing the number of restraints. The use of multiple models generated a greater number of results and thus better statistics to draw reliable conclusions.

Table 5.1 shows results of the interaction motif searches with the five different models of the carbonyl with a selection of donor groups. In general, the

Table 5.1 Results from interaction searching with five carbonyl models (with decreased restraints and specificity left to right) and 11 donor group descriptions. Reprinted with permission from L. K. Mapp, S. J. Coles and S. Aitipamula, *Cryst. Growth Des.*, 2017, 17, 163–174, Copyright 2017 American Chemical Society.

C=O model	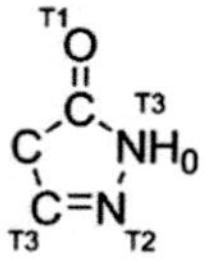		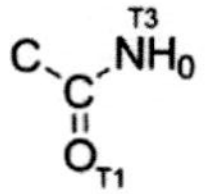		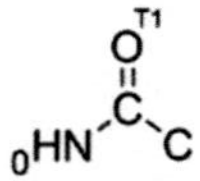		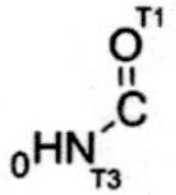		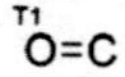	
Donor	No. structures	% Frequency	No. structures	% Frequency	No. structures	% Frequency	No. structures	% Frequency	No. structures	% Frequency
T3NH$_2$	15	45.5	529	47.3	771	49.6	1009	52.6	3302	58.6
T3NH$_1$_cycllic	14	34.1	1234	35.6	1589	37.2	2412	40	11 708	56.1
T3NH$_1$_acycllic	18	8.5	740	28.7	1343	31.3	918	29.9	2271	31.9
T3NH$_1$	32	13	1967	34.2	2918	36	3412	38.3	14 580	52.3
T2NH$_1$	0	0	0	0	1	33.3	8	16.7	20	14.7
CO$_2$H	1	20	282	29.7	389	30.9	354	30.2	1653	32.2
Aliphatic CO$_2$H	0	0	196	34.4	271	33.2	254	35.4	995	40.6
Aromatic CO$_2$H	1	50	68	24.3	94	28.6	77	23.8	386	30.9
OH_general	21	31.3	2571	49.4	3463	49.8	3379	48.4	14 175	51.9
OH_cyclic	10	25.6	1378	51	1755	51.4	1782	48.4	8775	50.2
OH_acyclic	11	35.5	1227	41.6	1750	42.8	1661	39	5802	43.7

statistics suggest that NH_2 is the most favoured donor; however, the isopropyl substituent in close proximity of the carbonyl in Propy could influence the interaction that an amine group can form with the carbonyl. In this regard, hydroxyl donors that contain less steric hindrance may be favoured over a primary amine group. Indeed, the hydroxyl groups resulted in similar statistics as amine group. Hydroxyl groups are more favoured over carboxylic acids with cyclic hydroxyls being more favourable compared to their acyclic counterparts. Cyclic NH donors provide the next best group; however, interactions with an acyclic NH are less common. As the restraints on the carbonyl fragment are released, NH_2 donors become the most favourable and there is a greater separation to the hydroxyl group results. In these instances, carboxylic acids also become more favourable, which would indicate that when restraints and constraints are irrelevant, the larger, bulkier donor groups become more favourable than when the acceptor environment is constrained. The analysis of intermolecular interactions using the combination of IsoStar and Motif contact searching tools revealed that aromatic amines and hydroxyls, aliphatic hydroxyls, and carboxylic acid have the greatest propensity of interaction with the carbonyl of Propy, and suggest that molecules containing these donor groups are the most suitable for co-crystal screening with Propy.

5.2.5.3 Experimental Screening

Experimental screening was conducted with co-formers that contain the main functional groups identified in the CSD analysis, either in isolation or in various combinations. Co-crystal screening was conducted using solid-state grinding techniques and supplemented with solvent-evaporative co-crystallization experiments. The screening resulted in eight co-crystals for which single crystals suitable for structural analysis were obtained and many samples for which no single crystals were produced. All the co-crystals were experimentally characterized by X-ray diffraction techniques. Interestingly, all the successful co-formers contain hydroxyl, carboxylic acid or both, and form the expected primary intermolecular interaction with the carbonyl group of the Propy (Figure 5.11). In the case of co-formers that contain both these functional groups, the hydroxyl group preferentially forms an O–H···O hydrogen bond with the carbonyl of Propy and the carboxylic acid forms an acid–acid dimer synthon, which is in accordance with Etter's rules of hydrogen bonding.[17]

The molecular complementary prediction methodology of Fábián[26] was also applied to Propy in order to compare the experimental findings with prediction results. Mercury (v3.8, build RC2) was used to generate descriptor values and to determine whether a combination of Propy and the selected co-formers was likely to form a co-crystal. The preloaded set of co-formers and the additional co-formers, which were found to result in a co-crystal with Propy, were utilized. The results of the prediction suggest that the eight co-formers that formed co-crystals in the current study were found in both the pass and fail lists (as predicted by the molecular complementarity method)

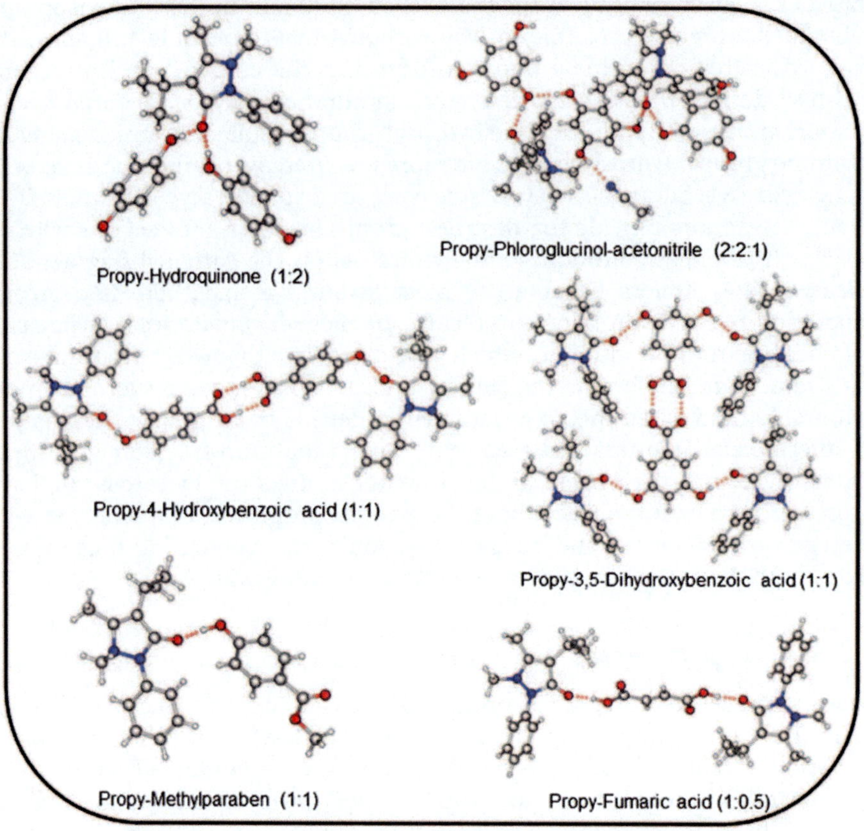

Figure 5.11 Crystal structures of the selected co-crystals of Propy. Adapted with permission from L. K. Mapp, S. J. Coles and S. Aitipamula, *Cryst. Growth Des.*, 2017, **17**, 163–174, Copyright 2017 American Chemical Society.

in an almost equal distribution. As the prediction methodology is primarily based on the polarity and presence of hydrogen bond donors and acceptors, the prediction model is perhaps not best suited to the molecules that are devoid of hydrogen bond donors. The current case study reaffirms the importance of complementary prediction tools, such as knowledge-based approaches, which aid in co-crystal screening for the selection of the most likely co-formers for a target molecule.

5.3 Process Development and Scale-up of Co-crystallization

Once a co-crystal has been determined for commercial production, a crystallization process must be developed for commercial production in large scale. While methods like evaporation and milling are useful for co-crystal preparation in lab-scale, they are not easily scalable. Hence solution crystallization is the preferred choice since it is amenable for industrial production, and existing manufacturing facilities for single component crystallization can be utilized. When developing a crystallization process, three important crystallization mechanisms need to be considered: nucleation, crystal growth, and agglomeration. They play different roles in defining the quality of the final crystal products. Below is a short summary of these mechanisms. Readers can refer to published books for more a rigorous description of fundamentals and mechanisms,[35] industrial crystallization techniques,[36] and pharmaceutical crystallization.[37]

Nucleation creates new solid surfaces from clear solution (primary nucleation), or in the presence of seeds (secondary nucleation). The number of nuclei and their residence time dictates the particle size distribution (PSD) of final product. Moreover, nucleation is often the limiting step in polymorphic crystallization.[38–40] Primary nucleation is inflicted with three problems. Firstly, it is stochastic in nature and the point of its occurrence changes from batch to batch, leading to quality variations. Secondly, relatively high supersaturation is necessary for primary nucleation to happen, which may cause fouling of crystallizer walls and internal fittings. Thirdly, oiling out, instead of nucleation, may take place in some crystallizing systems. Therefore, seeding is often employed to prevent primary nucleation.

Crystal growth competes with nucleation to transfer solute molecules from liquid to solid phase. A growth-dominated process produces much larger crystals than a nucleation-dominated one. The crystal growth rate is less sensitive to supersaturation than nucleation. Consequently, a relatively low supersaturation level is preferred to promote crystal growth over nucleation. In addition, it is the relative growth rate of different crystal faces that defines crystal habit. Changes in solvent and operating conditions may result in habit modification.[41]

Agglomeration leads to particle enlargement. It is envisioned as a two-steps process. In the first step, particles collide with each other and stick together, followed by cementing at the contacting points in the second step.[35] In most cases it is not desirable due to entrapment of impurities in the void space between primary particles. The agglomeration rate is mainly influenced by the interactions between crystal surfaces and solvent, and agitation intensity and supersaturation level. Particle collision increases with higher agitation intensity, leading to more aggregates. Meanwhile, aggregates will fall apart under higher shear rate. There is a dynamic equilibrium between aggregation and disassembling. Cementing at contacting points of aggregates is achieved *via* crystal growth and therefore agglomeration rises with increasing supersaturation.

It can be seen that supersaturation drives these three mechanisms. In crystallization development, it is worthwhile keeping its dynamics in mind. For practical purposes, it is calculated as a concentration difference:

$$\Delta c = c - c^{*} \tag{5.1}$$

where c and c^* is the transient solute concentration and solubility at a certain temperature, respectively. Another definition in frequent use is the supersaturation ratio:

$$S_c = c/c^*. \tag{5.2}$$

5.3.1 General Crystallization Development Procedure

The three mechanisms are the key to process understanding during crystallization development. Process understanding is one of the principles in the QbD[42] approach advocated by regulatory bodies. Other principles include understanding the target product quality profile, functional relationship between process parameters and quality, application of design of experiment (DoE), conduction of risk assessment, *etc.* A general crystallization development procedure will be described following a brief introduction to these principles. This development procedure will be applied to a case study of co-crystallization in the next section.

5.3.1.1 *Understanding Target Quality Attributes*

Target quality attributes of crystals must be clearly defined at the start. Among them are purity, impurity profile, solvent residual, particles size, crystal morphology, and solid form. All these attributes are to be within their respective specifications. However, one or more of them is more restrictive than others in specific cases, called critical quality attributes.

Crystallization development of a new compound must be preceded by a comprehensive polymorph and solvate screening to prevent the unexpected appearance of new solid forms. The conversion relationship between polymorphs are to be determined, if any. Usually a thermodynamically stable polymorph is chosen for development. When a meta-stable polymorph has to be chosen, the risks of polymorphic transformation must be assessed thoroughly. If the solvent for crystallization can form solvate with the compound, its critical activity at different temperatures is to be probed to delineate the safe operating region. When hydrates emerge in the landscape, determination of critical water activity and the humidity of the purging gas for drying comes into the agenda.

5.3.1.2 *Solvent Screening*

Solvent screening is an important step for process development. Solvent effects have been widely reported on crystal habit and polymorphism. A lot of effort has been made to rationalize and predict solvent effects on solubility, polymorphism, crystal habit, *etc.* to a certain degree of success. However, experimental screening remains the most reliable and applicable way in practice up to now. Solvents to be screened can be found in International Conference of Harmonization (ICH) guidelines (confined to classes II and III).

5.3.1.2.1 Preliminary Estimation of Solubility. Solvent screening starts with preliminary estimation of solubility at room temperature. It can be measured by stepwise addition of solvent/solvent mixtures to a given amount of solids. As a rule of thumb, a solubility of 10–30 mg ml^{-1} at room temperature often indicates a promising candidate solvent. Solubility in a given solvent often doubles with every increment of 10 °C in temperature. An increment of 30 °C above room temperature boosts solubility eight-fold, which gives a satisfactory solvent efficiency when the solubility at room temperature is 20 mg ml^{-1}. When the solute is practically not soluble in single solvents, solvent mixtures will have to be explored. There are a large number of possible combinations of single solvents and the composition of solvent mixtures can vary within 0–100%, making the screening task seem very daunting. Advice from synthesis personnel may help to narrow down the range. The measured solubility gives an idea of solvent efficiency, crystal yield and type of crystallization (cooling, antisolvent, reactive, or evaporative crystallization).

5.3.1.2.2 Growth Experiment. The number of solvents/solvent mixtures as potential candidates for the crystallization medium will be greatly reduced after preliminary solubility measurement. The crystal growth experiment will then be performed using the shortlisted solvents/solvent mixtures. The following quality attributes of resulting crystal products will be assessed in the growth experiment:

a. Particle size or growth tendency.
b. Crystal habit and crystallinity.
c. Impurity rejection.
d. Chemical stability.

The growth experiment is usually conducted by heating–cooling cycles. A saturated solution can be prepared at a relatively high temperature such as 50 °C. Seeds are added to the solution and the suspension is cooled to a relatively low temperature, for example room temperature. After a few cycles, the suspension is filtered, and solids samples are taken for characterization. Growth experiments can be conducted in small vials placed on a shaker for agitation purpose. Magnetic stirrer bars should be avoided for possible grinding effects.

It is possible that new solid forms such as solvates or polymorphs may appear during solvent screening. Once they appear, their relative stability must be understood by appropriate experiments such as competitive slurry experiments or solubility measurement.

5.3.1.3 *Measurement of Solubility and Meta-stable Zone Width (MSZW)*

The outcome of growth experiments will allow one to decide on the final solvent/solvent mixture for process development. More accurate solubility data is needed, such as by HPLC assay of liquid samples taken from slurries that have been stirred sufficiently. Solubility data points are often fitted with polynomial or exponential functions for interpolation or extrapolation. A van't Hoff-type expression is often used as well:

$$\ln(c) = \frac{\Delta H_f}{R}\left(\frac{1}{T_f} - \frac{\Delta}{T}\right) \tag{5.3}$$

Where ΔH_f is the enthalpy of melting, T_f is the melting point, c is the solubility in [g g^{-1} solvent] and T is the temperature in Kelvin. Eqn (5.3) is derived from the thermodynamic functions of ideal solutions. The trend of the solubility–temperature curve of many compounds, instead of absolute values, agrees well with it. This equation is often used to determine the transition temperature between enantiotropic polymorphs.

The meta-stable limit is defined as the maximum supersaturation that the solution can be brought to at a certain rate without triggering massive nucleation, as shown in Figure 5.12 for cooling crystallization. It is not a thermodynamic property of the system, but depends on many operating factors, *e.g.*, the volume of crystallizers, agitation intensity and generation rate of supersaturation.

The meta-stable zone width (MSZW) is the gap between meta-stable limit and solubility curve. Its measurement in cooling crystallization is conducted by slowly cooling a clear solution of a given concentration (0.1 °C min^{-1}).

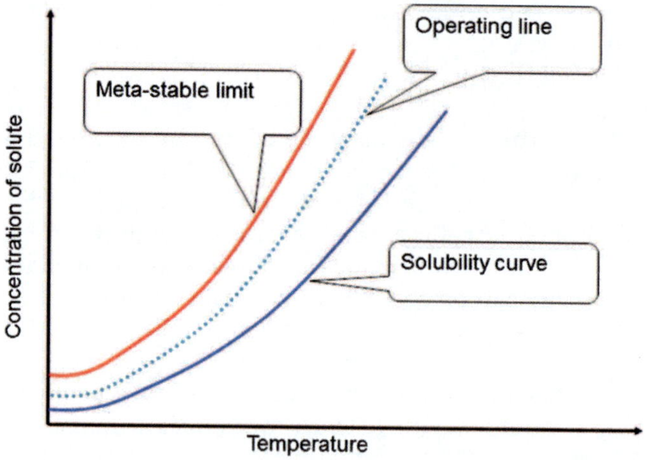

Figure 5.12 Meta-stable zone width in cooling crystallization.

It can be done together with measurement of dynamic solubility in small vials that are equipped with a turbidity meter and temperature control (such as Crystal 16® or Crystalline®). Similarly, MSZW measurement in anti-solvent crystallization is carried out by slow addition of anti-solvent to a clear solution.

The significance of MSZW measurement lies in the feasibility assessment of the seeding operation and the determination of the seeding point. With a narrow MSZW, seeding operation may not be necessary in the plant. For crystallizing systems with a wide MSZW, seeding is often effective to maintain quality consistency.

Related to MSZW is induction time, which is the length of time needed for nuclei to appear in a clear solution under a given supersaturation. It is measured isothermally. The induction time, when compared with the mixing time, serves as an indicator of the sensitivity of nucleation to scale-up.

5.3.1.4 Effects of Process Parameters and Process Optimization

The next stage will involve the determination of the process parameters and eventually process optimization to ensure efficient and robust operation of the crystallization process. The potential process parameters include cooling rate, antisolvent addition rate, temperature profile, seed mass, seed addition point, *etc.* Due to constraints in time and manpower, complete investigation of all process parameters may not be possible. One strategy is to conduct the first batch under the conditions that favour the most the target quality attributes and then relax some process parameters in the next batches to study their effects on the quality attributes. After a systematic assessment of the effect of the different operating parameters, the process parameters that most influence the critical quality attributes, termed the critical process parameters (CPPs), and the optimal operating conditions can be identified.

5.3.1.4.1 Potential Critical Process Parameters and Their Effects.

- Seeding protocol (unmilled seeds, 1%), including seed type (milled or pristine), seed loading, seed size, application method (dry or slurry) and seeding point. Seed loading depends on the purpose of seeding, to induce or to subdue secondary nucleation. The seed loading and size determine the total surface area of seeds. Seeds are to be applied when the solution is in its meta-stable zone.
- Cooling profile (0.2 $^{\circ}C\,min^{-1}$). The cooling profile defines the generation rate of supersaturation. It should be commensurate with the seeding protocol to maintain a desired supersaturation level. Ageing periods may be necessary in the profile.
- The addition rate of the anti-solvent or the second reactant (4–8 h) also determines the generation rate of supersaturation.

- Composition of the anti-solvent. When solubility is very sensitive to solvent composition, spontaneous nucleation may take place near the entry point of neat anti-solvent. Local nucleation can be avoided by using the solvent mixture as anti-solvent.
- Addition order of anti-solvent. High supersaturation is generated by reverse addition, *i.e.* API solution is added to anti-solvent. It can be employed to produce meta-stable polymorph or small particles.
- Agitation intensity or tip speed of the impeller (0.5–1 $m\,min^{-1}$). Agitation affords suspension and carrying of crystals across the crystallizer. It causes collision of particles and aggregation. At the same time, crystal breakage and de-agglomeration happens under intense agitation. Splashing of solution on crystallizer walls should be avoided.

Values in parenthesis are recommended settings of process parameters for the first few batches. They can be adjusted accordingly in the next batches based on the results of previous batches.

5.3.1.5 *Robustness Study*

An optimized crystallization process has taken shape from bench scale experiments at this time. Nevertheless, this process may turn out to be problematic when scale-up, due to the inherent irreproducibility of mixing conditions across scales, experimental errors at bench scale and operational variations at larger scale. A robustness study addresses these issues and makes necessary modifications to the optimal operating conditions.

The mixing conditions at larger scales are very different from those at bench scales. They have great impact on heat and mass transfer and thus crystallization kinetics. They should be evaluated during the robustness study in order to minimize possible scale-up effects. For example, the temperature drop between the wall and bulk solution is greater in a bigger crystallizer tank than in a smaller one when operating at the same cooling rate. Consequently, the local supersaturation is higher and may bring about changes in quality attributes. The wall temperature in a larger crystallizer can be simulated with computational fluid dynamics software if measurements are not available. This scenario can be imitated by a series of experiments that run at a faster cooling rate than the putative optimal operating conditions. The temperature drop at larger scales should be tolerated in the finalized operating conditions.

The Statistical Design of Experiment (DoE) is very useful to analyse the risks associated with experimental errors and operational variations. Regression of experimental data generates an empirical process model that quantifies the effects of CPPs. This empirical process model can be used to assess the risks of quality failure due to operational fluctuations and experimental errors on bench *via* process simulation. A design space can be obtained from this analysis.

5.3.1.6 Scale-up

Scale-up is often carried out in an existing stirred tank with fixed geometry and heating/cooling characteristics. Stirrer speed may be the only process parameter that can be adjusted in this case. The following information should be collected and kept in mind for pilot or commercial crystallizers at the start of crystallization development. It should be taken into account during experimental design on bench scale in order to avoid scale-up effects.

- Geometrical design of crystallizers and impellers. Is the starting volume sufficient to reach the impeller in anti-solvent crystallization for effective agitation?
- Cooling capacity and the temperature drop between wall and bulk solution. Will primary nucleation take place at the cooled wall? If yes, how can it be prevented?
- Mixing characteristics, such as mixing time, power number, *etc.* The ratio of induction time (or reaction time) to mixing time of the crystallizer is an indicator of the sensitivity of crystallization to mixing. Mixing time can be calculated from experimental correlations or computational fluid dynamics.
- Seed preparation and seeding facility. Specification of seeds applied at bench scale and seeding method (dry or slurry) must be achievable in pilot or commercial plant.
- Control accuracy of temperature and anti-solvent addition. The temperature may deviate from its target by around 3 °C in a commercial plant. It is common that reagent addition in commercial production is controlled manually.

Adjustments in stirrer speed have two conflicting impacts on crystallization. An increase in stirrer speed will enhance heat and mass transfer on the one hand, but it also brings about crystal breakage or attrition on the other hand. Three criteria are commonly employed to determine the stirrer speed upon scale-up, *i.e.*, equivalent tip speed of the impeller, equivalent energy dissipation rate and equivalent solid suspension state.

The crystallization process development procedure described above will be illustrated in a case study of caffeine–glutaric acid co-crystal. Not all elements in the procedure are included. A few process analytical technology (PAT) sensors were employed in the development, including PVM® (particle vision measurement by Mettler-Toledo), ATR-FTIR (attenuated total reflectance Fourier transform infrared spectroscopy, Thermo Analytics), and FBRM® (focused beam reflectance measurement).

Reviews of PAT have been made by several research authors.[43,44] Here, a brief introduction to the abovementioned sensors suffices. PVM is a probe-based high-speed camera that can be inserted into the crystallizer to capture and record crystal images, providing useful information on particle size, crystal habit and agglomeration. ATR-FTIR affords information in the liquid phase. The spectra of ATR-FTIR can be correlated with the concentrations of multiple components *via* chemometrics. The penetration depth of infrared waves is shorter than the boundary layer surrounding the ATR crystal. Therefore, its measurement is not affected by the presence of solids in normal operating conditions. FBRM offers information on solids concentration and particle size in the slurry. It measures the chord lengths of particles, instead of conventional particle size. The control interface of FBRM provides a few statistics of chord length distribution (CLD) such as total counts of chord lengths, square-weighted chord lengths, *etc.* The relationship between CLD and PSD was reviewed by Yu *et al.*[45] In the following case study, a sharp increase in total counts of chord lengths indicated the occurrence of massive nucleation.

5.3.2 Case Study: Development of Caffeine–Glutaric Acid Co-crystallization

Caffeine (CA) and glutaric acid (GA) can form a 1 : 1 co-crystal that has two polymorphs. Form I is the metastable form with a needle-like shape and form II is more stable with a prismatic shape. They can be differentiated by *in situ* imaging sensors due to their distinct shape. It is desired to obtain pure form II consistently. It was found that CA–GA co-crystal (form II) can grow reasonably fast in acetonitrile. Acetonitrile was chosen as the solvent for cooling co-crystallization without extensive solvent screening.

5.3.2.1 Construction of Phase Diagram

To construct the phase diagram for a co-crystal system, not only the concentration of each component, but also the composition of the corresponding solid phase in slurries needs to be determined. Depending on liquid phase composition, the solid phase may change from pure co-crystal to mixtures of co-crystal and crystals of single components. Usually the composition of the liquid phase is assayed by HPLC and the solid phase sample is analyzed by off-line characterization techniques such as PXRD. This method involves a lot of sample preparation and waiting time. An analysis of the generalized phase diagram of co-crystal systems, coupled with PAT application, allows us to propose a more efficient method for phase diagram construction.[46]

5.3.2.1.1 Generalized Phase Diagram and Relative Stability of Different Crystals. A generalized solubility curve of a 1 : 1 co-crystal at a certain temperature is shown in Figure 5.13. The *x*-axis refers to the equilibrium concentration of component A and the *y*-axis to component B. S_A on the *x*-axis and S_B on the *y*-axis denotes the solubility of pure components A and B, respectively. *E2* and *E1* are the eutectic points on the solubility curve of co-crystal with the concentration of component A at *E1* equal to C_A and

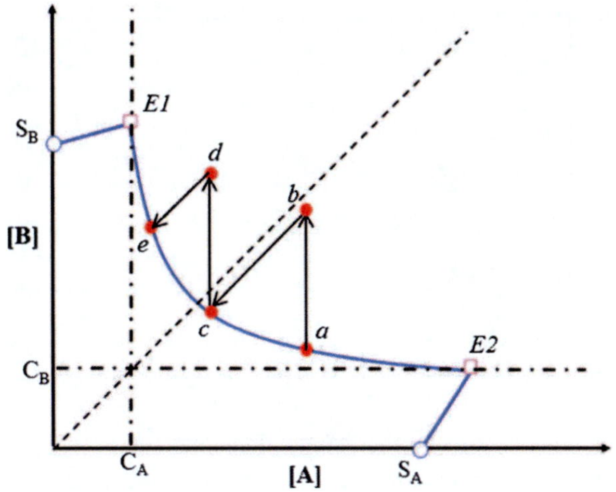

Figure 5.13 A generalized solubility curve of 1 : 1 co-crystal at a certain temperature. Reprinted with permission from Z. Q. Yu, P. S. Chow and R. B. H. Tan, *Cryst. Growth Des.*, 2010, **10**, 2382–2387, Copyright 2010 American Chemical Society.

the concentration of component B at *E2* equal to C_B. The dashed line bisects the coordinate system, representing the stoichiometry of 1 : 1.

Suppose we start from point *a* on the solubility curve. If excess B is added, the system state moves to point *b*. Nucleation of co-crystal may occur spontaneously or growth of co-crystal can be initiated by seeding in this zone. The system state then moves along a pathway parallel to the bisecting line. Eventually co-crystallization stops when the system reaches point *c* on the solubility curve. If more solid B is added, the system state moves to point *d* and then to point *e* on the solubility curve due to co-crystallization. If sufficient solid B is put in, the system state will eventually move to the eutectic point *E1* wherein solid B and A–B co-crystal coexist in the suspension and both are in equilibrium with the liquid phase. Likewise, eutectic point *E2* can be located by successively adding in excess solid A. The process can be monitored by ATR-FTIR to measure the equilibrium concentration of both components and to ascertain the eutectic points.

5.3.2.1.2 **Phase Diagram of Caffeine–Glutaric Acid in Acetonitrile.** The equilibrium concentration curves at different temperatures were obtained using ATR-FTIR and are presented in Figure 5.14. Notice that different scales are adopted for *x*- and *y*-axes to make variations in caffeine concentration more visible. Equilibrium concentrations at each temperature are joined together by a dotted line. The solubility data of pure caffeine and pure glutaric acid lie on the *y*-axis and *x*-axis, respectively.

Eutectic points for caffeine/co-crystal and for glutaric acid/co-crystal at different temperatures are linked together by the two red bold lines, which mark the two boundaries of the stability zone of caffeine–glutaric acid

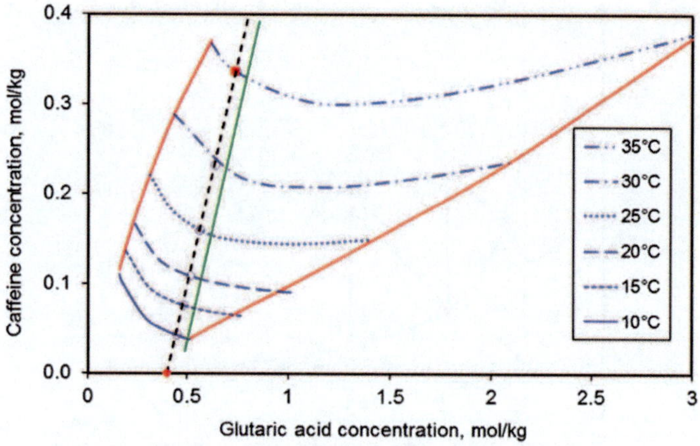

Figure 5.14 Equilibrium concentration of caffeine and glutaric acid at different temperatures.
Reprinted with permission from Z. Q. Yu, P. S. Chow and R. B. H. Tan, *Cryst. Growth Des.*, 2010, **10**, 2382–2387, Copyright 2010 American Chemical Society.

co-crystal. On its left is the stability zone for caffeine and on its right is that for glutaric acid.

With a complete phase diagram in hand, the operating region for co-crystal production can be delineated by considering the stoichiometry of the co-crystal system, and the stability zones at the initial and final temperatures for cooling crystallization. A relatively narrow operating region is obtained if the initial and final temperatures are set at 35 °C and 10 °C, respectively, as shown in Figure 5.14. The operating region is sandwiched between the left boundary of the stability zone of the co-crystal and thc bold straight line (green) that passes through the eutectic point for glutaric acid/co-crystal at 10 °C. The slope of the straight line corresponds to the stoichiometry of caffeine–glutaric acid co-crystal. The composition of starting solutions must be within this narrow region to obtain pure co-crystal.

5.3.2.1.3 Definition of Solubility and Supersaturation in Co-crystallization.[47]

In co-crystallization, the concentrations of both components change with time and supersaturation has been defined by Rodriguez-Hornedo *et al.* as a ratio:

$$\sigma = \left(\frac{C_1 C_2}{K_{SP}}\right)^{\frac{1}{2}} \tag{5.4}$$

where K_{SP} is the solubility product, and C_1 and C_2 are the concentrations of each solute component. In this definition, it is assumed that the equilibrium concentration curve of the co-crystal can be described by the solubility product in the liquid phase (ideal solution). Eqn (5.4) is not suitable for most real solutions. Therefore, a new definition of solubility and supersaturation for co-crystallization development will be given as follows.

Within the region in which CA–GA co-crystal is in the stable phase, CA and GA always come out of solution simultaneously as a 1 : 1 co-crystal. This means that for each mole of co-crystal that crystallized out, one mole each of CA and GA must be transferred from the liquid phase to the solid phase. Therefore, the solute concentration evolution follows the 1 : 1 stoichiometry line drawn from the starting concentration. As an example, the composition of a starting solution is marked with a solid square on the equilibrium concentration curve at 35 °C in Figure 5.14. A dashed straight line with a slope of 1 : 1 is drawn through this point and the crossing points with various equilibrium concentration curves are marked with open circles. It crosses the *x*-axis at point B1 (marked by a solid diamond). The CA molality is zero and the molality of the CA–GA pair is also zero at point B1. Using B1 as the reference, the molality of an imagined CA–GA pair at each circle along the straight line is defined, which is the solubility of co-crystal at the corresponding temperature. It should be noted that the solubility of co-crystal depends on GA concentration of the starting solution, which dictates the position of B1 on the *x*-axis.

The resulting solubility of co-crystal is plotted against temperature and fitted in Figure 5.15.

The transient molality of the CA–GA pair during a co-crystallization run is the same as that of CA measured by ATR-FTIR, and the difference between the transient concentration and solubility is defined as supersaturation.

With the new definition of supersaturation and the solubility curve of CA–GA co-crystal, the process can be optimized with the objective of producing pure form II of CA–GA co-crystal.

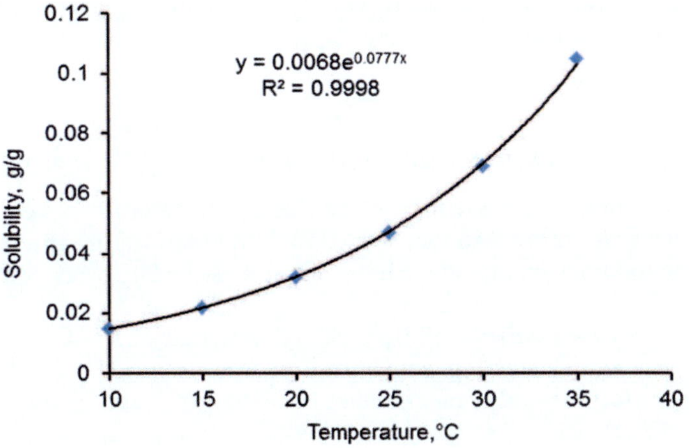

Figure 5.15 Solubility curve of CA–GA pair in acetonitrile.
Reprinted with permission from Z. Q. Yu, P. S. Chow, R. B. H. Tan and W. H. Ang, *Cryst. Growth Des.*, 2011, **11**, 4525–4532, Copyright 2011 American Chemical Society.

5.3.2.2 *Process Optimization*

During the measurement of the MSZW at a cooling rate of 0.1 °C min^{-1}, it was found that the meta-stable form I nucleated out first and then converted to the stable form II. Therefore, it was decided to apply seeds of form II to avoid the appearance of form I. There exists a supersaturation threshold below which the stable form can be obtained consistently in seeded co-crystallization. Therefore, control over polymorphic purity can be translated into locating the supersaturation threshold and the corresponding settings of process parameters.

In batch crystallization, supersaturation goes through a maximum with time and it is the peak supersaturation that should not exceed a certain threshold. Supersaturation dynamics are affected by five process parameters, including cooling profile, seed loading, seeding temperature, seed particle size distribution and starting concentration. The optimal operating conditions can be identified by the univariate design of the experiment. When the metastable form I appears in one batch, it can be concluded that the corresponding supersaturation level is too high. Then the settings of process parameters can be adjusted in the next batch according to the qualitative relationship between these process parameters and supersaturation. The operating range of each process parameter can be determined experimentally but it can be very time-consuming to find the operating ranges of all these process parameters by experiments alone. This can be expedited by coupling experiments with process modelling.[48,49]

The effects of the five process parameters are accounted for in a semi-mechanistic process model that describes the dynamics of supersaturation and the first three moments of PSD.[49] A successive design of experiment aided by process modelling was used to identify the supersaturation threshold. The supersaturation threshold for producing form II of CA–GA co-crystal was determined to be 0.015 g g^{-1} after only six crystallization experiments at bench scale.

5.3.2.3 *Risk Assessment and Determination of Operating Ranges*

In lab experiments, the settings of process parameters can be controlled tightly. However, there will be operational fluctuations in them at larger scales. The risks associated with them can be assessed by process modelling.

5.3.2.3.1 Risk Assessment of Operational Fluctuations. As an example, variability in actual starting concentration was assumed to follow a normal distribution around a mean value of 0.105 g g^{-1} with a standard deviation of 0.0025 g g^{-1} (±2.5%). The combined impacts of variability in starting concentration and model parameter uncertainty were assessed by Monte Carlo simulation, as shown in Figure 5.16. The red dashed line denotes the supersaturation threshold. The blue curve is from a process simulation with neither model parameter uncertainty nor variability in

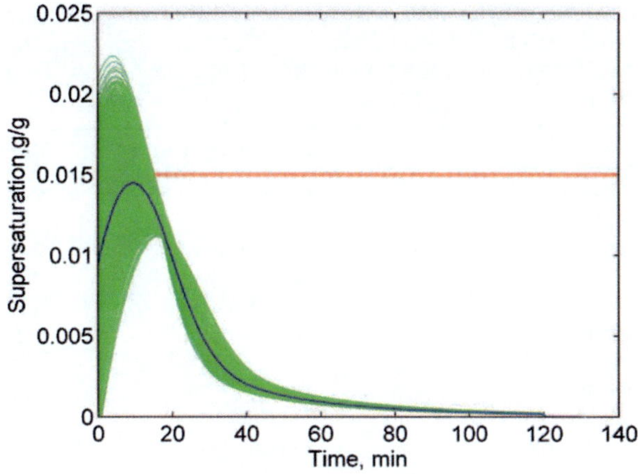

Figure 5.16 Simulated supersaturation trajectories without (blue curve) and with (green curves) model parameter uncertainty or operational variability in starting concentration.
Reprinted with permission from Z. Q. Yu, P. S. Chow and R. B. H. Tan, *Cryst. Growth Des.*, 2014, **14**, 3949–3957, Copyright 2014 American Chemical Society.

starting concentration. Green curves are from 10^5 simulations wherein model parameters and starting concentration are randomly sampled in their respective distributions. Only 63.4% of the green curves have a peak value below the supersaturation threshold, while the whole blue curve is below it. Operational fluctuations lead to significant risk of producing the meta-stable form I.

Similarly, the effects of operation fluctuations in temperature and seed loading can be investigated by Monte Carlo simulation. For detailed results, readers are referred to the report by Yu *et al.*[50]

5.3.2.3.2 **Determination of Operating Ranges.** The supersaturation threshold for producing form II is believed to be valid for all scales. The threshold should be translated to the operating ranges of each process parameter. This task can be neatly completed by process modelling. During translation, operational fluctuations of each parameter must be considered. In other word, the resulting operating ranges of process parameters should be robust towards operational fluctuations. The operating ranges of process parameters were determined and are shown in Table 5.2.

5.3.2.4 Scale-up

The operating ranges represented in Table 5.2 were verified with a scale-up experiment in a 10 l crystallizer. The stirrer speed in the 10 l crystallizer was determined according to an equivalent dissipation energy rate to the bench scale crystallizer. Simulated and experimental supersaturation trajectories are shown in Figure 5.17 along with the temperature profile. Seeds were

Table 5.2 Operating ranges and assumed variability of process parameters. Reprinted with permission from Z. Q. Yu, P. S. Chow and R. B. H. Tan, *Cryst. Growth Des.*, 2014, **14**, 3949–3957, Copyright 2014 American Chemical Society.

Process parameters	Operating range	Standard deviation
Starting concentration, $g\,g^{-1}$	0.105	0.00250
D-values of seeds		
D10, μm	≤67.0	0
D50, μm	≤113	0
D90, μm	≤151	0
Seed loading, wt%	≥5.00	0.125
Seeding temperature, °C	33.9–34.6	1.00
Temperature profile		
Age time after seeding, min	≥30.0	0
Linear cooling time, min	≥100	0
Age time at final temperature, min	≥10.0	0

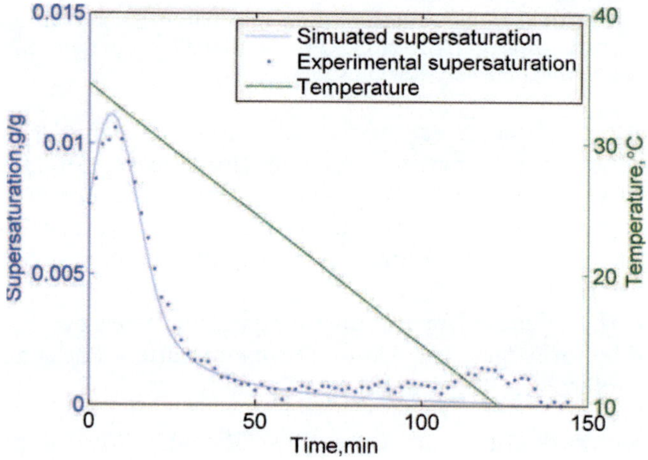

Figure 5.17 Simulated and experimental supersaturation trajectories and temperature profile in the scale-up experiment. Seeds were applied at time zero. Reprinted with permission from Z. Q. Yu, P. S. Chow and R. B. H. Tan, *Cryst. Growth Des.*, 2014, **14**, 3949–3957, Copyright 2014 American Chemical Society.

applied at time zero and there was no ageing period after seeding because the variability of process parameters on this scale is still negligible. It can be seen that the model agrees well with scale-up results. Meta-stable co-crystals did not appear throughout the scale-up batch.

5.3.2.5 *Feedback Control of Supersaturation*

With the use of the new definition of supersaturation for the co-crystal system described earlier, feedback control of supersaturation could also be implemented on a co-crystallization process.[47] During feedback control, the supersaturation is controlled by manipulating the transient temperature in the crystallizer such that the supersaturation remains below the supersaturation threshold. Shown in Figure 5.18 is a two-level feedback control structure. In the first level, an interface on the control computer is responsible for the supersaturation calculation based on spectral data from ATR-FTIR and transient temperature from the crystallizer. It compares actual supersaturation with its set point and decides the new set point for crystallizer temperature to keep supersaturation constant. The second level is actually the PID controller integrated in the circulator. It receives and executes commands from the first level.

The set-point of supersaturation was set at 0.08 $g\,g^{-1}$, lower than the supersaturation threshold obtained before. In addition, the control algorithm includes two constraints: (1) the final temperature is 15 °C, (2) the highest allowable cooling rate is 0.4 $°C\,min^{-1}$. The feedback control loop was activated when supersaturation was about 0.006 $g\,g^{-1}$ (crystallizer temperature was 34 °C). 0.5 g of milled seeds was applied at the same time. The resulting supersaturation and temperature profiles are plotted in Figure 5.19.

It can be seen that supersaturation decreased a little at the beginning because seeds started growing and incorporating solute molecules. Cooling was started because supersaturation was below its set point. In a few minutes, the temperature decreased from 34 °C to 32.6 °C and supersaturation overshot to 10 $g\,kg^{-1}$. Subsequently, cooling was suspended for a short while and supersaturation decreased slowly to its set point due to co-crystal growth. During this period, the temperature changed little. When supersaturation dropped below its set point, cooling was sped up and the temperature profile became steeper. Supersaturation stayed slightly below its set

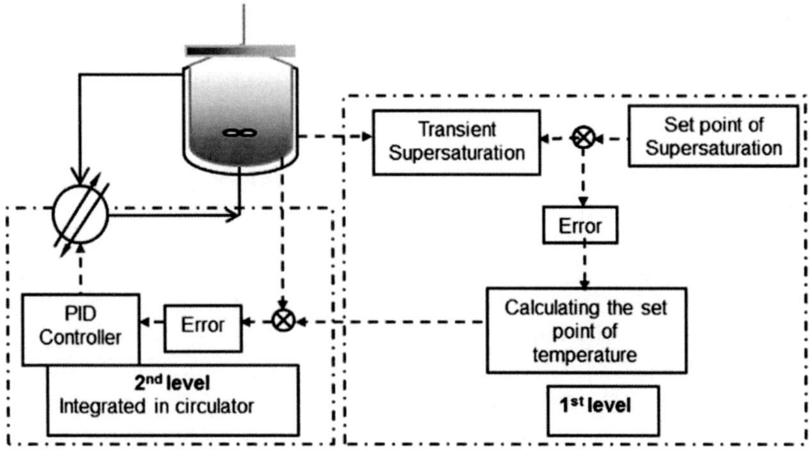

Figure 5.18 Structure of the feedback control system.
Reprinted with permission from Z. Q. Yu, P. S. Chow, R. B. H. Tan and W. H. Ang, *Cryst. Growth Des.*, 2011, **11**, 4525–4532, Copyright 2011 American Chemical Society.

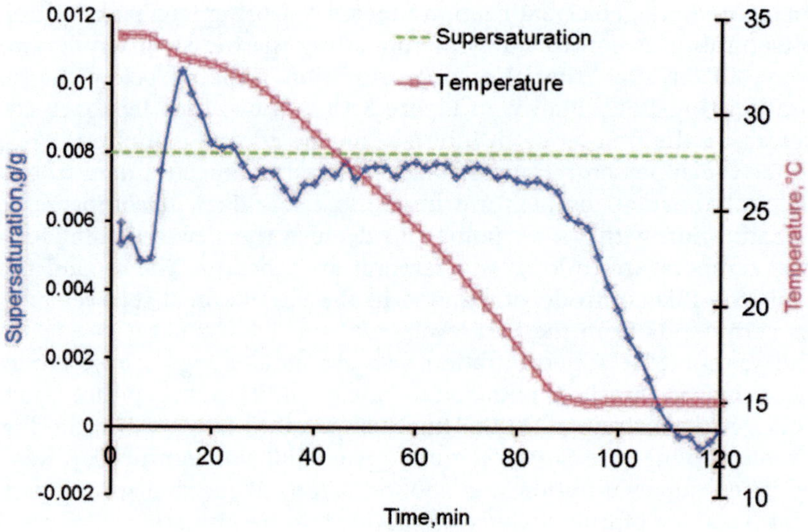

Figure 5.19 The trajectories of temperature and supersaturation during feedback control.
Reprinted with permission from Z. Q. Yu, P. S. Chow, R. B. H. Tan and W. H. Ang, *Cryst. Growth Des.*, 2011, **11**, 4525–4532, Copyright 2011 American Chemical Society.

point until the temperature reached 15 °C, at which time it was kept constant at 15 °C and supersaturation was consumed along the way by crystal growth. With this feedback control, form II of CA–GA was consistently produced.

5.4 Conclusions and Outlook

The emergence of pharmaceutical co-crystals as alternatives to long-known salts, solvates/hydrates, polymorphs, *etc.* has provided additional means of fine-tuning physicochemical properties of drug molecules. In this regard, strategies for systematic design of co-crystals have received paramount importance. Design of co-crystals demands prior knowledge of intermolecular interactions, which can aid in judicious selection of co-formers. A majority of co-crystal design strategies have relied on strong hydrogen bonds or synthons and proved effective in the design of co-crystals for a wide variety of APIs. The molecules that are devoid of potential hydrogen bonding sites, or those that contain only hydrogen bond acceptor groups, present challenges for co-crystal design and require a greater knowledge and control over weak intermolecular interactions. Synthon-based co-crystal design strategies offer limited assistance in these cases and necessitate alternative approaches for understanding weaker intermolecular interactions. Knowledge-based methods, which incorporate structural, geometrical, and chemical information pertaining to the API and co-formers, promise potential applications in co-crystal design. As co-crystals increasingly become valuable in the development of novel drug formulations, we hope that applying knowledge-based co-crystal design principles in the selection of co-formers reduces the effort, time, and consumption of resources. The recent developments in the co-crystal design methodologies could pave the way for the design of co-crystals for molecules of higher complexity with multiple hydrogen-bonding sites. Robust co-crystallization processes can be developed by following the principles of Quality-by-Design (QbD). The stability zone of co-crystals in the phase diagram must be delineated to assure the purity of co-crystal. Supersaturation manipulation is very important for process optimization. Many process parameters affect supersaturation dynamics and thus have a bearing on quality attributes. A new definition of solubility and supersaturation has been proposed and employed for co-crystallization development. Process analytical technology (PAT) can be employed in various stages of process development by providing on-line information of liquid and solid phases. Process modelling can speed up process optimization and determination of design space of co-crystallization. Possible scale-up effects should be considered and assessed during bench scale experiments to make technology transfer from lab to plant as smooth as possible.

Acknowledgements

We gratefully acknowledge the financial support from the Science and Engineering Research Council of A*STAR (Agency for Science, Technology and Research), Singapore. Part of the work described in this chapter was conducted by our former student, Lucy K. Mapp. It is our pleasure to acknowledge her dedication and hardwork.

References

1. D. P. Kale, S. S. Zode and A. K. Bansal, *J. Pharm. Sci.*, 2017, **106**, 457–470.
2. S. R. Byrn, *Solid State Chemistry of Drugs*, Academic Press, New York, 1982.
3. J. Bernstein, *Cryst. Growth Des.*, 2011, **11**, 632–650.
4. S. Aitipamula, P. S. Chow and R. B. H. Tan, *CrystEngComm*, 2014, **16**, 3451–3465.
5. S. R. Chemburkar, J. Bauer, K. Deming, H. Spiwek, K. Patel, J. Morris, R. Henry, S. Spanton, W. Dziki, W. Porter, J. Quick, P. Bauer, J. Donaubauer, B. A. Narayanan, M. Soldani, D. Riley and K. McFarland, *Org. Process Res. Dev.*, 2000, **4**, 413–417.
6. *Pharmaceutical Salts and Co-crystals. [In: RSC Drug Discovery Ser., 2012; 16]*, ed. J. Wouters and L. Quere, Royal Society of Chemistry, 2012.
7. S. Aitipamula, R. Banerjee, A. K. Bansal, K. Biradha, M. L. Cheney, A. R. Choudhury, G. R. Desiraju, A. G. Dikundwar, R. Dubey, N. Duggirala, P. P. Ghogale, S. Ghosh, P. K. Goswami, N. R. Goud, R. R. K. R. Jetti, P. Karpinski, P. Kaushik, D. Kumar, V. Kumar, B. Moulton, A. Mukherjee, G. Mukherjee, A. S. Myerson, V. Puri, A. Ramanan, T. Rajamannar,

C. M. Reddy, N. Rodriguez-Hornedo, R. D. Rogers, T. N. G. Row, P. Sanphui, N. Shan, G. Shete, A. Singh, C. C. Sun, J. A. Swift, R. Thaimattam, T. S. Thakur, R. Kumar Thaper, S. P. Thomas, S. Tothadi, V. R. Vangala, N. Variankaval, P. Vishweshwar, D. R. Weyna and M. J. Zaworotko, *Cryst. Growth Des.*, 2012, **12**, 2147–2152.
8. *Discovering and Developing Molecules with Optimal Drug-like Properties*, Springer-Verlag, New York, 1 edn, 2015.
9. J. R. P. Khadka, H. Kim, I. Kim, J. T. Kim, H. Kim, J. M. Cho, G. Yun and J. Lee, *Asian J. Pharm. Sci.*, 2014, **9**, 304–316.
10. N. K. Duggirala, M. L. Perry, O. Almarsson and M. J. Zaworotko, *Chem. Commun.*, 2016, **52**, 640–655.
11. G. Bolla and A. Nangia, *Chem. Commun.*, 2016, **52**, 8342–8360.
12. A. V. Trask, *Mol. Pharmaceutics*, 2007, **4**, 301–309.
13. *Guidance for Industry: Regulatory Classification of Pharmaceutical Co-crystals*, Food and Drug Administration, Silver Spring, MD, December, 2011.
14. *EMA Reflection Paper on the Use of Co-crystals and other Solid State Forms of Active Substances in Medicinal Products*, European Medicines Agency, 2014.
15. http://www.fda.gov/downloads/drugs/guidancecomplianceregulatoryinformation/guidances/ucm516813.pdf.
16. P. A. Wood, N. Feeder, M. Furlow, P. T. A. Galek, C. R. Groom and E. Pidcock, *CrystEngComm*, 2014, **16**, 5839–5848.
17. M. C. Etter, *Acc. Chem. Res.*, 1990, **23**, 120–126.
18. G. R. Desiraju, *Crystal Engineering: The Design of Organic Solids*, Elsevier, Amsterdam, 1989.
19. C. R. Groom, I. J. Bruno, M. P. Lightfoot and S. C. Ward, *Acta Crystallogr., Sect. B: Struct. Sci., Cryst. Eng. Mater.*, 2016, **72**, 171–179.
20. G. R. Desiraju, *Angew. Chem., Int. Ed. Engl.*, 1995, **34**, 2311–2327.
21. S. L. Childs, P. A. Wood, N. Rodriguez-Hornedo, L. S. Reddy and K. I. Hardcastle, *Cryst. Growth Des.*, 2009, **9**, 1869–1888.
22. A. Planavila, R. Rodriguez-Calvo, A. F. de Arriba, R. M. Sanchez, J. C. Laguna, M. Merlos and M. Vazquez-Carrera, *Mol. Pharmacol.*, 2006, **69**, 1174–1181.
23. S. Aitipamula, L. K. Mapp, A. B. H. Wong, P. S. Chow and R. B. H. Tan, *CrystEngComm*, 2015, **17**, 9323–9335.
24. S. Aitipamula, V. R. Vangala, P. S. Chow and R. B. H. Tan, *Cryst. Growth Des.*, 2012, **12**, 5858–5863.
25. N. Takata, R. Takano, H. Uekusa, Y. Hayashi and K. Terada, *Cryst. Growth Des.*, 2010, **10**, 2116–2122.
26. L. Fábián, *Cryst. Growth Des.*, 2009, **9**, 1436–1443.
27. S. Karki, T. Friscic, L. Fabian and W. Jones, *CrystEngComm*, 2010, **12**, 4038–4041.
28. I. J. Bruno, J. C. Cole, J. P. M. Lommerse, R. S. Rowland, R. Taylor and M. L. Verdonk, *J. Comput.-Aided Mol. Des.*, 1997, **11**, 525–537.

29. I. J. Bruno, J. C. Cole, P. R. Edgington, M. Kessler, C. F. Macrae, P. McCabe, J. Pearson and R. Taylor, *Acta Crystallogr., Sect. B: Struct. Sci.*, 2002, **58**, 389–397.
30. C. F. Macrae, I. J. Bruno, J. A. Chisholm, P. R. Edgington, P. McCabe, E. Pidcock, L. Rodriguez-Monge, R. Taylor, J. van de Streek and P. A. Wood, *J. Appl. Crystallogr.*, 2008, **41**, 466–470.
31. L. K. Mapp, S. J. Coles and S. Aitipamula, *Cryst. Growth Des.*, 2017, **17**, 163–174.
32. A. Lemmerer, J. Bernstein, U. J. Griesser, V. Kahlenberg, D. M. Toebbens, S. H. Lapidus, P. W. Stephens and C. Esterhuysen, *Chem. - Eur. J.*, 2011, **17**, 13445–13460. S13445/13441–S13445/13448.
33. M. Veverka, T. Dubaj, J. Gallovič, E. Švajdlenka, B. Mel'uchová, V. Jorík and P. Šimon, *Monatsh. Chem.*, 2013, **144**, 1335–1349.
34. N. Issa, S. A. Barnett, S. Mohamed, D. E. Braun, R. C. B. Copley, D. A. Tocher and S. L. Price, *CrystEngComm*, 2012, **14**, 2454–2464.
35. J. W. Mullin, *Crystallization*, Butterworth-Heinemann, Oxford, 4 edn, 2001.
36. *Handbook of Industrial Crystallization*, ed. A. S. Myerson, Butterworth-Heinemann, New York, 2002.
37. H.-H. Tung, E. L. Paul, M. Midler and J. A. McCauley, *Crystallization of Organic Compounds: An Industrial Perspective*, Wiley, 2009.
38. D. Mangin, F. Puel and S. Veesler, *Org. Process Res. Dev.*, 2009, **13**, 1241–1253.
39. M. Kitamura, *CrystEngComm*, 2009, **11**, 949–964.
40. M. Muller, U. Meier, D. Wieckhusen, R. Beck, S. Pfeffer-Hennig and R. Schneeberger, *Cryst. Growth Des.*, 2006, **6**, 946–954.
41. *Molecular Modeling Applications in Crystallization*, ed. A. S. Myerson, Cambridge University Press, 1999.
42. *ICH Q8(R2). Pharmaceutical Development in ICH Harmonized Tripartite Guideline*, 2009.
43. L. L. Simon, H. Pataki, G. Marosi, F. Meemken, K. Hungerbuhler, A. Baiker, S. Tummala, B. Glennon, M. Kuentz, G. Steele, H. J. M. Kramer, J. W. Rydzak, Z. P. Chen, J. Morris, F. Kjell, R. Singh, R. Gani, K. V. Gernaey, M. Louhi-Kultanen, J. O'Reilly, N. Sandler, O. Antikainen, J. Yliruusi, P. Frohberg, J. Ulrich, R. D. Braatz, T. Leyssens, M. von Stosch, R. Oliveira, R. B. H. Tan, H. Q. Wu, M. Khan, D. O'Grady, A. Pandey, R. Westra, E. Delle-Case, D. Pape, D. Angelosante, Y. Maret, O. Steiger, M. Lenner, K. Abbou-Oucherif, Z. K. Nagy, J. D. Litster, V. K. Kamaraju and M. S. Chiu, *Org. Process Res. Dev.*, 2015, **19**, 3–62.
44. S. Bordawekar, A. Chanda, A. M. Daly, A. W. Garrett, J. P. Higgins, M. A. LaPack, T. D. Maloney, J. Morgado, S. Mukherjee, J. D. Orr, G. L. Reid, B. S. Yang and H. W. Ward, *Org. Process Res. Dev.*, 2015, **19**, 1174–1185.
45. Z. Q. Yu, P. S. Chow and R. B. H. Tan, *Org. Process Res. Dev.*, 2008, **12**, 646–654.

46. Z. Q. Yu, P. S. Chow and R. B. H. Tan, *Cryst. Growth Des.*, 2010, **10**, 2382–2387.
47. Z. Q. Yu, P. S. Chow, R. B. H. Tan and W. H. Ang, *Cryst. Growth Des.*, 2011, **11**, 4525–4532.
48. S. B. Brueggemeier, E. A. Reiff, O. K. Lyngberg, L. A. Hobson and J. E. Tabora, *Org. Process Res. Dev.*, 2011, **16**, 567–576.
49. Z. Q. Yu, P. S. Chow, R. B. H. Tan and W. H. Ang, *Org. Process Res. Dev.*, 2013, **17**, 549–556.
50. Z. Q. Yu, P. S. Chow and R. B. H. Tan, *Cryst. Growth Des.*, 2014, **14**, 3949–3957.

CHAPTER 6

Co-crystallization of Energetic Materials

S. R. KENNEDY AND C. R. PULHAM*

EaStCHEM School of Chemistry, University of Edinburgh, Joseph Black Building, David Brewster Road, Edinburgh, Scotland, EH9 3FJ, UK, Email: Stuart.Kennedy2@ed.ac.uk
*Email: C.R.Pulham@ed.ac.uk

6.1 Introduction to Energetic Materials

Energetic materials are defined as those that release heat and/or gaseous products at a high rate upon stimulus by heat, impact, shock, spark, *etc.* They have widespread military and civilian uses that include: munitions, mining, quarrying, demolition, emergency signalling, breathing apparatus, automotive safety, space exploration, and entertainment. They can be broadly classified as explosives, propellants, gas generators, and pyrotechnics. Explosives can be further classified as (a) primary explosives for which a mild impetus leads to a short, strong shock wave, and (b) secondary explosives for which a strong impetus leads to a long-duration shock wave.[1] Primary explosives result in shock waves that have a strong local effect, but which do much less damage over longer distances and so their main function is to act as an initiator for the detonation of a secondary explosive. Examples of commonly used primary explosives include lead azide [$Pb(N_3)_2$], lead styphnate (lead 2,4,6-trinitroresorcinate), and mercury fulminate [$Hg(ONC)_2$]. Once initiated, a secondary explosive liberates a large amount of energy and hence a sustained shock wave that causes more damage at a distance. Examples of widely used secondary

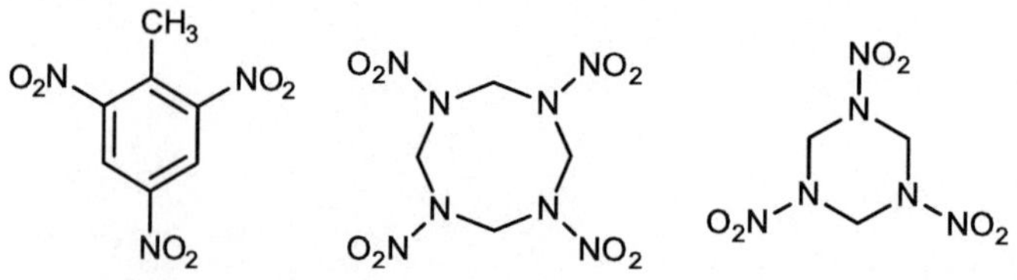

Figure 6.1 Molecular structures of TNT, HMX and RDX.

explosives include trinitrotoluene (TNT), HMX (1,3,5,7-tetranitro-1,3,5,7-tetrazacyclooctane), and RDX (1,3,5-trinitrohexahydro-s-1,3,5-triazine). The molecular structures of these compounds are shown in Figure 6.1.

The defining feature of these molecules is an appropriate balance of oxidizer (*e.g.* nitro-groups) in close proximity to readily oxidizable fuel (*e.g.* aromatic hydrocarbon rings). This results in a high energy density and on initiation a very rapid release of this stored chemical energy. Important properties of energetic materials include: detonation velocity, detonation pressure, oxygen balance, density, crystal morphology, number and type of crystal defects, thermal and long-term chemical stability, melting point, and sensitivity to initiation through mechanical impact, friction, spark, and shock. The detonation velocity and pressure refer to the shock wave that moves through the material during detonation and these properties depend on both the density of the compound and on its oxygen balance (OB or OB%)—an expression that is used to indicate the degree to which an explosive can be oxidized. Molecules that contain just enough oxygen to form carbon dioxide and water have a zero oxygen balance; those that have a positive oxygen balance contain more oxygen than is needed; those that have a negative oxygen balance contain less oxygen than is needed for complete combustion.[1] Perhaps the most important property is sensitivity to initiation and so a major priority for both military and civilian applications is the development of less-sensitive formulations and compositions that are less prone to accidental initiation. This is particularly true for munitions used under battle-field conditions, which may experience impact by bullets and shrapnel, and exposure to fire and shock. For this reason, substantial effort is expended in the development of compositions that are less sensitive to initiation, namely so-called "insensitive munitions". Conventional methods of reducing sensitivity include: (i) control of particle size of the energetic material, (ii) dispersing the polycrystalline energetic material into a polymer binder to produce a polymer-bonded explosive (PBX), in which the polymer binder plays roles in dissipating the energy of impact and reducing the effects of friction, and (iii) the development of new energetic compounds that are intrinsically less sensitive to initiation. Some examples of the latter approach include the development of triaminotrinitrobenzene (TATB) and diaminodinitroethane (DADNE or FOX-7), both of which are significantly less sensitive than HMX (Figure 6.2).

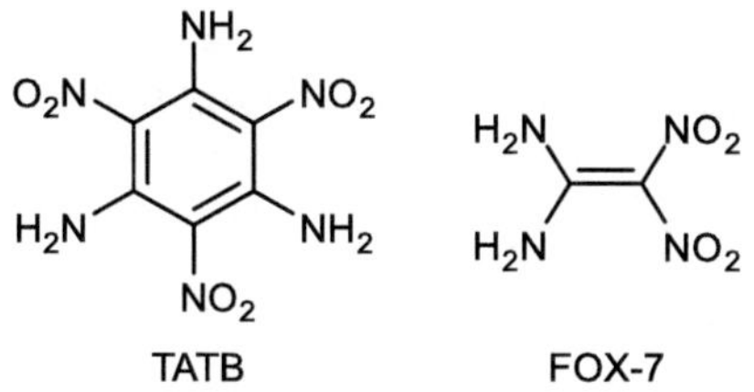

Figure 6.2 Molecular structures of TATB and FOX-7.

However, this approach requires substantial R&D effort and the scale-up required to produce industrial quantities of these new compounds can be both time-consuming and expensive. A relatively new concept for altering the physicochemical properties of energetic materials that is currently attracting significant interest is *co-crystallization*. Although now established within the pharmaceutical industry as a means of modifying the physicochemical properties of active pharmaceutical ingredients to meet specific criteria (*e.g.* increased solubility, dissolution rates, and chemical stability), co-crystallization is currently at an earlier stage of development within the field of energetic materials, but is beginning to show substantial promise. This approach circumvents additional synthetic steps and instead combines established energetic materials in innovative ways to produce new, solid-state materials with tailored properties. A widely accepted and useful definition of a co-crystal is a crystal in which two or more neutral molecular components interact non-covalently, typically in a defined stoichiometric ratio.[2–4] In principle, this can include solvates and for the purposes of this review chapter, solvates will also be included on the basis that solvates of energetic materials can substantially modify their physicochemical properties and performance. There are several reasons why co-crystallization of energetic materials is less well developed than co-crystallization of pharmaceuticals. In part, this reflects the much less widespread research effort into energetic materials conducted within academic institutions because of the often-stringent licensing, safety and security requirements associated with acquiring, storing, and handling these materials. An additional factor, however, is that the structures of energetic molecules are generally defined by the presence of relatively non-polar, weakly interacting nitro-groups that are often reluctant to form hydrogen bonds, rather than the much wider array of often very polar functional groups found in pharmaceutical molecules. Nevertheless, this position is changing, and it is now clear that energetic co-crystals can be obtained that feature C–H···nitro, π–π stacking, and even halogen-bonding interactions. It is also noteworthy that although the emergence of the field of crystal engineering is relatively recent, as early as 1988, whilst exploring the interaction of solvents (dopants) with HMX to prepare solvates, Brill *et al.* commented that "*In fact, the physical and chemical properties of HMX might be tailored systematically by such dopants.*"[5]

The aim of this chapter is therefore to provide a current perspective of research into energetic co-crystals that highlights recent progress in tailoring the properties of these materials, whilst identifying some of the remaining challenges in the field. We have chosen to order this chapter by specific energetic material and its associated co-crystals. This should allow readers to gain a more comprehensive view of the current status of co-crystals of some energetic materials. We have also attempted to provide some historical context of how the field has progressed. Readers are also directed to a recently published highlight by Zhang and Shreeve that also covers some of the issues associated with energetic co-crystals.[6]

6.2 Co-crystals of TNT (2,4,6-Trinitrotoluene)

The first structurally characterized co-crystal of TNT was in fact reported as early as 1987, although the authors did not comment on the significance of this 1 : 1 TNT–pyrene "complex" (shown in Figure 6.3) in terms of its energetic properties.[7]

A fluorescence-based chemo-sensor has been designed to detect trinitro aromatic compounds.[8] The molecular scaffold of this molecule comprises a calixarene framework with pyrene functionality. Low concentrations of TNT in acetonitrile were shown to reliably quench the fluorescence of the calixarene molecule. This behaviour is attributed to a charge transfer event between the pyrene subunit and the TNT molecule. Figure 6.4 shows the crystal structure of the co-crystal in which π-stacking interactions akin to the first TNT/pyrene co-crystal (Figure 6.3) are clearly evident, thus highlighting the formation of this type of non-covalent interaction between TNT and other π-systems.[7]

In 2010 Landenberger and Matzger published their seminal paper that reported 17 co-crystals of TNT and introduced some of the concepts associated with energetic co-crystals and their properties.[9] The structures of the

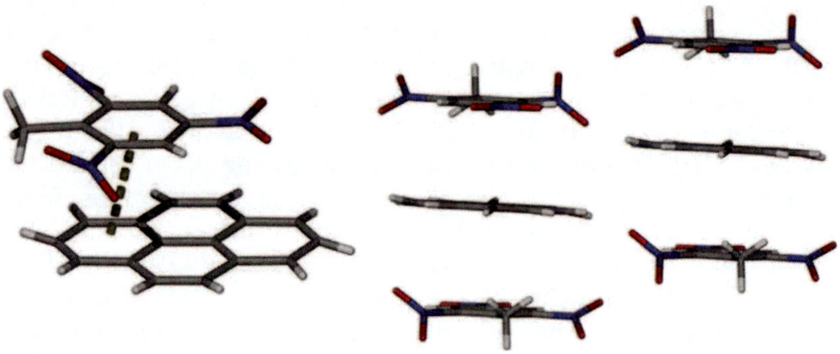

Figure 6.3 Structure of the TNT pyrene co-crystal where the image on the left highlights the π-stacking (dashed line) between both molecules. The image on the right shows the layered packing observed throughout the structure.[7]

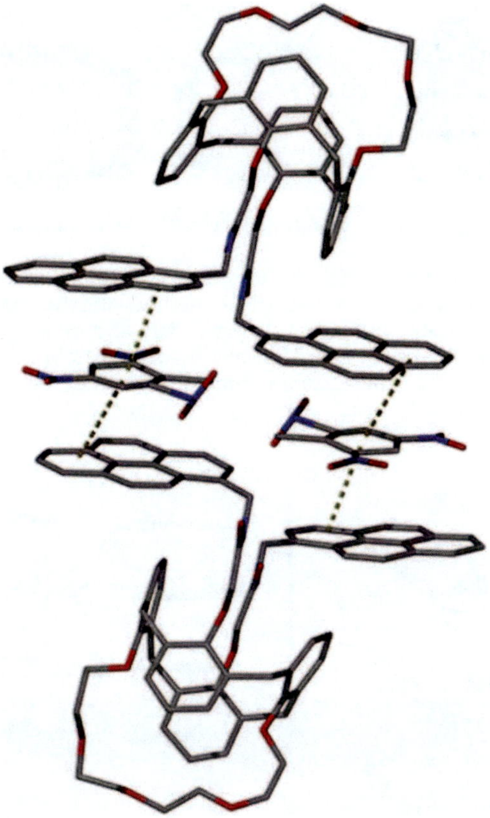

Figure 6.4 A calixarene/TNT co-crystal with a similar hydrogen-bonding motif to that found in the TNT/pyrene co-crystal.[8]

co-crystals formed between TNT and polycylic aromatic hydrocarbons such as naphthalene, anthracene, and phenanthrene highlight the importance of π–π interactions in stabilizing these new structures. Due to the strong electron-withdrawing effects of the nitro-groups, TNT has an electron-poor π-system, whereas the co-crystal formers all possess relatively electron-rich aromatic rings, thereby promoting favourable interactions between the electron-poor and electron-rich π-systems. This is illustrated by the calculated electrostatic potential surfaces of TNT and selected co-crystal formers, as shown in Figure 6.5.

As a result, face-to-face π-stacking is observed in all the co-crystals, in contrast to the edge-to-face interactions found in TNT itself. Representative examples of both packing motifs are shown in Figure 6.6.

Also highlighted was the importance of amine–nitro interactions for TNT co-crystals containing amines such as 1,2-phenylenediamine, and 4-aminobenzoic acid. One significant result of co-crystallizing TNT was that significant changes in the density of the co-crystal were achieved. The density of an energetic material is particularly important as it affects both

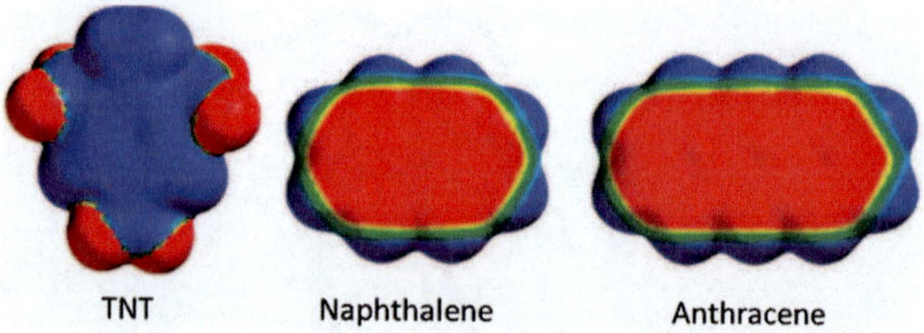

Figure 6.5 Electrostatic potential surfaces of TNT and the co-crystal formers calculated using the semi-empirical method and the AM1 model. Red and blue surfaces represent electron rich and poor regions, respectively. Surfaces are normalized between −25 and 25 kJ mol^{-1}.
Reprinted with permission from K. B. Landenberger and A. J. Matzger, *Cryst. Growth Des.*, 2010, **10**, 12, 5341, Copyright 2010 American Chemical Society.

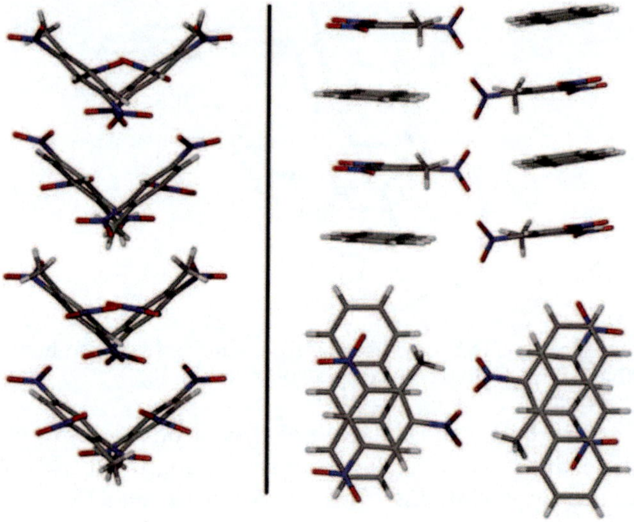

Figure 6.6 Comparison of the herringbone packing of monoclinic TNT (left) and the face-to-face π-stacking of the TNT/anthracene co-crystal.
Adapted from K. B. Landenberger and A. J. Matzger, *Cryst. Growth Des.*, 2010, **10**, 12, 5341, Copyright 2010 American Chemical Society.

the detonation velocity and detonation pressure—denser materials exhibit higher detonation velocities and pressures, and are therefore more powerful explosives. Furthermore, many applications of energetic materials are volume-limited, and it is advantageous to use higher density materials. Many of the TNT co-crystals were found to have a density in between those of the pure components, but this was not always the case and the TNT/tetrathiafulvalene co-crystal was found to have a density that was higher than those of two of the reported forms of tetrathiafulvalene and the orthorhombic form of TNT. Even though many of the TNT co-crystals do not show such high densities as

TNT, the packing coefficient of TNT in many of the co-crystals was observed to increase substantially, indicating strong, complementary relationships between molecules. The authors also investigated some of the physico-chemical properties of the co-crystals. For example, the melting points of the TNT co-crystals exhibit a wide range of melting points—the melting point could be lowered by as much as 36 °C or increased by 94 °C compared to that of TNT. The decomposition temperatures of the co-crystals were also modified, with substantial increases (compared to TNT) observed for co-crystals containing phenanthrene, naphthalene, and dibenzothiophene. Even though the energetic component in these co-crystals was substantially diluted by the non-energetic component, DSC measurements demonstrated that violent, exothermic decomposition was still achievable.

In 2011, Bolton and Matzger addressed the issue of dilution of the energetic component by co-crystallizing two energetic co-formers TNT and CL-20 in a 1 : 1 ratio.[10] CL-20 (or HNIW) is a relatively new energetic compound developed by the United States Navy and exhibits a high density, a high detonation velocity, and a favourable oxygen balance.[11] CL-20 is the most powerful explosive in current use, although some concerns remain over its sensitivity to detonation and its high cost of production. These issues have hindered the introduction of CL-20 into explosive applications. By contrast, TNT has a lower density and more modest detonation properties, but features lower sensitivity and manufacturing costs. The 1 : 1 CL-20/TNT co-crystal was formed by crystallization from ethanol and its crystal structure shows the capability of aliphatic nitramines to co-crystallize, despite the lack of strong interactions and π–π stacking (Figure 6.7). In fact, this co-crystal shows a series of weak C–H···O hydrogen-bonding interactions, together with interactions between two nitro-groups of CL-20 and the electron-poor

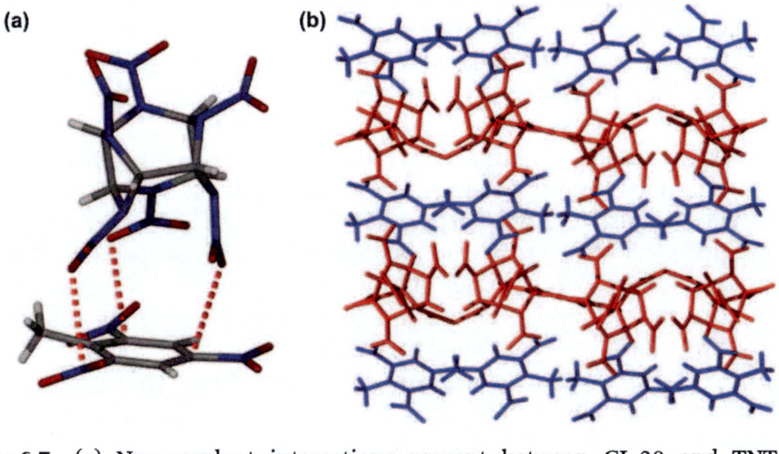

Figure 6.7 (a) Non-covalent interactions present between CL-20 and TNT. Red dashed lines are used to highlight nitro–aromatic and nitro–nitro interactions. (b) Extended structure of the co-crystal where CL-20 and TNT are shown as red and blue, respectively.

ring of TNT. A third nitro-group of the CL-20 molecule facing the TNT molecule forms a weak interaction with the nitrogen atom of the 2-nitro-group of the TNT molecule. A subsequent detailed analysis of the crystal structure by Zhang *et al.* showed that there was little molecular deformation of the CL-20 molecule apart from some changes in the torsional angles of some of the nitro-groups, and that several computational approaches confirm that the co-crystal is thermodynamically favoured compared to the individual components.[12]

Bolton and Matzger showed that the density of the co-crystal at room temperature was 1.84 $g\,cm^{-3}$, lying midway between the densities of the two pure components, thereby offering a respectable explosive performance.[10] The most important observation was that the co-crystal was substantially (~50%) less sensitive to impact than CL-20, thereby exemplifying the practical merits of energetic co-crystallization—the co-crystal combines the lower sensitivity of TNT with the density and power of CL-20 to give a homogenous energetic material with high explosive power and excellent insensitivity. A further observation that has potential implications for the safer transport of energetic materials was the effect of heating the co-crystal. At 136 °C, the co-crystal melted to form liquid TNT and either the β- or γ-form of CL-20, depending on the degree of heating at temperatures up to 150 °C. Both of these polymorphs of CL-20 are more sensitive to impact than the co-crystal and the ε-form of CL-20. The authors highlight that this co-crystallization approach could be used to generate a smart energetic material that could be stored and transported in a stable, less-sensitive form and then converted into an activated, more sensitive form for deployment. The thermal decomposition of nano-sized co-crystal CL-20/TNT has been studied in more detail using a combination of DSC, ultrafast chip calorimetry, and high-resolution X-ray powder diffraction.[13] These *in situ* measurements showed that the onset of melting of the co-crystal occurred at 130 °C followed immediately by the crystallization of the γ-form of CL-20 from the melt. A computational study using reactive molecular dynamics simulations has compared the thermal decomposition of the TNT/CL-20 co-crystal with that of pure crystals of TNT and CL-20, and with a simple physical mixture of TNT and CL-20.[14] The results showed that the co-crystal has a lower decomposition rate than CL-20, but higher than TNT, which is consistent with experimental observations. The formation of carbon clusters arising from TNT, a carbon-rich molecule, was shown to play an important role in the thermal decomposition process. At low temperature and in the early stages of chemical reactions under high temperature, the co-crystal released energy more slowly than the simple mixture of CL-20 and TNT. According to the authors, this explained the decrease in sensitivity of the co-crystal, but there appears to be some confusion in the paper about thermal behaviour and sensitivity to impact—these properties are quite distinct and are not directly related. A further study on this system has focused on both the intra- and intermolecular interactions within the co-crystal.[15] Work on the thermodynamics of the formation of a series of energetic–energetic co-crystals has been investigated in more depth.[16]

Rather surprisingly, there have been relatively few further reports describing co-crystals of TNT, although it seems likely that TNT should readily form co-crystals with electron-rich, aromatic π-systems. One example of note is the 1 : 1 co-crystal formed with 1-nitronaphthalene.[17] The authors report that the impact sensitivity and thermal stability of this co-crystal was greatly improved compared to that of pure TNT, and the co-crystal had a lower melting point than that of TNT. Another example is provided by the co-crystallization of TNT, picric acid, and 1,3-dinitrobenzene with the electron-rich polyaromatic compound 1,3,5-tris(4′-aminophenyl)benzene (TAPB) as the co-former.[18] The crystal structures show that strong, intermolecular π–π interactions direct the intercalation of the polynitro–aromatic molecules between the layers of TAPB molecules, leading to the formation of vertically overlapped –A–B–A–B– types of π-stacks. The crystal lattices are further stabilized through inter-stack hydrogen bonds (N–H···N and N–H···O) between amino groups of TAPB and the nitro-groups of the polynitro–aromatic molecules. A final example is of a 1 : 1 co-crystal formed with 1,3,5-trinitrobenzene (TNB).[19] Various non-covalent interactions are present in the co-crystal including C–H···O hydrogen bonding, π-stacking and p–π interactions between the TNT nitro groups and the benzene ring of TNB. The authors highlight that the co-crystal has a reduced sensitivity compared to both TNT and TNB.

6.3 Co-crystals of Benzotrifuroxan (BTF)

BTF is a hydrogen-free explosive that contains only C, N and O atoms and can be used as an igniter powder and high explosive—see Figure 6.8.

Nevertheless, it suffers from high sensitivity to initiation and this limits its uses. This prompted a study by Zhang *et al.* to explore whether the energetic properties of BTF could be improved by co-crystallization.[20] Five co-crystals were prepared and structurally characterized, comprising the following co-formers: 1,3,5-trinitrobenzene (TNB), TNT, 2,4,6-trinitroaniline (TNA), methylamino-2,4,6-trinitrobenzene (MATNB), and 1,3,3-trinitroazetidine (TNAZ) shown in Figure 6.9.

The structures of the co-crystals feature $-NO_2$ interactions with the relatively electron-poor, aromatic π-system of BTF molecules in neighbouring layers.

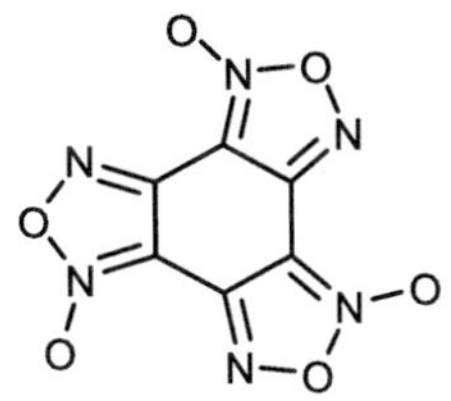

BTF

Figure 6.8 Molecular structure of BTF.

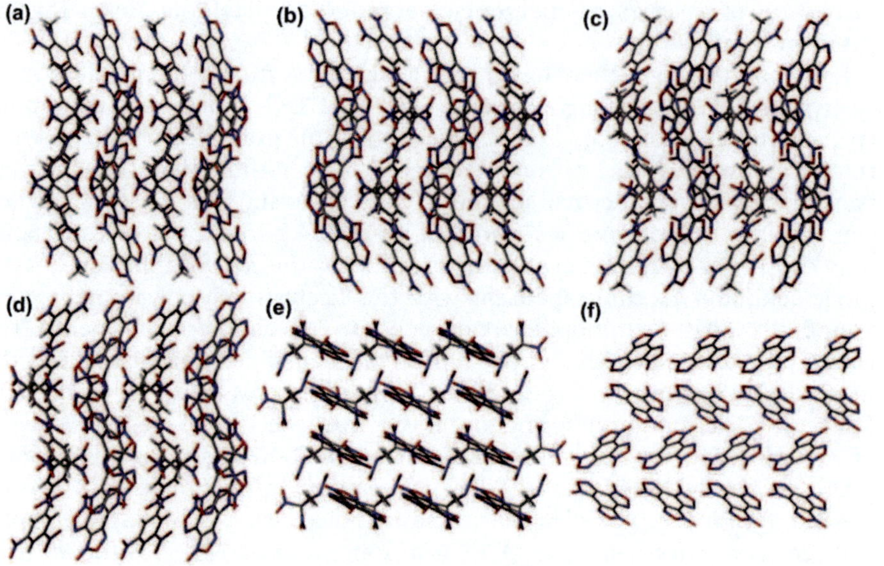

Figure 6.9 Crystal structures of BTF/MATNB (a), BTF/TNA (b), BTF/NT (c), BTF/TNB (d), and BTF/TNAZ (e) co-crystals and BTF (f) viewed along the *c*-axes. Adapted with permission from H. Zhang, C. Guo, X. Wang, J. Xu, X. He, Y. Liu, X. Liu, H. Huang and J. Sun, *Cryst. Growth Des.*, 2013, **13**, 679, Copyright 2013 American Chemical Society.

Hydrogen-bonding and π-stacking interactions were also found to be prominent. The densities of the co-crystals lie in the range 1.804–1.884 $g\,cm^{-3}$ and hence are slightly lower than that of BTF, but nevertheless are comparable with the density of RDX. The two co-crystals formed with TNT and TNB were found to be less impact-sensitive than BTF, but the other three were all more sensitive than their component co-formers. The authors speculated that this may be the result of the strong hydrogen bond between BTF and the co-former thereby weakening some of the bonds within the BTF molecule. This work was followed up by a study of the properties of the 1 : 1 co-crystal formed between BTF and 1,3-dinitrobenzene.[21] DSC studies showed that the co-crystal has a melting point of 130 °C (38 °C higher than pure DNB) and a similar decomposition temperature to that of pure BTF. The co-crystal was also shown to be substantially less sensitive to impact than BTF. Its measured density of 1.736 $g\,cm^{-3}$ lies midway between the densities of the two co-formers and leads to a predicted detonation velocity of 7373 $m\,s^{-1}$. (See section on CL-20 co-crystals for 1 : 1 co-crystal between BTF and CL-20.)

6.4 Co-crystals of HMX and RDX

HMX (1,3,5,7-tetranitro–1,3,5,7-tetrazacyclooctane or cyclotetramethylene tetranitramine or octogen) is one of the most widely used secondary explosives. Under ambient conditions, three polymorphs (α, β, and δ) and a hemihydrate

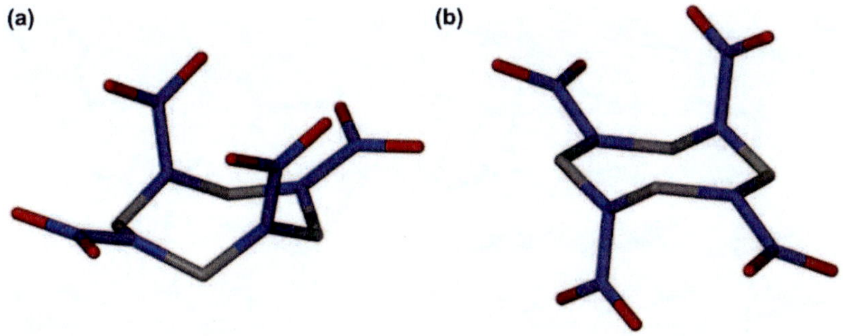

Figure 6.10 Representation of the two conformations of HMX, the chair–chair (a) and the chair (b) conformations.

(usually referred to as γ-HMX) are known and have been structurally characterized.[22–24] Under ambient conditions, the β-form is the most thermodynamically stable and densest form, and the shock- and impact-sensitivities of the four forms follows the order $\delta > \gamma > \alpha > \beta$.[25,26] For these reasons β-HMX is used in energetic applications. Two conformations of the HMX molecule can be observed: the chair–chair and chair conformations (Figure 6.10). The chair conformation is the more energetically favourable and is observed in the crystal structure of β-HMX. In contrast, the α-, δ-, and γ-forms all adopt the chair–chair conformation—the repulsion between the four closely spaced nitro-groups makes this conformation less stable.

The first reports of solvates of HMX date back to the early 1960s. For example, Cady *et al.* prepared 1 : 1 solvates of HMX with cyclopentanone, *N*-methyl-2-pyrrolidinone, and *N*,*N*-dimethylacetamide. All of the solvates were shown to be less impact-sensitive than HMX.[27] Structural characterization of a 1 : 1 solvate of HMX with DMF was first reported in 1975[28] and again in 1983.[29] Brill *et al.* structurally characterized a 1 : 1 solvate of *N*-methyl-2-pyrrolidinone and highlighted how "*the physical and chemical properties of HMX might be tailored systematically by such dopants.*"[5] Despite this potential avenue, up until 2012 only two solvates of HMX had entries in the Cambridge Structural Database—HMX/DMF[28,29] and HMX/2,4-dinitro-2,4-diazapentane.[30] Matzger *et al.* subsequently structurally characterized seven co-crystals of HMX and identified at least three distinct structural motifs, each dictated by the electronic character and size of the co-crystal former.[31] These are:

(i) The chair–chair/layered motif observed with co-crystal formers consisting of electron-rich aromatic rings such as 1,2-phenylenediamine and 2-bromoaniline in which homogeneous layers of HMX molecules in the chair–chair conformation alternate with layers of co-crystal former (Figure 6.11).
(ii) The chair–chair/pocket motif (observed with *N*-methyl-2-pyrrolidone, 2-pyrrolidone, 4-picoline-*N*-oxide, and DMF) in which HMX molecules adopt the chair–chair conformation, with layers consisting of three

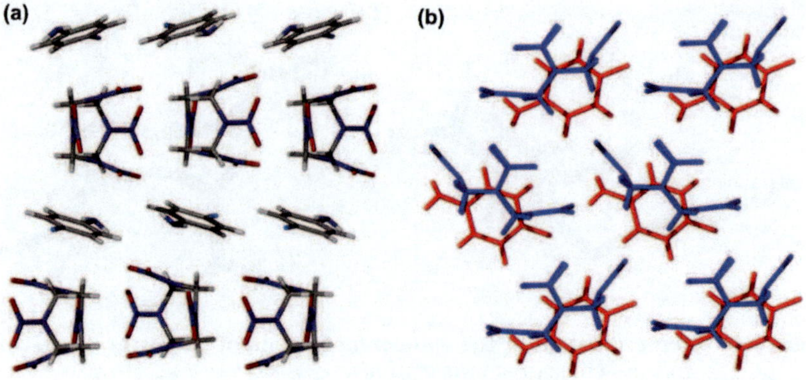

Figure 6.11 The chair–chair conformation adopting a layered packing motif observed in the HMX/4-fluoroaniline co-crystal (a). A view down the *c*-axis where HMX is blue and 4-fluoroaniline is red (b).
Adapted with permission from K. B. Landenberger and A. J. Matzger, *Cryst. Growth Des.*, 2012, **12**, 3603, Copyright 2012 American Chemical Society.

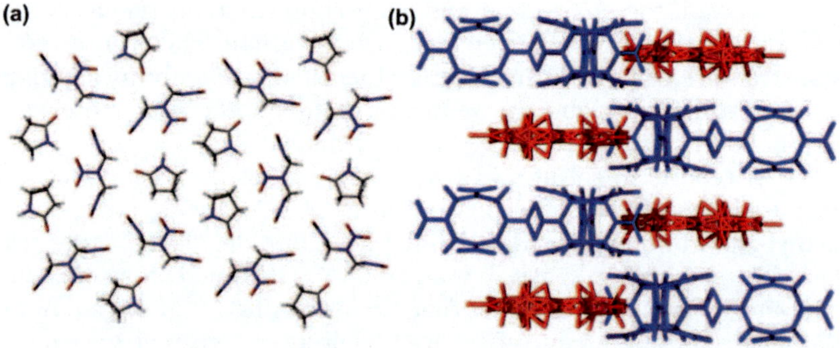

Figure 6.12 The chair–chair pocket packing motif observed in the HMX/2-pyrrolidone co-crystal viewed down the *c*-axis (a). Alternative view of the packing where HMX is blue and 2-pyrrolidone is red (b). The 2-pyrrolidone molecules are disordered over two positions.
Adapted with permission from K. B. Landenberger and A. J. Matzger, *Cryst. Growth Des.*, 2012, **12**, 3603, Copyright 2012 American Chemical Society.

co-crystal former molecules surrounded by six HMX molecules, three of which have nitro-groups facing toward the co-crystal former, and three of which have nitro-groups facing away (Figure 6.12). The co-crystal formers appear to serve as interchangeable guest molecules that stabilize the network of HMX molecules, suggesting that the co-crystal formers need only fulfil requirements of size, shape, and electronic factors to fill the interstitial spaces and stabilize the crystal. This effect, combined with the pocket-like nature of this motif results in a high propensity for the co-former molecules to be disordered.

(iii) The chair/layered motif (observed with 1,4-piperazinedicarboxaldehye, 2-picoline-*N*-oxide, and 2,4-dinitro-2,4-diazapentane) contains layers of HMX molecules in the chair conformation alternating with layers of the co-crystal former (Figure 6.13). In contrast to the other motifs, the interactions between HMX molecules are relatively distant and the interactions between HMX and the various co-crystal formers in this motif are shorter and more numerous. In general, the co-crystal formers generating this motif display greater variation in size and functionality than the other two motifs, and the authors suggest that this motif holds the greatest promise for accommodating other energetic compounds. Other researchers have looked at the binding energies of the HMX/2-picoline-*N*-oxide co-crystal using molecular dynamics in greater detail.[32]

All of the HMX co-crystals were found to be significantly less sensitive to impact than either β- or δ-HMX and it was proposed that this was because the layered structures found for all of these motifs introduce slip planes into the structure—other authors have suggested that the presence of such slip planes reduces the shock sensitivity of explosives.[33,34] The thermal behaviour of these HMX co-crystals was also investigated and it was observed that all of the co-crystals dissociated over a range of temperatures (125 to 179 °C), well below the decomposition temperature of HMX, and in all cases the less stable and more impact-sensitive δ-form of HMX was produced. This was a particularly interesting result as it demonstrated the concept of exploiting the insensitivity of HMX co-crystals, thus providing a means to

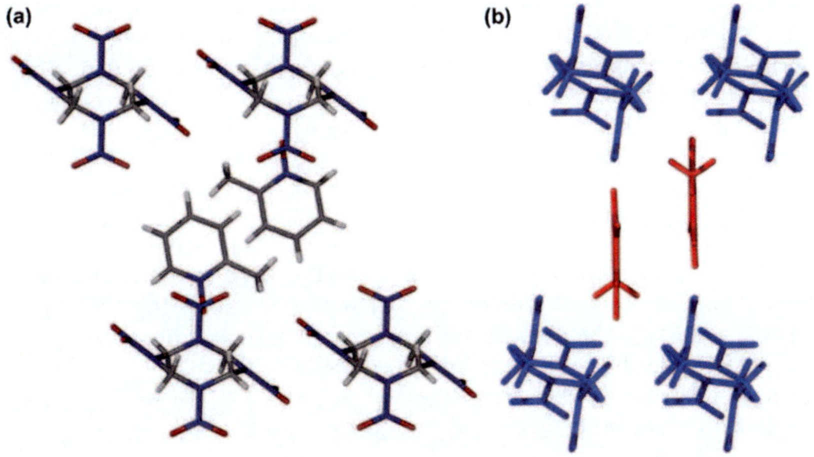

Figure 6.13 The chair layered packing motif observed in the HMX/2-picoline-*N*-oxide co-crystal (a). Alternative view of the packing down the *b*-axis where HMX is blue and 2-picoline-*N*-oxide is red (b).
Adapted with permission from K. B. Landenberger and A. J. Matzger, *Cryst. Growth Des.*, 2012, **12**, 3603, Copyright 2012 American Chemical Society.

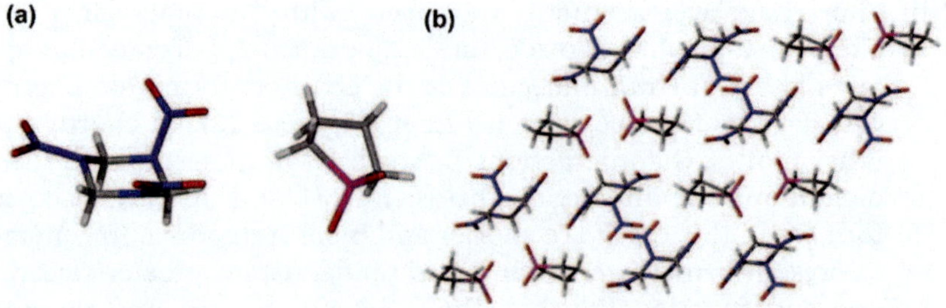

Figure 6.14 Crystal structure of the only known solvate of RDX from sulfolane.

safely transport HMX, combined with the ability to regenerate δ-HMX on demand.

Other co-crystals of HMX that have been reported include the 1 : 1 solvate formed with 1,3-dimethyl-2-imidazolidinone (DMI).[35] The experimental crystal structure, combined with further analysis using DFT computational techniques, showed that hydrogen bonding between methylene groups of HMX molecules and oxygen atoms of DMI molecules represent the main intermolecular interactions.

In marked contrast to the propensity of HMX to form co-crystals, there has only been one well-documented account of a co-crystal formed by the closely related molecule RDX. Crystallization of RDX from sulfolane (2,3,4,5-tetrahydrothiophene-1,1-dioxide) gave a 1 : 1 solvate (Figure 6.14)[36] in which each of the three nitro-groups of the RDX molecule adopt different orientations (axial, equatorial, and planar), resulting in differences in the N–N bond lengths, but no sensitivity data have been reported for this material.

The contrast between RDX and HMX in their propensity to form co-crystals presumably reflects the much greater conformational flexibility of HMX compared to RDX—in all four of the structurally characterized forms of RDX the six-membered ring adopts a chair conformation.[37–44]

6.5 Co-crystals of CL-20

On account of the favourable thermal stability, high density, and excellent explosive performance of CL-20, it is perhaps not surprising that this is an energetic material that is attracting substantial attention. As described earlier, one issue that hampers its wider use is its relatively high sensitivity to impact and shock. For this reason, there has been substantial effort expended on co-crystallization studies of CL-20 with the aim of developing CL-20 co-crystals that maintain high levels of performance, but which exhibit reduced sensitivity. Four crystalline forms of CL-20 are known to exist at ambient pressure: the ε-form is the most stable and has the highest density; the β-form can be crystallized from hot benzene; the γ-form can be obtained by heating the ε-form above 125 °C; and the α-form was first obtained by recrystallization from nitric acid and shown to be a hemihydrate.[45]

Millar *et al.* reported the preparation and structural characterization of four co-crystals (solvates) of CL-20 with *N,N*-dimethylformamide (DMF), 1,4-dioxane, hexamethylphosphoramide (HMPA), and γ-butyrolactone.[46] This study demonstrated that in the absence of traditional crystal-engineering motifs such as hydrogen bonding or π–π stacking, co-crystal formation appears to be directed by much weaker C–H···O and P=O···H intermolecular interactions (Figure 6.15). Thermally induced desolvation of CL-20/dioxane, CL-20/DMF and CL-20:butyrolactone, in which the CL-20 molecules adopt the same conformation as the γ-form of CL-20, resulted in the formation of the γ-form. Desolvation of CL-20/HMPA, in which the CL-20 molecules adopt the same conformation as the β-form of CL-20, resulted in the formation of the β-form. Hence, in all cases, the molecular conformation of the CL-20 molecules in the co-crystals appeared to direct which polymorph was produced on desolvation. Although no sensitivity measurements were made on these co-crystals, the authors suggested, based on the layered motifs observed for three of the structures, that they might display improved stability to friction or shock through dissipation of energy through the layers.

Further studies have focused on the intermolecular hydrogen-bonding interactions within co-crystals of CL-20/*p*-benzoquinone and CL-20/1,4-naphthoquinone.[47] The authors reported a substantial reduction in sensitivity to impact over that for ε-CL-20.

A very unusual 1:5 co-crystal of CL-20 with caprolactam has been reported and structurally characterized.[48] It could only be prepared under dry conditions, as in the presence of atmospheric moisture it transformed to α-CL-20. The crystal structure shows CL-20 molecules forming hydrogen-bonding interactions with caprolactam molecules (Figure 6.16), and not surprisingly the co-crystal has a substantially reduced density compared to CL-20. The material was found to be substantially less sensitive to impact than CL-20.

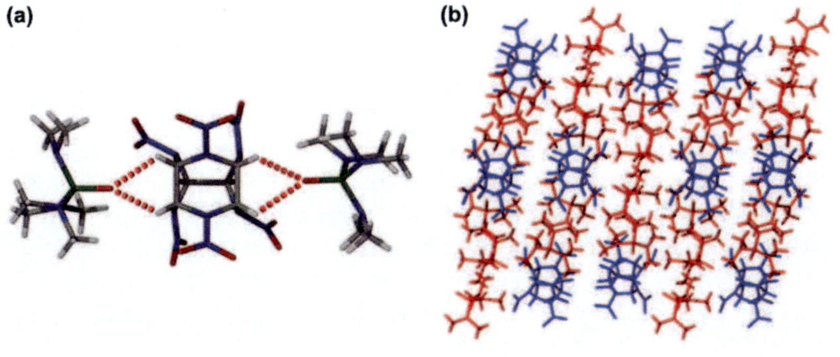

Figure 6.15 (a) Highlighting the P=O···H intermolecular hydrogen-bonding interactions observed in the CL-20/HMPA co-crystal. Hydrogen-bonding interactions shown as red dashed lines. (b) Crystal packing of the CL-20/HMPA co-crystal showing a layered packing motif. CL-20 is highlighted in blue and HMPA in red.
Adapted from ref. 46 with permission from the Royal Society of Chemistry.

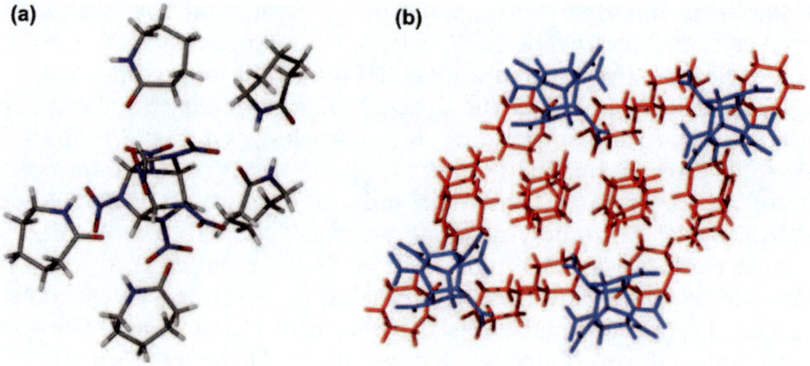

Figure 6.16 (a) Asymmetric unit of the co-crystal. (b) Packing observed in the CL-20/caprolactam co-crystal where CL-20 is red and caprolactam is blue.

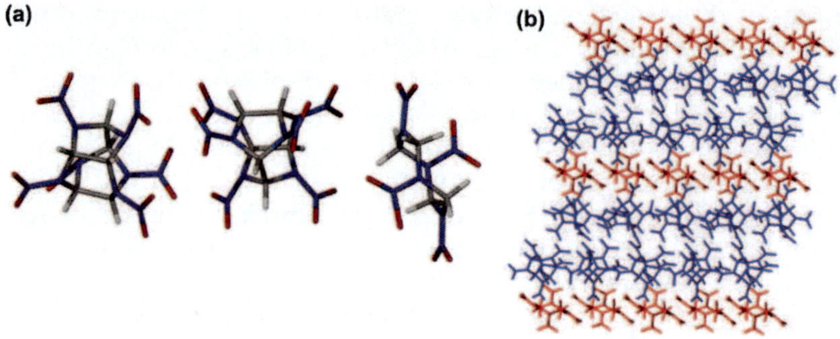

Figure 6.17 (a) Part of the asymmetric unit of the 2 : 1 CL-20/HMX co-crystal. (b) Crystal packing observed in the CL-20-HMX co-crystal highlighting the layered packing motif. CL-20 is shown as blue and HMX is red.[49]

A 2 : 1 CL-20/HMX co-crystal was first reported and structurally characterized by Bolton *et al.* and, based on its high density of 1.945 $g\,cm^{-3}$, this was predicted to have a detonation velocity of 9480 $m\,s^{-1}$, representing a power increase of ~20% compared to β-HMX (Figure 6.17).[49] Furthermore, in small-scale impact drop tests the co-crystal was reported to exhibit sensitivity that was indistinguishable from that of β-HMX and significantly lower than that of ε-CL-20. The authors suggested that this surprisingly low sensitivity was caused, in part, by dilution of the more sensitive component, but also because of an increased degree of C–H···O hydrogen-bonding interactions observed in the structure of the co-crystal compared to those of pure HMX and CL-20. A molecular dynamics (MD) simulation offered another possible explanation in which the co-crystal was shown to have shorter $N{-}NO_2$ bond lengths than either a composite of CL-20 and HMX or pure CL-20, and that the cohesive energy density of the co-crystal is higher than that of the composite at the same temperature.[50] The authors highlight that the $N{-}NO_2$ bonds are "trigger" bonds, *i.e.* these are the bonds that are broken on initiation. In a subsequent experimental study by Anderson *et al.*, in which the co-crystal was subjected to small-scale sensitivity testing (impact, friction, and electrostatic discharge), the authors noted that the sensitivity data did not correlate with the previously reported values in the literature, and that the co-crystal was in fact more sensitive to impact than ε-CL-20.[51] The authors highlighted this as a potential safety concern, and this exemplifies one of the issues associated with sensitivity testing. It is commonly accepted that there are many factors (*e.g.* different operating procedures, particle size, presence of adsorbed water and solvent) that can lead to variations in the determination of impact sensitivity and that comparison of data between laboratories should be done with care.[52]

The thermal decomposition of $(CL\text{-}20)_2$/HMX has been studied using a combination of techniques including DSC, hot-stage microscopy, ultrafast chip calorimetry, and high-resolution X-ray powder diffraction.[13,49,51,53,54] These studies showed that the co-crystal does not melt, but like pure CL-20 it begins to decompose at *ca.* 235 °C followed by detonation at *ca.* 240 °C.

Co-crystallization of CL-20 with BTF (benzotrioxofuran) from ethanol has been demonstrated to give a 1:1 co-crystal that has been structurally characterized.[55] In the crystal structure, the closest interaction is the edge-to-face π–π stacking of BTF rings, while the second closest interaction is the unique C–H···N hydrogen bonds between the two co-formers (Figure 6.18). The CL-20/BTF co-crystal has a crystallographic density of 1.918 g cm^{-3} at

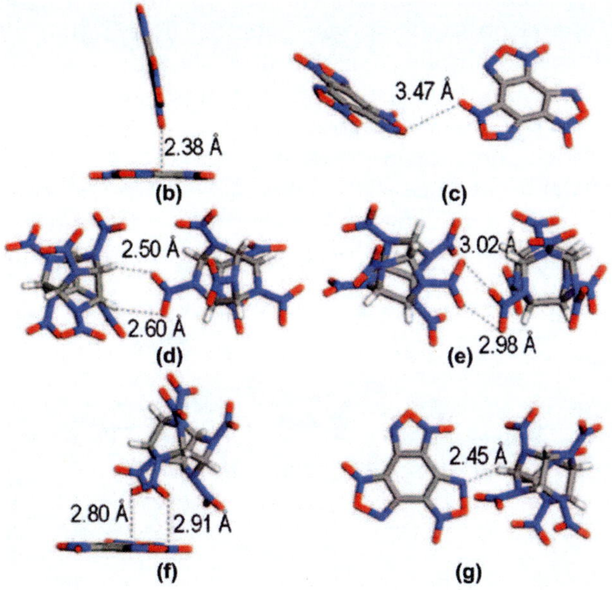

Figure 6.18 Highlighting the various hydrogen-bonding interactions observed in the CL-20/BTF co-crystal.[55]
Reprinted with permission from Z. Yang, H. Li, X. Zhou, C. Zhang, H. Huang, J. Li and F. Nie, *Cryst. Growth Des.*, 2012, **12**, 5155. Copyright 2012 American Chemical Society.

room temperature, which, although somewhat higher than that of BTF (1.901 g cm^{-3}), is lower than that of ε-CL-20 (2.035 g cm^{-3}). The predicted detonation performance of the co-crystal indicates that the detonation pressure and detonation velocity relative to BTF were increased by 14.0% and 6.5%, respectively.

This experimental approach of stabilizing hydrogen-free energetic molecules such as BTF by crystallizing with other hydrogen-containing energetic molecules has also been verified by theoretical work that highlights the role of hydrogen-bonding in encouraging co-crystal formation and potentially conferring additional stability.[56]

Another approach has been demonstrated in the synthesis and characterization of a 1 : 1 co-crystal of CL-20 with 1-methyl-3,4,5-trinitropyrazole (MTNP) by X-ray diffraction.[57] Instead of co-crystallization with carbon-rich aromatic molecules such as TNT, the authors selected the more nitrogen-rich heterocycle, MTNP, which is itself a potential candidate for replacement of TNT as a melt-cast explosive (Figure 6.19).[58,59]

CL-20/MTNP has a high crystal density of 1.932 g cm^{-3} at 293 K and superior detonation performance due to its high heat of formation, nitrogen content, and oxygen balance. Moreover, measured impact and friction sensitivities show that it is less sensitive than CL-20, with a sensitivity close to those of RDX and HMX. In order to analyse the intermolecular interaction of CL-20/MTNP, a series of theoretical analyses was employed including Hirshfeld surface analysis, non-covalent interaction plots, interaction energy calculations, and electrostatic surface potential distributions. The physicochemical performance implied that CL-20/MTNP could serve as a promising energetic material, and the use of methyl-substituted low melting point compounds acting as donor–acceptors could be a new strategy for preparing novel nitrogen-rich energetic co-crystals

A 1 : 1 co-crystal between CL-20 1-methyl-3,5-dinitro-1,2,4-triazole (MDNT) has recently been reported by Anderson *et al.* and has been denoted as NEX-1.[60] MDNT is a potential melt-cast explosive and the authors surmised,

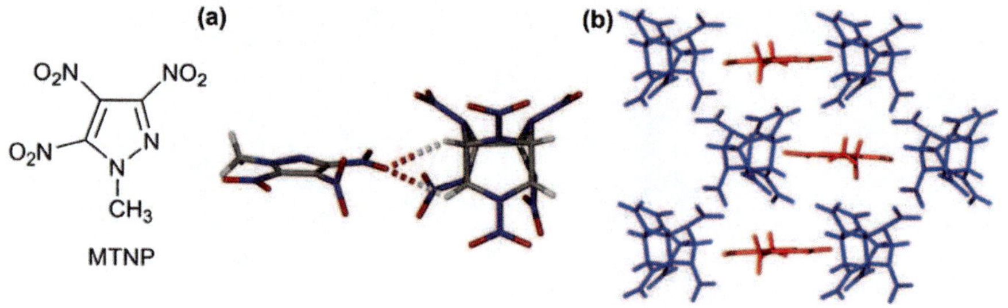

Figure 6.19 Left: molecular structure of MTNP. (a) Part of the asymmetric unit showing the intermolecular CH···O hydrogen-bonding interactions as dashed lines. (b) Packing observed in the CL-20/MTNP co-crystal where CL-10 is highlighted in blue and MTNP in red.
Adapted from ref. 57 with permission from The Royal Society of Chemistry.

based on the results of excess enthalpy calculations, that this would be a good candidate material for co-crystallization with CL-20. This technique, implemented in the program COSMOtherm[61] compared the excess enthalpy of CL-20 when combined in variable mole fractions with a selection of energetic co-formers (HMX, TNT, and MDNT). A comparison of the predicted excess enthalpy values showed a significant negative excess enthalpy at a mole fraction of CL-20 of 0.40, suggesting that there would be an increased stabilization offered by the co-crystallization of CL-20 and MDNT in a stoichiometric ratio.

The crystal structure of NEX-1 shows a nitro-group interacting with the π-system of the triazole ring with distances in the range 2.96–3.22 Å. As both components of the co-crystal are rich in nitro-groups, but lack hydrogen-bond donors, the crystal packing is driven by electrostatic interactions generated by the polar nitro-groups. The abundance of such anion–π interactions between adjacent molecules of both components results in a high density of the co-crystal (1.883 g cm^{-3} at 95 K) that lies midway between the densities of the two co-formers. The authors note that the conformation of the CL-20 molecule resembles that of the α-form of CL-20 (a hydrate) rather than that of the denser ε-form. The lower density of the co-crystal is attributed to this conformational difference as it creates small pockets of inaccessible space around every CL-20 molecule.

Unlike MDNT, the co-crystal did not melt, but instead showed the onset of decomposition at 200 °C with what appeared to be a greater energy release per mole than CL-20 (2166 kJ mol^{-1} for NEX-1 *vs.* 2019 kJ mol^{-1} for CL-20). The co-crystal showed a similar sensitivity to impact to that of ε-CL20, but was less sensitive to friction and electrostatic discharge.

Bennion and co-workers recently reported two hydrogen peroxide solvates of CL-20.[62] Crystallization of CL-20 from a 1 : 1 mixture of acetonitrile and hydrogen peroxide resulted in the formation of two different crystal habits that were shown to be polymorphic solvates both with the stoichiometry 2 : 1 CL-20/H_2O_2. One of the solvates was found to be isostructural with α-CL-20—the hemihydrate of CL-20. This solvate was found to have a higher crystallographic density than α-CL-20 (2.033 g cm^{-3} compared to 1.970 g cm^{-3}). The inclusion of hydrogen peroxide within the structure results in an improved oxygen balance in comparison to α-CL-20 (−8.79% for the solvates and −0.84% for α-CL-20).

A co-crystal of CL-20 and RDX has recently been reported by slow evaporation of the two components from a solution of acetonitrile.[63] The material was characterized by vibrational spectroscopy, powder X-ray diffraction, and DSC. The authors also measured the impact sensitivity of the material and reported that it was less sensitive than CL-20. Whilst this represents a potentially intriguing result, the evidence presented for the formation of a genuine co-crystal is not strong, *i.e.* no characterization by single-crystal diffraction techniques. Further discussion about the potential pitfalls that can arise in the identification of energetic co-crystals is presented later in this review.

6.6 Co-crystals of Azole Derivatives (NTO, DNBT, and DNPP)

3-Nitro-1,2,3-triazole-5-one (NTO) was first synthesized in the 1980s and was initially regarded as a replacement for RDX. It has a calculated detonation velocity and pressure similar to those of RDX, but is much less sensitive.[64] NTO exists as one of two polymorphs, the α-form or β-form.[65] Due to its high acidity, the use of NTO can be problematic. To date there has only been one reported co-crystal of NTO.[66] In this study NTO was co-crystallized with the nitrogen-rich compound 5,6,7,8-tetrahydrote-trazolo[1,5-*b*] [1,2,4]-triazine (TZTN). This co-former was chosen as it is a weak base and it could potentially neutralize the acidity of NTO. Strong intermolecular NH···N and NH···O hydrogen-bonding interactions are the primary driving force in the formation of the NTO/TZTN energetic–energetic co-crystal (Figure 6.20). The co-crystal has a crystallographic density of 1.665 $g\,cm^{-3}$, which is higher than TZTN, but lower than NTO. Both the detonation velocity and detonation pressure were found to be slightly lower than those of NTO.

Azole-based energetic materials are of significant interest for co-crystallization studies as they are relatively good hydrogen bond donor or acceptor molecules. Examples of such molecules include 5,5′-dinitro-3,3′-bi-1,2,4-triazole (DNBT) and 3,6-dinitropyrazolo[4,3-*c*]pyrazole (DNPP) (Figure 6.21).[6]

DNBT was first developed in the 1970s and has a relatively high density (1.89 $g\,cm^{-3}$); however, DNBT is readily found in hydrate form and has a significantly lower density (1.74 $g\,cm^{-3}$).[67] Recently Matzger *et al.* reported the formation of a series of co-crystals with DNBT.[68] Three co-crystals were reported with co-formers DNPP, 5-amino-3-nitro-1*H*-1,2,4-triazole (ANTA) and 3,4-dinitropyrazolo (3,4-DNP). The DNBT/DNPP co-crystal was formed in a 1 : 1 ratio whereas the other two formed in a 1 : 2 ratio of DNBT to co-former (Figure 6.22).

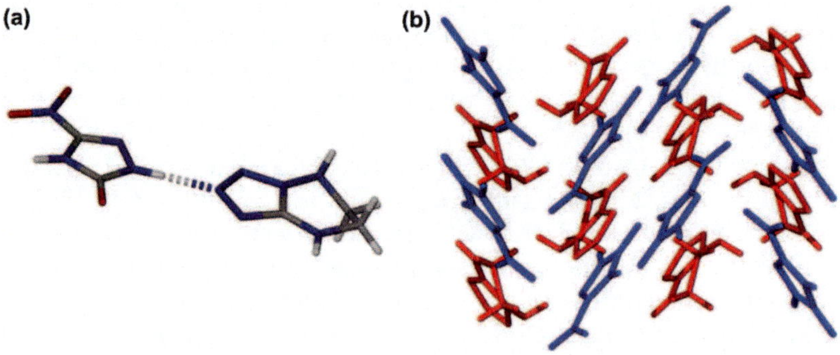

Figure 6.20 (a) The asymmetric unit found in the NTO/TZTN co-crystal with the intermolecular NH···N hydrogen-bonding interaction highlighted as a mixed colour dashed line. (b) Packing motif observed in the co-crystal where NTO is highlighted in blue and TZTN is red.

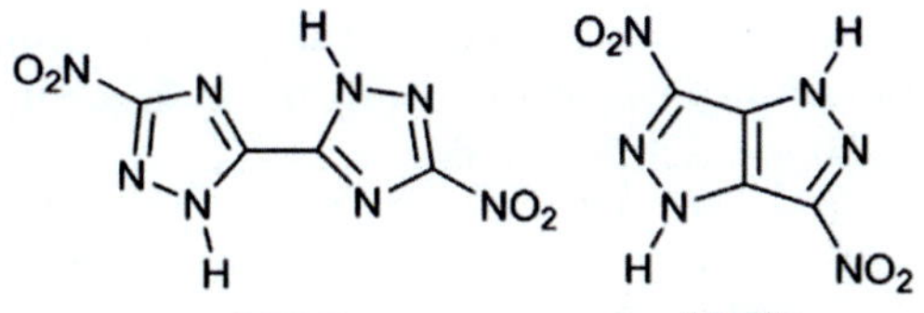

Figure 6.21 Molecular structure of DNBT and DNPP.

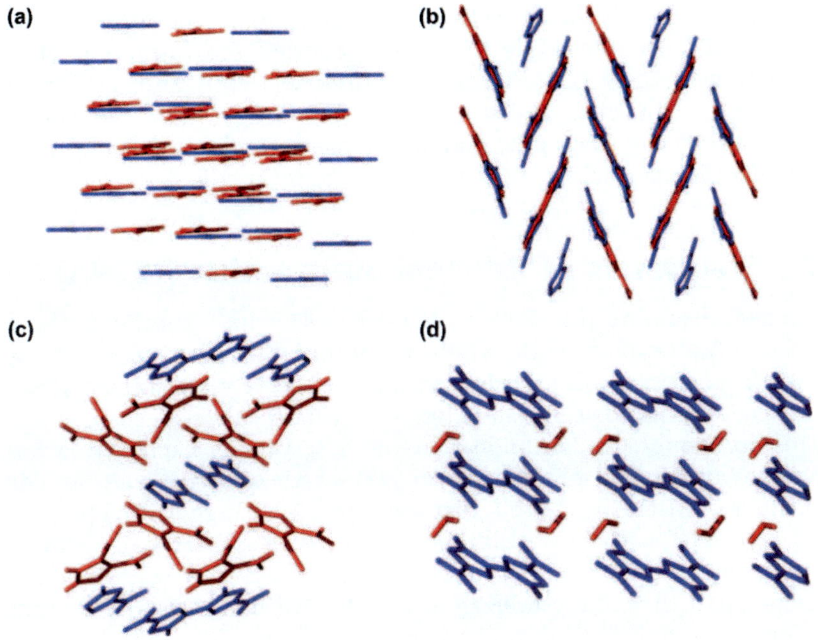

Figure 6.22 Packing motifs observed in the co-crystals of DNBT. (a) Layered packing observed in DNBT/ANTA. (b) Herringbone packing in DNBT/DNPP. (c) Alternating layered system in DNBT/3,4-DNP. (d) Layered packing found in hydrated DNBT. In all images DNBT is blue and the co-formers are red.

All three co-crystals are held together through a series of NH···N and NH···O hydrogen-bonding interactions. DNTB/DNPP and DNTB/ANTA were found to have a calculated detonation velocity lower than those of their individual co-formers whereas the DNTB/3,4-DNP co-crystal has a detonation velocity comparable to 3,4-DNP. Sensitivity testing of the co-crystals revealed both DNTB/DNPP and DNTB/ANTA can be classified as insensitive explosives. However, DNTB/3,4-DNP was found to be less sensitive than 3,4-DNP but more sensitive than DNTB.

DNPP is a strong acid that would typically form energetic salts[69] when reacted with bases. However, Shreeve *et al.* reported the formation of energetic co-crystals of DNPP with the basic co-formers 3-amino-1,2,4-triazole

(3-AT) and 4-amino-1,2,4-triazole (4-AT).[70] Co-crystallization with weak bases can neutralize the acidity of DNPP resulting in a neutral co-crystal. The DNPP/3-AT and DNPP/4-AT co-crystals exists in a 1:2 ratio. The room-temperature densities of DNPP/3-AT (1.72 $g\,cm^{-3}$) and DNPP/4-AT (1.68 $g\,cm^{-3}$) were found to lie between those of DNPP and their respective co-formers. Extensive intermolecular hydrogen-bonding interactions are observed throughout the co-crystals. This could be directly influencing the sensitivity of the co-crystals to impact as they were found to be less sensitive than DNPP and were thus easier to manipulate.

A further example of azole-based energetic materials was reported by Zhang *et al.*[71] In this study the group reported the formation of a 1:2 co-crystal of 1*H*,1′*H*-5,5′-bitetrazole-1,1′-diolate (BTO) and 1-amino-1,2,3-triazole (ATZ). This co-crystal was reported as having a greatly improved heat of formation over that of common explosives and an enhanced impact sensitivity thereby making this material safer to handle.

6.7 Co-crystals of Ethylenedinitramine (EDNA)

Ethylenedinitramine (EDNA) also known as Haleite[72] combines the properties of a high explosive like TNT and an initiating agent like mercury fulminate; it possesses a high brisance, a comparatively low impact sensitivity and a high decomposition temperature (Figure 6.23).

This compound finds applications as a secondary explosive, a booster explosive, and in the preparation of melt-cast explosives such as Ednatol (a mixture of EDNA and TN). It displays a relatively high acidity ($pK_{a1} = 5.31$ and $pK_{a2} = 6.64$) with the result that it is corrosive and can react with metals to form salts, thereby limiting its more widespread use. Aakeröy *et al.* have demonstrated that co-crystallization of EDNA with six selected compounds containing suitable hydrogen-bond acceptor groups reduced its acidity while retaining other valuable characteristics.[73] For example, the co-crystal of EDNA and pyrazine *N,N′*-dioxide was shown to be less corrosive towards metallic copper than EDNA, and this was rationalized by the introduction of a "supramolecular protecting group" through co-crystal formation that reduced the effective acidity of the protons on the EDNA molecule. The co-crystal of EDNA and 1,2-bis(4-pyridyl)ethylene was found to be more thermally stable than EDNA, while the co-crystals EDNA:4,4′-bipyridine and EDNA/pyrazine *N,N′*-dioxide were found to be about 50% less impact-sensitive than EDNA. The authors also highlighted that changing the

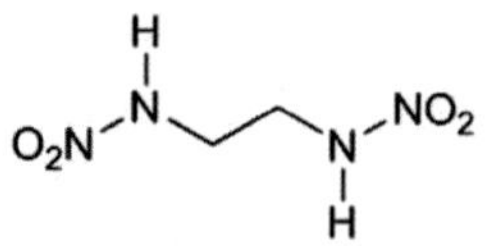

Figure 6.23 Molecular structure of EDNA.

co-former from the parent *N*-heterocycle to the corresponding *N*-oxide serves two purposes in the design and synthesis of energetic co-crystals: not only is the oxygen balance enhanced, but the density can also improve. Such a strategy has also been employed in the design of single component energetic materials based on *N*-oxides of substituted tetrazines.[74]

6.8 Co-crystals of Diacetone Diperoxide (DADP)

Although known for over 100 years the crystal structure of diacetone diperoxide (DADP) has only recently been determined.[75] Belonging to the family of compounds related to triacetone triperoxide (TATP), DADP is the less common of the two (Figure 6.24) and both compounds are extremely sensitive to impact and friction. One advantage DADP holds over TATP is that the peroxide oxygen atoms are accessible for potential co-crystal formation through non-covalent interactions whereas the peroxide oxygen atoms in TATP are shielded by the methyl groups making it less likely to form co-crystals.[76]

Matzger and co-workers investigated the formation of a series of co-crystals containing DADP trinitrotrihalogenated benzene co-crystals.[76,77] 1 : 1 DADP co-crystals were formed with 1,3,5-trichloro-2,4,6-trinitrobenzene (TCTNB), 1,3,5-tribromo-2,4,6-trinitrobenzene (TBTNB) and 1,3,5-triiodo-2,4,6-trinitrobenzene (TITNB). The TCTNB and TBTNB co-crystals were found to be isostructural where the main interaction with DADP was through a peroxide–aromatic ring interaction (Figure 6.25). The TITNB co-crystal differs from the other two as there is a unique halogen-bonding interaction with the peroxide (Figure 6.26). This co-crystal has a better oxygen balance and a higher density than the other two co-crystals. Sensitivity tests of DADP, DADP/TCTNB and DADP/TITNB gave some surprising results in that the DADP/TCTNB co-crystal was found to be only slightly less sensitive than DADP, thereby demonstrating that co-crystallization can produce a material with improved oxygen balance and density whilst retaining a similar sensitivity. In contrast, the DADP/TITNB co-crystal was found to be much less sensitive than both DADP and TITNB, demonstrating that combining two materials with relatively high sensitivities can result in a much less-sensitive co-crystal.[77]

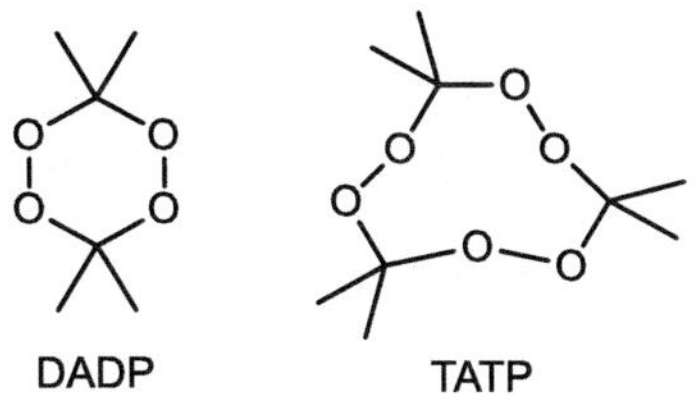

Figure 6.24 Molecular structures of DADP and TATP.

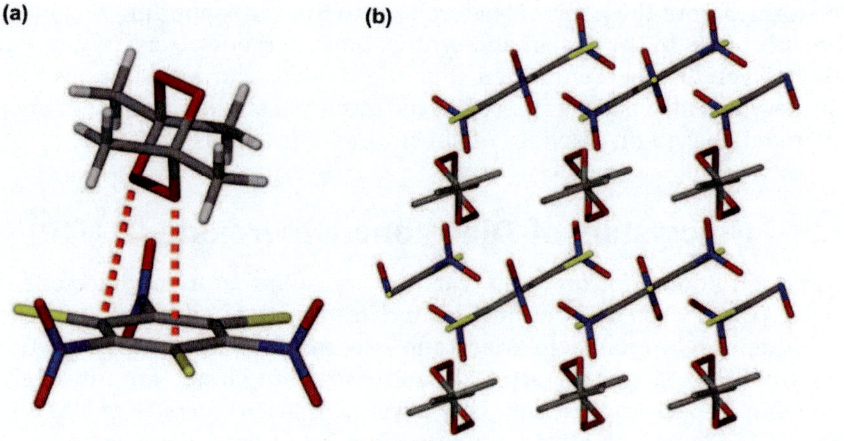

Figure 6.25 (a) 1 : 1 co-crystal of DADP/TBTNB showing the peroxide aromatic interaction (red dashed lines). (b) The extended structure of the co-crystal.

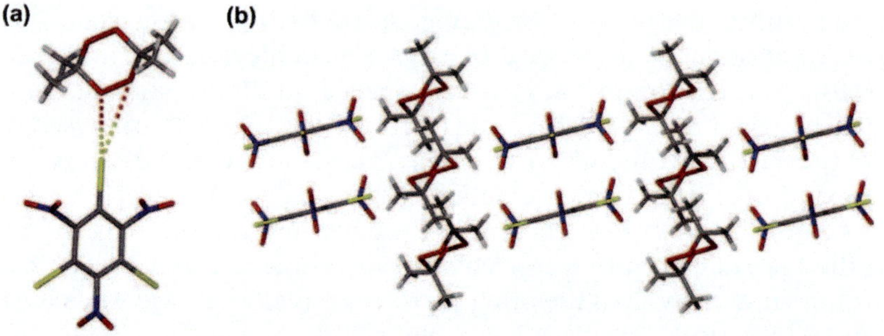

Figure 6.26 (a) 1 : 1 co-crystal of DADP/TITNB showing the peroxide halogen interaction (mixed colour dashed lines). (b) The extended structure of the co-crystal.
Adapted with permission from K. B. Landenberger, O. Bolton and A. J. Matzger, *J. Am. Chem. Soc.*, 2015, **137**, 5074, Copyright 2015 American Chemical Society.

6.9 Stabilization of High-nitrogen Compounds

There is significant interest within the energetics community in the synthesis of new molecules that are rich in nitrogen, and which therefore decompose very exothermically to gaseous dinitrogen. As would be expected, these compounds are often highly sensitive to initiation, displaying impact sensitivities of less than 1 J. Hence, in order to stabilize the highly sensitive salt 1,1′-(triaz-1-ene-1,3-diyl)bis(1*H*-tetrazol-5-amine) chloride (abbreviated to N11-Cl), the compound was successfully isolated in a co-crystal form combined with two molecules of 5-aminotetrazole (ATE) (Figure 6.27). This N11-Cl/ATE co-crystal exhibited a much lower impact sensitivity of 10 J,

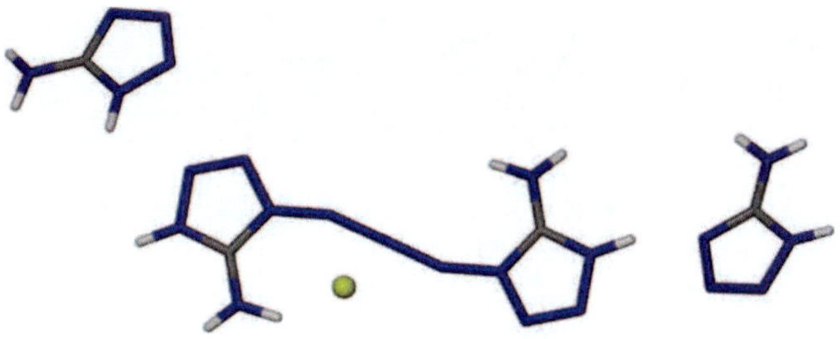

Figure 6.27 Structure of the N11-Cl/ATE co-crystal.

compared to the reported N10 compound, 1,1′-azobis(tetrazole) with an impact sensitivity of ≪1 J.[78]

6.10 A Melt-castable Co-crystal

There are two main methods for the loading of energetic materials into munitions. The first involves hydraulic pressing and subsequent milling of a pure or formulated energetic material to give the required dimensions, whilst the second involves melt-casting of the liquid material or mixture directly into a munition. Melt-casting offers several advantages over hydraulic pressing—these include: elimination of voids, leading to the attainment of the theoretical maximum density; greater processing efficiency; and the ability to fill more complex geometric shapes. Desirable properties of a melt-castable material include: a low melting point (between 70 and 120 °C), a large range between the melting point and the onset of thermal decomposition, and a high loading density. Relatively few energetic materials fit these criteria and traditional melt-cast technology has focused on materials such as TNT, 2,4-dinitroanisole (DNAN), and 1,3,3-trinitroazetidine (TNAZ). Hence attention has turned towards co-crystallization as a potential method to expand the range of high-performance melt-castable formulations. A great example of this approach was recently reported by Bennion *et al.* in which the authors prepared and characterized a low-melting 1 : 1 co-crystal formed between 3,4-diaminofurazan (DAF) and 4-amino-3,5-dinitropyrazole (ADNP) (Figure 6.28).[79]

ADNP possesses a high density (1.90 g cm^{-3}) and low sensitivity to impact, but has poor thermal stability and melts with decomposition near 180 °C. The authors proposed that the poor thermal properties of ADNP could be improved by the addition of a co-former that could alter the hydrogen-bonding network, thereby stabilizing ADNP in the melt. With a relatively low density (1.61 g cm^{-3}) and an oxygen balance (OB = −80) similar to that of TNT, DAF exhibits only a modest performance. However, it has a melting point of 180 °C, decomposes at 250 °C, and has the capacity to form hydrogen bonds. These factors therefore make it an attractive potential co-former for combination with ADNP. Co-crystals of DAF:ADNP were obtained either from acetonitrile solution, or from an isopropanol slurry or from the melt, and the crystal structure shows the formation of hydrogen bonds

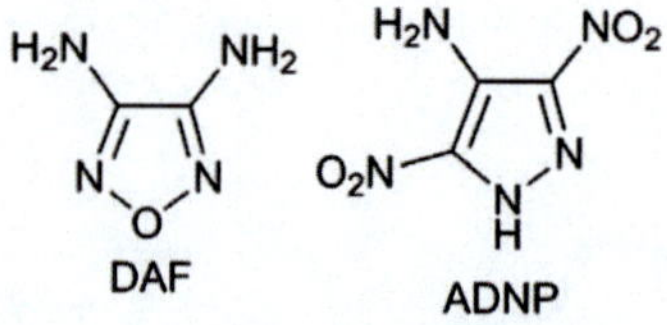

Figure 6.28 Molecular structure of DAF and ADNP.

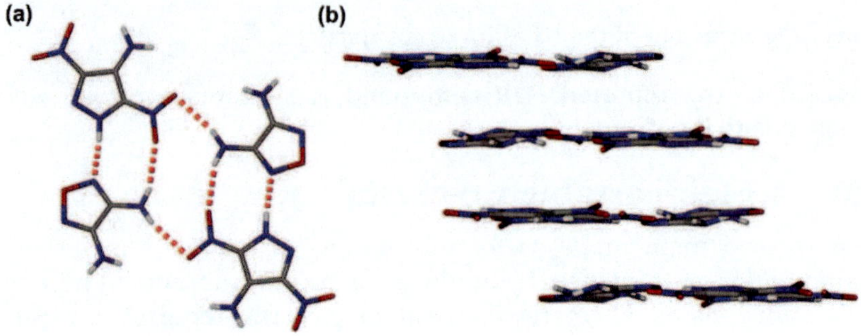

Figure 6.29 (a) Hydrogen-bonding interactions observed in DAF/ADNP. (b) Layered packing observed in the extended structure.

between the amine and nitro groups (2.01 and 2.22 Å) as well as short hydrogen-bonding interactions between DAF and ADNP involving the amine and the furazan (1.77 Å). The shortest hydrogen-bonding interaction in the co-crystal is significantly shorter than that of any hydrogen-bonding interactions in either of the pure components (Figure 6.29).

Raman spectroscopic studies showed that the strong hydrogen bonding between the two components persisted on melting and this leads to a significantly increased thermal stability compared to DAF, *i.e.* the co-crystal melted at 162 °C and decomposed near 212 °C.

6.11 Co-crystals or Physical Mixtures?

The excitement generated by co-crystallization as a potential route for the discovery of new energetic materials has undoubtedly resulted in increasing levels of research activity in this field. The co-crystallization of two energetic compounds is often challenging to achieve and, in the absence of unambiguous data from a single crystal X-ray diffraction study, can present significant challenges for characterization. The differentiation between physical mixtures and true co-crystalline materials is challenging due to the propensity for energetic compounds to form solvates and/or polymorphs. These challenges have been highlighted and discussed in a study by Wiscons and Matzger[80] in which they re-investigated six previously published materials that claimed to be co-crystals (HMX/TATB, RDX/LLM-105, HMX/LLM-105, CL-20/LLM-105, HMX/AP, and CL-20/TATB),[81–84] but for which no structures obtained from single-crystal X-ray diffraction measurements had been presented. It was found that in all six cases the data originally published were consistent with crystallization of *physical mixtures* of the two starting components; in three of these cases the crystallization procedures were reproduced, and the characterization data supported the presence of physical mixtures. The authors highlight that characterization studies that rely solely on vibrational spectroscopy, crystal morphology, and powder X-ray diffraction can be misleading. For example, if polycrystalline samples are not ground to give finely divided powders (for understandable reasons, given their potential to be very sensitive), then the effects of preferred orientation of the sample can be severe, resulting in widely varying relative intensities of Bragg peaks in the PXRD pattern such that expected reflections appear to be very weak or absent and low-intensity peaks may be unduly prominent. This can result in PXRD patterns that appear to be very different from those expected for the two components and lead to the erroneous conclusion that a co-crystal has been formed.

6.12 Novel Preparative Methods of Energetic Co-crystal RAM Mixing and Vacuum Freeze-drying

Currently, the most common technique used to produce co-crystals is by evaporation of a solution containing a mixture of co-formers. Although extremely useful for the creation of crystalline material, this method can often be slow, requires large ratios of solvent volume to solutes, and is often difficult to scale up. A recent development for the large-scale formation of co-crystals is through the use of resonant acoustic mixing (RAM).[51,85] Low-frequency, high-intensity acoustic energy results in very effective mixing of the material within the reaction vessel. This enables larger scale co-crystallization experiments to be performed over a much shorter time-scale. Initial co-crystallization experiments demonstrated proof-of-concept using non-energetic systems,[85] but this was soon followed by an energetic co-crystal experiment (see CL-20/HMX section 1.5).[51] This method for the production of energetic co-crystals on a larger scale offers a safer option for co-crystal formation compared to grinding and milling methods for which there is a substantial risk of initiation of sensitive materials. An additional disadvantage of the solvent evaporation method is that it is difficult to produce nanoscale particles. Qiu *et al.* recently reported a novel way of producing nanoscale CL-20/HMX by bead milling an aqueous suspension of the two components and monitoring the PXRD patterns of the resulting solids as a function of time.[86] Complete conversion to the co-crystalline material was reported over a period of 60 min, possibly as a result of a solvent-mediated mechanism. Scanning electron microscopy showed that the final co-crystal particles exhibited rounded morphology were less than 200 nm in diameter. Another technique that has been reported for the large-scale production of nano-sized energetic co-crystals is spray flash evaporation.[87,88] This continuous manufacturing approach can be used to prepare crystalline composites, partially crystalline mixtures, and co-crystals at rates of kg h^{-1}.

Gao and co-workers have recently reported a novel method for the preparation of a co-crystal of CL-20 and nitroguanidine (NQ) by a vacuum freeze-drying process.[89] Vacuum freeze-drying involves freezing a solvent within a material and then allowing the frozen solvent to sublime under vacuum. Although no single crystal data were reported, the authors claimed that nano-sized CL-20/NQ co-crystals could be obtained and claimed that the impact-sensitivity of this material is lower than that of CL-20. This reduced sensitivity is attributed to hydrogen bonding within the co-crystal.

6.13 Summary and Future Directions

This review has demonstrated that co-crystallization represents a powerful technique for the modification of the important properties of energetic materials—see Table 6.1. Examples have been presented that show how the following properties can be modified: density (CL-20/TNT);[10] impact sensitivity (DADP/TITNB);[77] thermal decomposition temperature (TNT/1-nitronaphthalene);[17] melting point (DAF/ADNP);[79] detonation velocity (CL-20/BTF);[55] switchable properties (HMX/2-picoline-*N*-oxide);[31] oxygen balance (CL-20/H_2O_2);[62] acidity and corrosive properties (NTO/TZTN and EDNA/*N*,*N*′-pyrazine-dioxide).[66,73] Hence there is substantial scope for the identification of new energetic materials with enhanced properties, *e.g.* reduced sensitivity, higher performance, increased damage resistance, tailored and/or switchable outputs. There is increasing recognition that co-crystallization of two components that have both already been developed and optimized for performance on their own may not in fact be the optimal strategy for the design of new energetic materials. Instead, an emerging strategy is to combine in a co-crystal two components that have not been optimized for performance when acting alone, but when combined in a co-crystal they act synergistically and result in a material with properties that are significantly enhanced, *i.e.* the combination of two components to give a product that is "more than the sum of its two parts". Hence this provides opportunities for re-visiting energetic compounds that had originally been rejected on the basis of their sub-optimal or undesirable properties, *e.g.* poor performance or high sensitivity, by combining them in a co-crystal. The example of ADNP/DAF highlights new opportunities for the discovery and development of new melt-castable materials based on co-crystals. At the same time, there are also likely to be opportunities for new propellants based on co-crystals, especially as there is increasing pressure to replace the widely used material ammonium perchlorate because of its environmental impact.

Nevertheless, despite the substantial progress in this research area over the last decade, there remain some substantial challenges. In particular, there is a key requirement to develop an enhanced understanding of how the crystal structures of energetic co-crystals influence key properties such as sensitivity

Table 6.1 Properties of some of the co-crystals discussed. All crystal densities are reported at room temperature unless otherwise stated.

Co-crystal	Tm (°C)[a]	Td (°C)[b]	D ($g\,cm^{-3}$)[c]	vD (ms^{-1})[d]	P (GPa)[e]	Ref.
TNT/naphthalene	95.9	304.2	1.487[f]	—	—	9
TNT/1-bromonaphthalene	73.3	296.4	1.737[f]	—	—	9
TNT/anthracene	108.5	293.5	1.515[f]	—	—	9
TNT/9-bromoanthracene	101.0	288.2	1.698[f]	—	—	9
TNT/phenanthrene	100.2	307.0	1.487[f]	—	—	9
TNT/perylene	151.5	268.6	1.531[f]	—	—	9
TNT/tetra thiafulvalene	108.8	202.9	1.707[f]	—	—	9
TNT/thieno[3,2-b]thiophene	83.7	291.9	1.677[f]	—	—	9
TNT/phenothiazine	107.2	237.3	1.561[f]	—	—	9
TNT/dibenzothiophene	118.8	312.8	1.545[f]	—	—	9
TNT/4,6-dimethyldibenzothiophene	116.3	299.1	1.496[f]	—	—	9
TNT/1,2-phenylenediamine	73.8	267.3	1.575[f]	—	—	9
TNT/1,4-dimethoxybenzene	45.2	289.5	1.501[f]	—	—	9
TNT/4-aminobenzoic acid 1:1	117.6	256.7	1.575[f]	—	—	9
TNT/4-aminobenzoic acid 1:2	175.2	251.8	1.509[f]	—	—	9
TNT/anthranilic acid 1:1	151.7	266.2	1.594[f]	—	—	9
TNT/anthranilic acid 1:2	139.3	258.4	1.544[f]	—	—	9
CL20/TNT	136	—	1.84	—	—	10
TNT/1-nitronaphthalene	62	240	1.593	5318.6	13.58	17
TNT/TAPB	170	241	1.442[g]	—	—	18
TNT/TNB	62.31	—	1.64	7070	20.93	19
BTF/MATNB	171.3	—	1.804[h]	—	—	20
BTF/TNA	205.8	—	1.884[i]	—	—	20
BTF/TNB	189.0	—	1.806	—	—	20
BTF/TNT	132.6	—	1.805[h]	—	—	20
BTF/TNAZ	164.5	—	1.834	—	—	20
BTF/1,3-DNB	130.0	286.0	1.735	7373	24.4	21

Table 6.1 (*Continued*)

Co-crystal	Tm (°C)[a]	Td (°C)[b]	D (g cm^{-3})[c]	vD (ms^{-1})[d]	P (GPa)[e]	Ref.
HMX/1,2-phenylenediamine	170	—	1.651[f]	—	—	31
HMX/2-bromoaniline	156	—	1.860[f]	—	—	31
HMX/3,4-diaminotoluene	148	—	1.657[f]	—	—	31
HMX/4fluoroaniline	128	—	1.687[f]	—	—	31
HMX/thieno[3,2-b]thiophene	179	—	1.765[f]	—	—	31
HMX/2-pyrrolidone	163	—	1.657[j]	—	—	31
HMX/4-picoline-N-oxide	125	—	1.749[f]	—	—	31
HMX/1,4-piperazinedicarboxaldehyde	151	—	1.630[f]	—	—	31
HMX/2-picoline-N-oxide	153	—	1.583[f]	—	—	31
CL-20/DMF	—	—	1.725[f]	—	—	46
CL20/1,4-dioxane	—	—	1.606[f]	—	—	46
CL-20/HMPA	—	—	1.436[k]	—	—	46
CL-20/γ-butyrolactone	—	—	1.856[k]	—	—	46
CL-20/p-benzoquinone	132	235	1.737	7800	27.4	47
CL-20/1,4-naphthoquinone	176	245	1.774	7000	22.4	47
CL-20/CPL	89	194	1.405[l]	—	—	48
CL-20/HMX	—	235	1.945	9484	—	49
BTF/CL-20	—	235	1.918	8969	39.1	55
CL-20/MTNP	215	222	1.932	9347	40.5	57
CL-20/MDNT	165	200	1.883[f]	—	—	60
EDNA/4,4′-bipy	146	189	1.452	6178	15.24	73
EDNA/4,4′-ethylenedipyridine	129	191	1.355	5967	13.70	73
EDNA/4,4′-vinylenedipyridine	—	198	1.434	6030	14.32	73
EDNA/4,4′-azobipyridine	154	165	1.529[m]	6504	17.21	73
EDNA/pyrazine-1,4-dioxide	156	172	1.64	7256	22.73	73
EDNA/azophenylpyridine	133	168	1.432[m]	5948	13.75	73

DADP/TCTNB	—	—	1.92	—	—	77
DADP/TBTNB	—	—	2.40	—	—	77
DADP/TITNB	—	—	2.82	—	—	77
NTO/TZTN	156.6	178	1.665	7458	23.5	66
DNBT/DNPP	—	—	1.802	8135	—	68
DNBT/ANTA	—	—	1.833	8097	—	68
DNBT/3,4-DNP	—	—	1.824	8203	—	68
DNPP/3-AT	252	282	1.70	8024	23.9	70
DNPP/4-AT	231	284	1.68	8234	25.6	70
BTO/ATZ	133.6	145.4	1.697	8088	28.1	71

[a]Melting point.
[b]Decomposition temperature.
[c]Density.
[d]Detonation velocity.
[e]Detonation pressure.
[f]95 K.
[g]150 K.
[h]145 K.
[i]135 K.
[j]85 K.
[k]100 K.
[l]143 K.
[m]120 K.

and energy output—structure–property relationships for these materials are often not well understood except at a very qualitative level. The developments of correlations of structure with sensitivity are further hampered by the variability of the methods of determining sensitivities from laboratory to laboratory, as well as the variability in sample preparation, *e.g.* particle size and shape, presence of adsorbed water and solvent. Furthermore, the heats of formation of most reported energetic co-crystals are theoretically determined and so if more accurate evaluation of detonation performance is to be achieved, then there is a requirement for improvements in the precision of these calculations. Energetic materials experience extremes of both pressure and temperature immediately prior to and during detonation. Hence there is a need to explore the effects of both static and dynamic compression on representative energetic co-crystals to discover how they respond to these conditions. The impact of possible phase transitions, separation of co-crystal components, or chemical reactions between the components may have significant effects on the performance and characteristics of these new materials. Whilst some progress is being made in the prediction of co-crystals, this is an area that requires further development both through the implementation of more effective techniques that use excess enthalpy methods, but also through *ab initio* crystal structure prediction (CSP) methods. In order to perform the extensive safety and performance tests that are typically required for qualification of an energetic composition, there is a need to scale up the production of energetic co-crystals to the multi-gram and multi-kilogram scale. As described earlier, traditional solution crystallization methods provide only a narrow scope for this degree of scale-up for co-crystals. Hence there will be a requirement to further develop mechanochemical methods such as resonant acoustic mixing. Given these remaining challenges and the observation that the field of energetic co-crystallization is still relatively young, especially when compared to pharmaceutical co-crystallization, it is perhaps important to reflect that it may yet be some time before an energetic co-crystal formulation is deployed in operational devices. Nevertheless, co-crystallization science will play an increasingly important role not only in assuring the development of next-generation energetic materials, but will also provide fascinating insights into molecular crystals.

References

1. J. Akhavan, *The Chemistry of Explosives*, 3rd edn, Royal Society of Chemistry, Cambridge, 2011.
2. N. Schultheiss and A. Newman, *Cryst. Growth Des.*, 2009, **9**(6), 2950.
3. G. P. Stahly, *Cryst. Growth Des.*, 2009, **9**(10), 4212.
4. S. Aitipamula, R. Banerjee, A. K. Bansal, K. Biradha, M. L. Cheney, A. R. Choudhury, G. R. Desiraju, A. G. Dikundwar, R. Dubey, N. Duggirala, P. P. Ghogale, S. Ghosh, P. K. Goswami, N. R. Goud, R. R. K. R. Jetti, P. Karpinski, P. Kaushik, D. Kumar, V. Kumar, B. Moulton, A. Mukherjee, G. Mukherjee, A. S. Myerson, V. Puri,

A. Ramanan, T. Rajamannar, C. M. Reddy, N. Rodriguez-Hornedo, R. D. Rogers, T. N. G. Row, P. Sanphui, N. Shan, G. Shete, A. Singh, C. C. Sun, J. A. Swift, R. Thaimattam, T. S. Thakur, R. K. Thaper, S. P. Thomas, S. Tothadi, V. R. Vangala, N. Variankaval, P. Vishweshwar, D. R. Weyna and M. J. Zaworotko, *Cryst. Growth Des.*, 2012, **12**(5), 2147.
5. T. M. Haller, A. L. Rheingold and T. B. Brill, *Acta Crystallogr.*, 1985, **C41**, 963.
6. J. Zhang and J. M. Shreeve, *CrystEngComm*, 2016, **18**, 6124.
7. J. C. Barnes and W. Golnazarians, *Acta Crystallogr.*, 1987, **C43**, 549.
8. Y. H. Lee, H. Liu, J. Y. Lee, S. H. Kim, S. K. Kim, J. L. Sessler, Y. Kim and J. S. Kim, *Chem. - Eur. J.*, 2010, **16**, 5895.
9. K. B. Landenberger and A. J. Matzger, *Cryst. Growth Des.*, 2010, **10**(12), 5341.
10. O. Bolton and A. J. Matzger, *Angew. Chem., Int. Ed.*, 2011, **50**, 8960.
11. K. J. Kraeutle, *JANNAF Proceedings. CPIA Publ.* 1988 (**October**), 498.
12. C. Zhang, X. Xue, Y. Cao, J. Zhou, A. Zhang, H. Li, Y. Zhou, R. Xu and T. Gao, *CrystEngComm*, 2014, **16**, 5905.
13. D. Doblas, M. Rosenthal, M. Burghammer, D. Cheryshov, D. Spitzer and D. A. Ivanov, *Cryst. Growth Des.*, 2016, **16**, 432.
14. D. Guo, Q. An, S. V. Zybin, W. A. Goddard III, F. Huang and B. Tang, *J. Mater. Chem. A*, 2015, **3**, 5409.
15. L. Shi, X.-H. Duan, L.-G. Zhu, X. Liu and C.-H. Pei, *J. Phys. Chem. A*, 2016, **120**(8), 1160.
16. X. Wei, A. Zhang, Y. Ma, X. Xue, J. Zhou, Y. Zhu and C. Zhang, *CrystEngComm*, 2015, **17**(47), 9037.
17. D. Hong, Y. Li, S. Zhu, L. Zhang and C. Pang, *Cent. Eur. J. Energ. Mater.*, 2015, **12**(1), 47.
18. P. Vishnoi, M. G. Walawalkar and R. Murugavel, *Cryst. Growth Des.*, 2014, **14**, 5668.
19. C. Guo, H. Zhang, X. Wang, X. Liu and J. Sun, *J. Mater. Sci.*, 2013, **48**, 1351.
20. H. Zhang, C. Guo, X. Wang, J. Xu, X. He, Y. Liu, X. Liu, H. Huang and J. Sun, *Cryst. Growth Des.*, 2013, **13**, 679.
21. Z. Yang, Y. Wang, J. Zhou, H. Li, H. Huang and F. Nie, *Propellants, Explos., Pyrotech.*, 2014, **39**, 9.
22. H. H. Cady, A. C. Larson and D. T. Cromer, *Acta Crystallogr.*, 1963, **16**, 617.
23. E. A. Zhurova, V. V. Zhurov and A. A. Pinkerton, *J. Am. Chem. Soc.*, 2007, **129**, 13887.
24. W. C. McCrone, *Anal. Chem.*, 1950, **22**, 1225.
25. G. R. Millar and A. N. Garroway, *Naval Research Laboratory*, NRL/MR6120-01-8585, 2001.
26. H. H. Cady and L. C. Smith, *Los Alamos Scientific Laboratory*, LAAMS-2652, 1962.
27. R. S. George, H. H. Cady, R. N. Rogers and R. K. Rohwer, *Ind. Eng. Chem. Prod. Res. Dev.*, 1965, **4**(3), 209.

28. R. E. Cobbledick and R. W. H. Small, *Acta Crystallogr.*, 1975, **B31**, 2805.
29. T. M. Haller, A. L. Rheingold and T. B. Brill, *Acta Crystallogr.*, 1983, **C39**, 1559.
30. I. V. Levakova, A. P. Korobko, S. V. Krasheninnikov and V. E. Zavodnik, *Kristallografiya*, 1996, **41**, 1015.
31. K. B. Landenberger and A. J. Matzger, *Cryst. Growth Des.*, 2012, **12**, 3603.
32. Z.-B. Xie, S.-Q. Hu and X. Xiong, *Mol. Phys.*, 2016, **114**(14), 2164.
33. A. R. Zamiri and S. De, *Propellants, Explos., Pyrotech.*, 2011, **36**, 247.
34. T.-T. Zhou, J.-F. Lou, H.-J. Song and F.-L. Huang, *Phys. Chem. Chem. Phys.*, 2015, **17**, 7924.
35. H. Lin, S.-G. Zhu, H.-Z. Li and X.-H. Peng, *J. Mol. Struct.*, 2013, **1048**, 339.
36. T. M. Haller, T. B. Brill and A. L. Rheingold, *Acta Crystallogr., Sect. C*, 1984, **C40**, 517.
37. W. C. McCrone, *Anal. Chem.*, 1950, **22**, 954.
38. R. J. Karpowicz and T. B. Brill, *J. Phys. Chem.*, 1984, **88**, 348.
39. C. S. Choi and E. Prince, *Acta Crystallogr.*, 1972, **B28**, 2857.
40. P. Hakey, W. Ouellette, J. Zubieta and T. Korter, *Acta Crystallogr.*, 2008, **E64**, o1428.
41. D. I. A. Millar, I. D. H. Oswald, D. J. Francis, W. G. Marshall, C. R. Pulham and A. S. Cumming, *Chem. Commun.*, 2009, 562.
42. D. I. A. Millar, I. D. H. Oswald, C. Barry, D. J. Francis, W. G. Marshall, C. R. Pulham and A. S. Cumming, *Chem. Commun.*, 2010, **46**, 5662.
43. A. J. Davidson, I. D. H. Oswald, D. J. Francis, A. R. Lennie, W. G. Marshall, D. I. A. Millar, C. R. Pulham, J. E. Warren and A. S. Cumming, *CrystEngComm*, 2008, **10**, 162.
44. J. A. Ciezak, T. A. Jenkins, Z. Liu and R. J. Hemley, *J. Phys. Chem. A*, 2007, **111**, 59.
45. A. T. Nielsen, A. P. Chafin, S. L. Christian, D. W. Moore, M. P. Nadler, R. A. Nissan, D. J. Vanderah, R. D. Gilardi, C. F. George and J. L. Flippen-Anderson, *Tetrahedron*, 1998, **54**, 11793.
46. D. I. A. Millar, H. E. Maynard-Casely, D. R. Allan, A. S. Cumming, A. R. Lennie, A. J. Mackay, I. D. H. Oswald, C. C. Tang and C. R. Pulham, *CrystEngComm*, 2012, **14**, 3742.
47. C. Zhang, Z. Yang, X. Zhou, C. Zhang, Y. Ma, J. Xu, Q. Zhang, F. Nie and H. Li, *Cryst. Growth Des.*, 2014, **14**(8), 3923.
48. C. Guo, H. Zhang, X. Wang, J. Xu, Y. Liu, X. Liu, H. Huang and J. Sun, *J. Mol. Struct.*, 2013, **1048**, 267.
49. O. Bolton, L. R. Simke, P. F. Pagoria and A. J. Matzger, *Cryst. Growth Des.*, 2012, **12**, 4311.
50. T. Sun, J. J. Xiao, Q. Liu, F. Zhao and H. M. Xiao, *J. Mater. Chem. A*, 2014, **2**, 13898.
51. S. R. Anderson, D. J. am Ende, J. S. Salan and P. Samuels, *Propellants, Explos., Pyrotech.*, 2014, **39**, 637.
52. R. M. Doherty and D. S. Watt, *Propellants, Explos., Pyrotech.*, 2008, **3**, 4.
53. B. Gao, D. Wang, J. Zhang, Y. Hu, J. Shen, J. Wang, B. Huang, Z. Qiao, H. Huang, F. Nie and G. Yang, *J. Mater. Chem. A*, 2014, **2**, 19969.

54. X. Xue, Y. Ma, Q. Zeng and C. Zhang, *J. Phys. Chem. C*, 2017, **121**(8), 4899.
55. Z. Yang, H. Li, X. Zhou, C. Zhang, H. Huang, J. Li and F. Nie, *Cryst. Growth Des.*, 2012, **12**, 5155.
56. X. Wei, Y. Ma, X. Long and C. Zhang, *CrystEngComm*, 2015, **17**, 7150.
57. Q. Ma, T. Jiang, Y. Chi, Y. Chen, J. Wang, J. Huang and F. Nie, *New J. Chem.*, 2017, **41**, 4165.
58. G. Hervé, C. Roussel and H. Graindorge, *Angew. Chem., Int. Ed.*, 2010, **49**, 3177.
59. Y. Zhang, Y. Guo, Y.-H. Joo, D. A. Parrish and J. M. Shreeve, *Chem. - Eur. J.*, 2010, **16**, 10778.
60. S. R. Anderson, P. Dubé, M. Krawiec, J. S. Salan, D. J. am Ende and P. Samuels, *Propellants, Explos., Pyrotech.*, 2016, **41**, 783.
61. Y. A. Abramov, C. Loschen and A. Klamt, *J. Pharm. Sci.*, 2012, **101**, 3687.
62. J. C. Bennion, N. Chowdhury, J. W. Kampf and A. J. Matzger, *Angew. Chem., Int. Ed.*, 2016, **55**, 13118.
63. H. Gao, W. Jiang, J. Liu, G. Hao, L. Xiao, X. Ke and T. Chen, *J. Energ. Mater.*, 2017, **35**, 490.
64. K. Y. Lee, L. B. Chapman and M. D. Cobur, *J. Energ. Mater.*, 1987, **5**, 27.
65. E. A. Zhurova and A. A. Pinkerton, *Acta Crystallogr., Sect. B.*, 2001, **B57**, 359.
66. J.-T. Wu, J.-G. Zhang, T. Li, Z.-M. Li and T.-L. Zhang, *RSC Adv.*, 2015, **5**, 28354.
67. E. V. Nikitina, G. L. Starova, O. V. Frank-Kamenetskaya and M. S. Pevzner, *Kristallografiya*, 1982, **27**, 485.
68. J. C. Bennion, A. McBain, S. F. Son and A. J. Matzger, *Cryst. Growth Des.*, 2015, **15**, 2545.
69. J. Zhang, D. A. Parrish and J. M. Shreeve, *Chem. - Asian J.*, 2014, **9**, 2953.
70. J. Zhang, D. A. Parrish and J. M. Shreeve, *Chem. Commun.*, 2015, **51**, 7337.
71. Z.-B. Zhang, T. Li, L. Yin, X. Yin and J.-G. Zhang, *RSC Adv.*, 2016, **6**(79), 76075.
72. G. C. Hale, *U. S. Pat.*, 2 .011.578, 1935, Dover, N.J.
73. C. B. Aakeröy, T. K. Wijethunga and J. Desper, *Chem. - Eur. J.*, 2015, **21**, 11029.
74. H. Wei, H. Gao and J. M. Shreeve, *Chem. - Eur. J.*, 2014, **20**, 16943.
75. F. G. Gelalcha, B. Schulze and P. Lönnecke, *Acta Crystallogr.*, 2004, **C60**, o180.
76. K. B. Landenberger, O. Bolton and A. J. Matzger, *Angew. Chem., Int. Ed.*, 2013, **52**, 6468.
77. K. B. Landenberger, O. Bolton and A. J. Matzger, *J. Am. Chem. Soc.*, 2015, **137**, 5074.
78. Y. Tang, H. Yang, B. Wu, X. Ju, C. Lu and G. Cheng, *Angew. Chem., Int. Ed.*, 2013, **52**, 4875.
79. J. C. Bennion, Z. R. Siddiqi and A. J. Matzger, *Chem. Commun.*, 2017, **53**, 6065.
80. R. A. Wiscons and A. J. Matzger, *Cryst. Growth Des.*, 2017, **17**, 901.

81. J. Shen, X. Duan, Q. Luo, Y. Zhou, Q. Boa, Y. Ma and C. Pei, *Cryst. Growth Des.*, 2011, **11**, 1759.
82. S. Zhang and H. Zhao, *Adv. Mater. Res.*, 2014, **900**, 251.
83. J. Chen, X. Duan and C. Pei, *Chin. J. Energ. Mater.*, 2013, **21**, 409.
84. H. Xu, X. Duan, H. Li and C. Pei, *RSC Adv.*, 2015, **5**, 95764.
85. D. J. am Ende, S. R. Anderson and J. S. Salan, *Org. Process Res. Dev.*, 2014, **18**, 331.
86. H. Qiu, R. B. Patel, R. S. Damavarapu and V. Stepanov, *CrystEngComm*, 2015, **17**, 4080–4083.
87. D. Spitzer, B. Risse, F. Schnell, V. Pichot, M. Klaumünzer and M. R. Schaefer, *Sci. Rep.*, 2014, **4**, 6575.
88. B. Gao, D. Wang, J. Zhang, Y. Hu, J. Shen, J. Wang, B. Huang, Z. Qian, H. Huang, F. Nie and G. Yang, *J. Mater. Chem. A*, 2014, **2**, 19969.
89. H. Gao, P. Du, X. Ke, J. Liu, T. Chen and W. Jiang, *Propellants, Explos., Pyrotech.*, 2017, **42**, 889.

CHAPTER 7

Paramagnetic Organic Co-crystals of Neutral or Ionic Radicals

MARC FOURMIGUÉ* AND IE-RANG JEON

Univ Rennes, CNRS, ISCR (Institut des Sciences Chimiques de Rennes) – UMR 6226, 35000 Rennes, France
*Email: marc.fourmigue@univ-rennes1.fr

7.1 Introduction

The definition of a co-crystal has recently raised a strong interest in the scientific community, following regulatory issues provided by the Food and Drug Administration (FDA) in the USA, in relation to the increasing number of solid forms of active pharmaceutical ingredients (APIs).[1] This classification of co-crystals as "dissociable API-excipient molecular complexes" was soon questioned as to whether salts should be included or not in that definition.[2] Indeed, a basic two-component AB crystal classification allows one to distinguish three broad classes defined as solvates (A solid, B liquid), co-crystals (A solid, B solid) and salts (A^+ solid, B^- either). At the overlap between the two latter classes stands what was later defined as co-crystal salt, involving then at least three components such as [A(solid), B^+C^-].[3] In the following, we will tentatively stick to this extended definition, albeit that the exact status of a charged molecule can often be difficult to decide in the so-called mixed-valence salts involving radical cation or radical anions. For the purpose of this review, we will concentrate on two classes of paramagnetic molecular compounds, (i) organic neutral radicals (Section 7.3); and

(ii) cation (or anion) radicals when associated with their neutral reduced (or oxidized) form (Section 7.4).

7.2 Radical Species and their Interactions

Neutral organic radical species that have been reported to be involved in co-crystal formation are depicted in Scheme 7.1. One finds nitroxides such as TEMPO (TEMPO = 2,2,6,6-tetramethylpiperidin-1-yloxyl) and analogs, isoindoline nitroxides such as TMIO (TMIO = nitroxide-1,1,3,3-tetramethylisoindolin-2-yloxyl), and 2-substituted tetramethylnitronylnitroxides 2-R-4,4,5,5-tetramethylimidazolidinyl-3-oxide-1-oxy noted R-NN, but also thiazyl radicals such as dithiadiazolyl (DTDA), and verdazyl radicals. We have excluded from this review all metal coordination complexes involving these N- or O-based radicals. The formation of co-crystals with these molecules involve their association in a crystalline form with other neutral, eventually also radical, co-formers. We will discuss these associations in Section 7.3, considering different situations such as: (i) embedding of the radical within an inclusion compound, (ii) associations stabilized through π–π interactions, (iii) associations stabilized through hydrogen bonding, and (iv) associations stabilized through halogen bonding.

In addition, other organic radical species can be also easily found from the oxidation (reduction) of organic donor (acceptor) molecules whose prototypical examples are tetrathiafulvalene (TTF) as the electron donor molecules and quinone derivatives (chloranil, TCNQ) as the acceptor molecules (Scheme 7.2). These radical species and their precursors are problematic when one wants to describe them in the context of radical co-crystals. Indeed, a donor molecule (D) like TTF can easily co-crystallize with a variety of electron acceptors (A), giving rise to *neutral* charge transfer complexes (DA), most often crystallizing into alternated stacks ...DADADA... with strong π–π interactions. Such compounds are not radical and therefore not included in the present review. On the other hand, when a full charge transfer occurs, as in $(TTF)^{+\bullet}$,Cl^-, we are faced here with a simple radical salt, also out of the scope of this review. The situation is much more complex if we consider mixed-valence salts such as $(TTF)_3(BF_4)_2$ or $(TMTTF)_2(BF_4)$ (TMTTF = tetramethyltetrathiafulvalene), where now a radical $TTF^{+\bullet}$ species can be considered as co-crystallizing with a neutral TTF^0 species, giving rise to the so-called co-crystal salt defined above in the

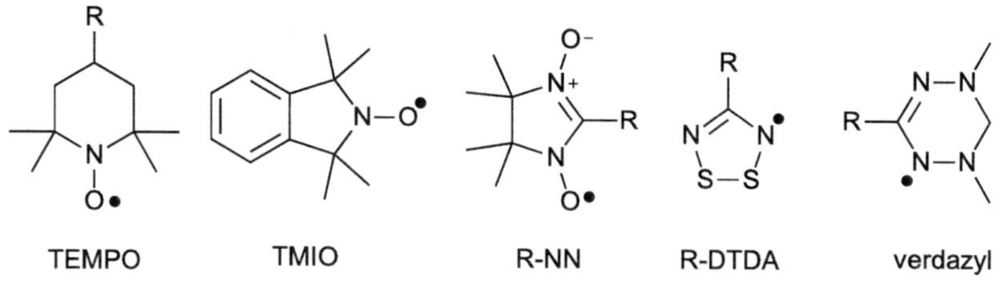

Scheme 7.1

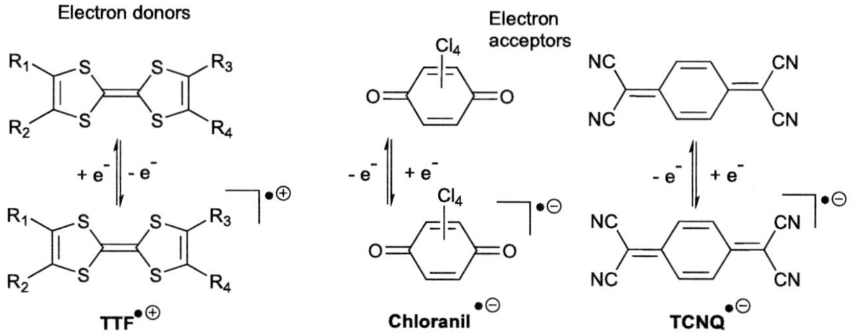

Scheme 7.2

[A(solid), B^+C^-] structure. These complex situations of paramagnetic organic co-crystals will be described in Section 7.4.

The paramagnetic susceptibility of non-interacting radical species follows a Curie law, which writes as $\chi = \frac{C}{T}$ with the Curie constant $C = \frac{Ng^2\mu_B^2 S(S+1)}{3k_B}$. In the presence of intermolecular interactions, the χT product is no longer constant. The increase in χT upon lowering the temperature indicates the presence of ferromagnetic interactions, while its decrease is the signature of antiferromagnetic interactions. The paramagnetic susceptibility can adopt different behaviors depending also on the dimensionality of these interactions, with typical situations such as uniform spin chains (also described as Bonner–Fisher systems) or dimeric associations (also described as singlet–triplet or Bleaney–Bower systems).[4] One can therefore anticipate that, unless there is a specific situation, the formation of co-crystals of radical species through their association with non-radical, closed-shell molecules, should in most cases decrease or suppress intermolecular interactions that could be potentially present within the crystals of the pure radical species. On the other hand, such associations in co-crystals can favor interactions in specific directions (and their associated magnetic response), which would not be found in the pure material. This balance between dilution/isolation effects and control of the interactions dimensionality is the salient feature of these radical co-crystals, as detailed in the following.

7.3 Neutral Radical Species in Co-crystals

7.3.1 Radical Co-crystals as Inclusion Adducts

Cucurbit[n]urils (CB[n]) are a family of synthetic macrocycles with a pumpkin shape and a hydrophobic cavity.[5,6] Their association into gels or crystals has been investigated from the EPR response of the included radical TEMPO derivatives with 4-amino (**1**),[7] or 4-methoxy (**2**) substituents (Figure 7.1, left).[8] Their organization into trimeric units is indeed revealed from the magnetic coupling of three identical nitroxide moieties, included in the CB[8]

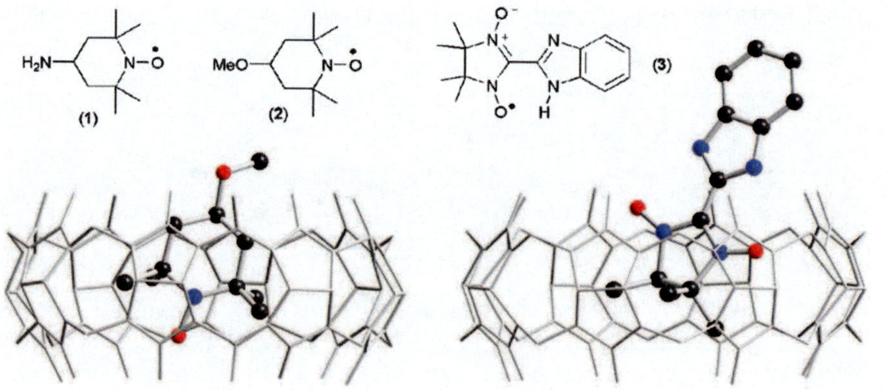

Figure 7.1 Structures of the cucurbit[8]urils (CB[8]) adducts of the radicals **1** (left) and **3**·HCl (right).

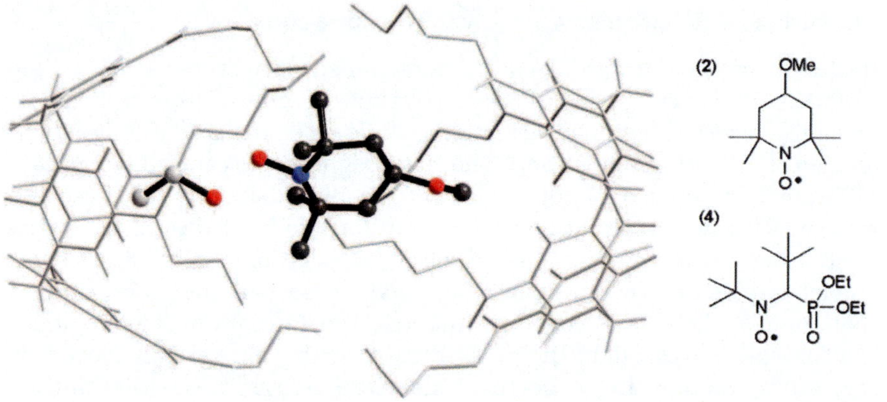

Figure 7.2 Structure of the calixarene adduct of (2) including one hydrogen-bonded ethanol molecule. Only one on the four disordered radicals has been shown for clarity.

molecule. Similarly, nitronylnitroxides such as 2-(2-benzimidazolyl)-4,4,5,5-tetramethylimidazolidinyl-3-oxide-1-oxy (**3**), as hydrochloride, was also included in a CB[8] host (Figure 7.1, right) and formulated as (**3**·HCl)@CB[8].[9] The adduct exhibits then a typical Curie-law of isolated spins, contrasting with the ferromagnetic interactions exhibited by the pure radical, but also by another non-crystalline inclusion compound formulated as $(\mathbf{3})_2$@$(CB[8])_3$, where ferromagnetic interactions within dimeric units are revealed ($J/k = +3.6(6)$ K, $g = 1.986$), the probable consequence of a face-to-face interaction between the outer benzimidazolyl parts of the radical.

Calixarenes provide another tool as they associate two-by-two to encapsulate radical species such as 4-methoxy-TEMPO (2) (Figure 7.2),[10] or the stable nitroxyl radical with bulky substituents: N-(^{t}butyl)-N-(1-diethylphosphono-2,2-^{t}butyl)-aminoxyl (**4**).[11] In the former, the extremely high g-tensor resolution of ESR at 360 GHz, allows us to distinguish between encapsulated nitroxide molecules forming a hydrogen bond between their O–(N) group and the OH group of the ethanol molecule occupying the same capsule and nitroxides missing this interaction.

In addition to these examples, radicals have been shown to be included in a variety of porous organic structures, for example those generated from tris(*o*-phenylenedioxy)cyclotriphosphazene (TPP),[12] or from the hydrogen bonded association of the guanidinium cation $C(NH_2)_3^+$ and organosulfonates $R–SO_3^-$.[13] Another remarkable example is provided by a dinitroxide biradical, which crystallizes itself by forming hexagonal open frameworks with one-dimensional corrugated channels filled with C_{60}.[14]

7.3.2 Radical Co-crystals Stabilized through π–π Interactions

Electron-rich aromatic molecules (D) are well known to co-crystallize with electron-poor molecules (A) such as chloranil or hexafluorobenzene to form charge-transfer complexes, where molecules stack on top of each other into alternated chains.[15] Such compounds raise a strong interest, either when the redox potentials of both D and A partners allows for a neutral-to-ionic transition upon cooling to give $(D^{+\bullet})(A^{-\bullet})$, or when it is used to control the proper alignment in the solid state of reactive double or triple bonds in stilbenes or diacetylenes for further cycloaddition reactions under irradiation.[16,17] A few examples of such strategies have been reported where the donor molecule D is actually a radical, as shown in Figure 7.3 with nitronylnitroxyde derivatives **5**[18,19] and **6**[20] interacting with either C_6F_6 or chloranil. It was shown that the trimeric entities $(\mathbf{5})_2(C_6F_6)$ further interact in the solid state through weak N–O···H–C contacts leading to antiferromagnetic 1D interactions with $g = 1.999$ and $J_{1D}/k_B = -1.716(3)$ K. Note

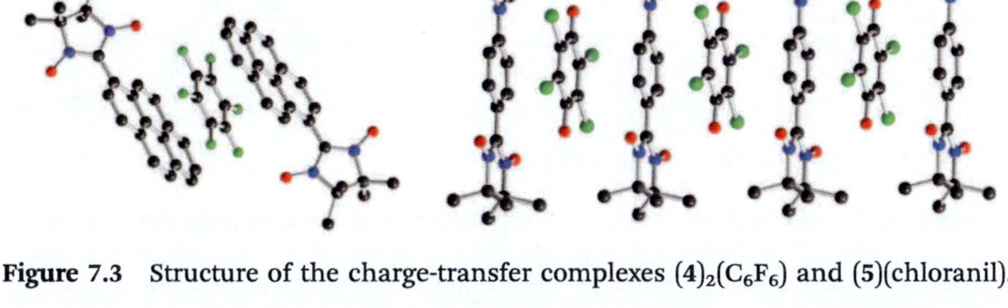

Figure 7.3 Structure of the charge-transfer complexes $(\mathbf{4})_2(C_6F_6)$ and (**5**)(chloranil).

that the pure compound **5** was dominated by a strong singlet–triplet behavior with $J_{ST}/k_B = -90/-100$ K. On the other hand, the 1:1 (**6**)(chloranil) charge transfer complex exhibits a Curie–Weiss behavior and an insulating character, indicating unambiguously the absence of electron transfer.

Another very attractive situation concerns the π–π interactions between two different radical species. As shown in Figure 7.4(a), the nitronylnitroxide radical **7** co-crystallizes with the bis(nitronylnitroxide) biradical **8** into alternating chains with an eclipsed face-to-face overlap between the nitrophenyl moiety in **7** and the phenyl moiety in **8**; another demonstration of the efficiency of this supramolecular strategy.[21] While the χT value at 250 K (1.13 emu K mol^{-1}) indicates that all spins behave paramagnetically at higher temperatures, the temperature dependence down to 10 K was analyzed as the combined contribution of an intramolecular ferromagnetic coupling (J_{intra}) with the meta-biradical **8** and an intermolecular antiferromagnetic coupling (J_{inter}) between **7** and **8**, with best fits giving $J_{intra}/k_B = +20$ K and $J_{inter}/k_B = -30$ K. Yet another illustration of these π–π interactions in radical co-crystals is provided by the two dithiadiazolyl radicals shown in Figure 7.4(b), bearing either a phenyl (**9**) or a pentafluorophenyl ring (**10**).[22] They co-crystallize into heteromeric (**9**·**10**) dyads with a face-to-face interaction between the electron-rich and electron-poor aryl moieties. The separation of the phenyl and perfluorophenyl ring centroids (3.642 Å) compares with those observed in the $C_6F_6 \cdot C_6H_6$ co-crystal and related phenyl–perfluorophenyl derivatives (3.4–3.8 Å).[23,24] The short intradimer S/S distances in the dyads, also observed in the homomeric

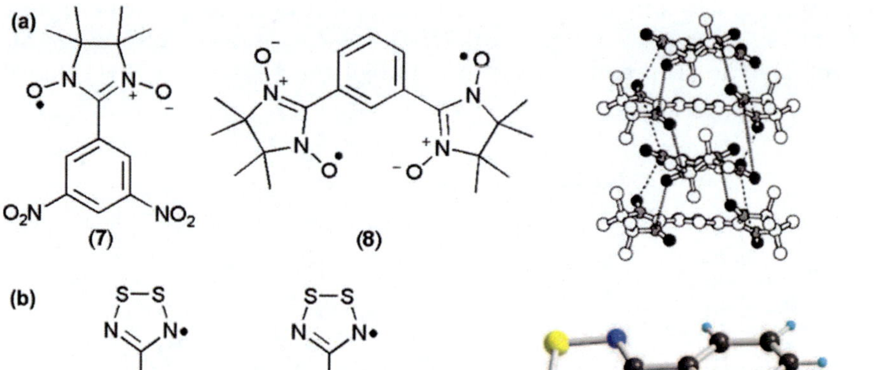

Figure 7.4 (a) Molecular structures of (**7**) and (**8**) and solid-state association in the bimolecular co-crystal (**7**)·(**8**). (b) Molecular structures of (**9**) and (**10**) and solid-state association in the bimolecular co-crystal (**9**)·(**10**).

systems lead to a strong antiferromagnetic interaction and an essentially diamagnetic behavior at room temperature.

7.3.3 Radical Co-crystals Stabilized through Hydrogen Bonding

Co-crystals of neutral radical species with hydrogen bond donors are found with OH-containing molecules such as hydroquinone or phenylboronic acid (Figure 7.5). The (4-pyridyl)-nitronylnitroxide radical **11** forms with hydroquinone a 2 : 1 adduct involving the oxygen atom of the N–O• group (rather than the pyridine nitrogen) as hydrogen bond acceptor, with a O···O distance of 2.728(5) Å (Figure 7.5(a)).[25] In the co-crystal, the radicals **11** stack on top of each other into head-to-tail uniform spin chains, isolated from each other by the hydroquinone. The temperature dependence of the magnetic susceptibility confirms this description, with a weak antiferromagnetic coupling estimated at $J_{chain}/k_B = -4.7$ K. The 1 : 1 adduct of the verdazyl radical **12** with hydroquinone (Figure 7.5(b)) adopts a different organization.[26] The hydroquinone form hydrogen bonded chains and the remaining hydrogen atom is linked to one nitrogen atom in **12**. Nevertheless, it also leads to a stacking of the neutral radicals **11** into weakly alternated spin chains, with an antiferromagnetic exchange interaction parameter $J_{chain}/k_B = -8.5$ K. A more complex structure (space group $P2_1/n$) is observed from the 1 : 1 association of the TEMPO nitronylnitroxide **13** and phenylboronic acid, which bridges two radicals by O–H···O–N hydrogen bonds, to form chains running along the 2_1 screw axis (Figure 7.5(c)).[27] Below $T = 220$ K, the compound undergoes a crystallographic phase transition to the $P\bar{1}$ space group, due to an ordering of the methyl groups in **13**. At variance with the previous examples, this specific geometry no longer allows face-to-face

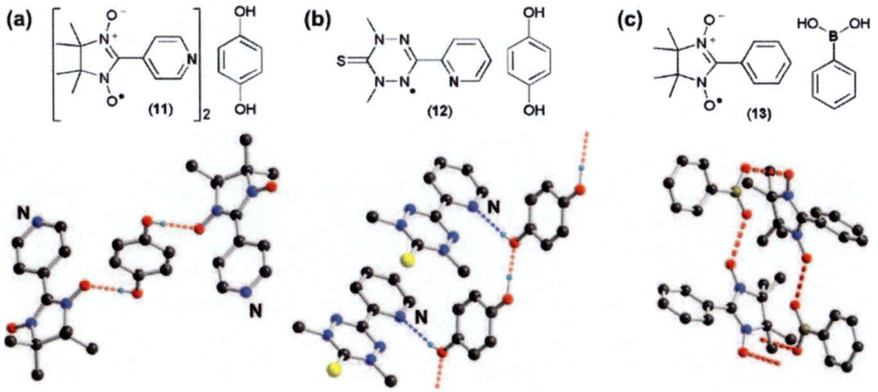

Figure 7.5 Hydrogen bonded adducts of (a) nitronylnitroxide radical **11** with hydroquinone (hq), $(\mathbf{11})_2$·hp; (b) verdazyl radical **12** with hydrquinone (**12**)·hp; (c) nitronylnitroxide radical **13** with phenylboronic acid. Hydrogen atoms of C–H groups have been omitted for clarity.

stacking of the radicals, allowing then for a ferromagnetic interaction with $J/k_B = +0.65$ K. In-depth theoretical calculations have shown that the spin density in **13**, essentially delocalized on the O–N–C–N–O moiety as usual, is actually dissymmetrized in the co-crystal, and significantly reduced on the oxygen atom involved in the shorter hydrogen bond with the $PhB(OH)_2$ molecule. Furthermore, a positive spin density contribution is also found on these two hydrogen atoms involved in the O–H···O–N hydrogen bonds.[28]

Other examples of hydrogen bonding with di-alcohols include the insertion of 4-amino-2,2,6,6-tetramethylpiperidin-1-yloxyl(4-amino-TEMPO) within host matrices composed of 9-hydroxy-fluorenyl-substituted polycondensed thiophenes.[29] Co-crystals involving stronger hydrogen bond donors are described with carboxylic acids such as fumaric,[21] squaric[21] and trichloroacetic acids,[30] all of them with the mentioned above 4-pyridylnitronylnitroxide **11**. As shown in Figure 7.6(c), the squaric acid adduct is actually a salt, with transfer of the two protons to the pyridine moieties, with, furthermore, short intermolecular distances between the NO moieties of neighboring radicals. The magnetic susceptibility of $(\mathbf{11})_2$(fumaric acid) follows a Curie–Weiss law with weak antiferromagnetic interactions ($\theta = -0.68$ K) while the intermolecular short NO···ON contacts in $(\mathbf{11})_2$(squaric acid) leads to a singlet–triplet magnetic behavior with $J_{ST}/k_B = -18.5$ K.

Last but not least, one should also mention the hydrogen-bonded associations between two different radical species. They were described with *para*-nitronylnitroxide benzoic acid (**14**) as the hydrogen bond donor,

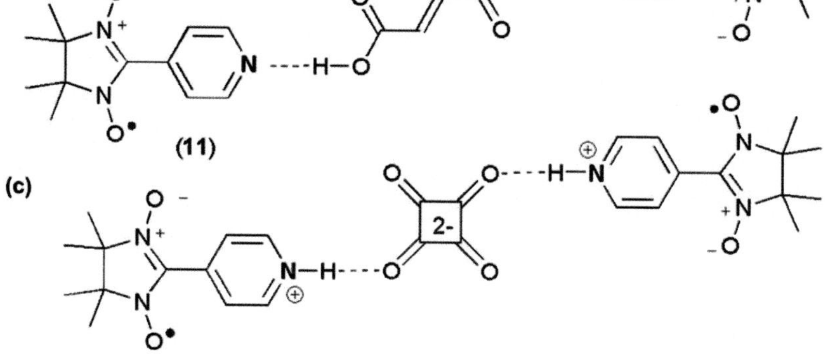

Figure 7.6 Schematic representation of the carboxylic acid adducts of the 4-pyridyl nitronylnitroxide radical **11** with (a) trichloroacetic acid, (b) fumaric acid, (c) squaric acid.

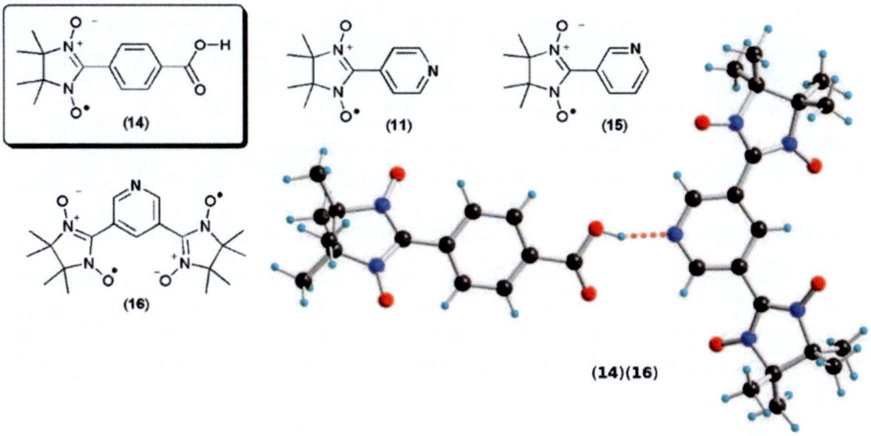

Figure 7.7 Detail of the structure of the bimolecular hydrogen-bonded adduct between *para*-nitronylnitroxide benzoic acid (**14**) and **16**.

co-crystallized with either the *para*-pyridyl-nitronylnitroxide (**11**),[31] the *meta*-pyridyl-nitronylnitroxide (**15**)[27] or the biradical **16**.[32] They all form 1 : 1 adducts with an O–H···N_{Py} hydrogen bond (Figure 7.7). The intermolecular O···N distances are 2.613(6) Å in (**14**)·(**11**), 2.642(5) Å in (**14**)·(**15**) and 2.656(4) in (**14**)·(**16**), indicating rather intense hydrogen bonds.

The magnetic properties of (**14**)·(**11**) and (**14**)·(**15**) are well interpreted in terms of the Curie–Weiss law with $\theta = -0.68$ K and a singlet–triplet model with $J_{ST}/k_B = -8.0$ K, respectively.[27] On the other hand, the χT value in (**14**)·(**16**) increases upon cooling and exceeds 1.38 emu K mol^{-1}, which is expected for a pair of an $S = 1$ (for **16**) and an $S = 1/2$ spin (for **14**). Short intermolecular contacts were identified as being responsible for these ferromagnetic interactions. A maximum of χT observed at 11 K, and a characteristic field dependence below 5K, unambiguously indicates the presence of a magnetic transition to an AF ground state below $T_{Néel} = 5$ K. As stated by the authors,[28] this "*complex ... is the first example of a hydrogen-bonded heterospin, heteromolecular complex exhibiting a magnetic phase transition. The present results indicate that the hydrogen bond between organic acids and bases is an efficient structure determining factor of forced co-crystallization of open-shell molecules with inequivalent spins, which leads to novel molecule-based magnetic materials such as ferrimagnets*".

7.3.4 Radical Co-crystals Stabilized through Halogen Bonding

Such systems involving halogen bonding[33] as the main tool for the formation of radical co-crystals were recently reviewed by one of us.[34] For the sake of completeness of this chapter, we will briefly mention the main results. The halogen bond interaction finds its origin in the electropositive area located on heavy halogen atoms (X = I, Br), in the prolongation of the Y–X bond, particularly when Y is an electron-withdrawing group (cyano, perfluoroaliphatic or perfluoroaromatic) or a π system (C≡C > C=C). The renewal of this interaction in the last twenty years is particularly attributable to the use of α,ω-diiodoperfluoroalkanes, as shown in Figure 7.8(a).[35] Diiodoperfluorobenzene is also a halogen bond donor of choice in that respect, as illustrated in Figure 7.8(b).[36] The reduction ratio (RR) is defined as the ratio between the actual interatomic distance relative to the sum of the van der Waals radii on the interacting atoms. RR values as low as 0.80 indicate a strong halogen bond interaction. Note here also two main modes of association, such as in Figure 7.8(a): a trimeric entity when co-crystallizing a monotopic XB acceptor (here the TEMPO radical) with a ditopic XB donor

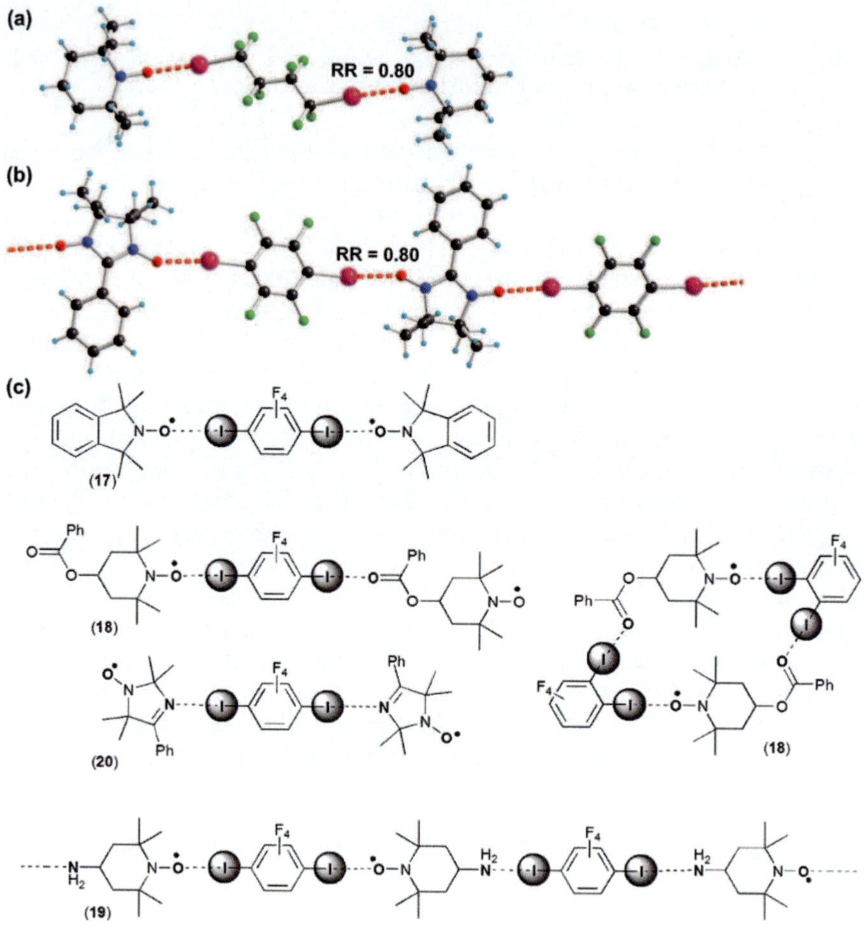

Figure 7.8 (a) Trimeric XB association between TEMPO radical and 1,4-diiodoperfluorobutane. (b) One-dimensional infinite structure from the XB association of nitronylnitroxide **13** and *para*-diiodoperfluorobenzene. (c) Other examples of XB co-crystals of neutral radicals.

(1,4-diiodoperfluorobutane), or, as in Figure 7.8(b), a chain-like structure when associating both the ditopic XB acceptor (the TEMPO nitronylnitroxide **13**) and the XB donor (*para*-diiodoperfluorobenzene). Other examples are collected in Figure 7.8(c) and involve isoindoline nitroxide **17**,[37] TEMPO derivatives such as **18**[38] or **19**,[39] or 3-imidazolin-1-yloxyl radical **20**.[35] In the nitronyl systems such as TEMPO, the setting of the halogen bond is associated with a shift of the spin density from the oxygen to the nitrogen atom in the N–O bond, revealed by an increase in the EPR a_N coupling constant.[31] In the solid state, most compounds exhibit weak antiferromagnetic interactions, demonstrating also that magnetic interactions are not transmitted through the halogen bond donor molecules.

7.4 Cation (Anion) Radical Species in Co-crystal Salts

7.4.1 Electron Transfer in TEMPO Radicals

In some instances, it appears that the co-crystal formation is also associated with electron transfer. The question then arises whether such systems should still be described as co-crystals, demonstrating the difficulty of rationalizing such a broad concept. An interesting example is provided by the association of TEMPO (**13**) or 4-azetinido-TEMPO (**21**) radicals with electron acceptors such as $TCNQF_4$ (Figure 7.9).[40,41] Indeed, in the co-crystal (TEMPO)($TCNQF_4$), the short N–O bond indicates the predominance of a cationic oxidized TEMPO structure with double bond character for the N=O bond, in accordance with intramolecular bond lengths within the $TCNQF_4$ molecule characteristic of a $TCNQF_4^{-\bullet}$ radical anion. On the other hand, the "normal" N–O bond in **21** indicates that its oxidation involves the azetidino group

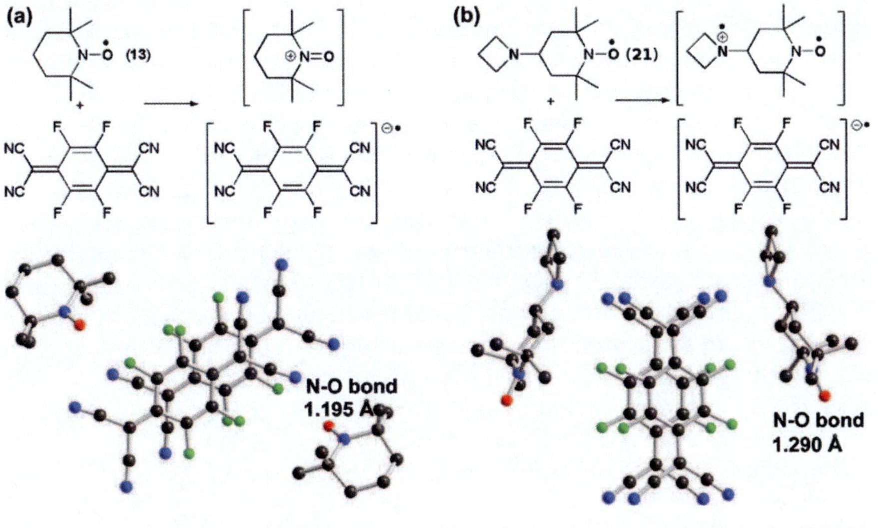

Figure 7.9 (a) Products of the co-crystallization of $TCNQF_4$ with (a) TEMPO, (b) azetidino-TEMPO (21).

rather than the N–O moiety. In (**13**)($TCNQF_4$), the $TCNQF_4^{-\bullet}$ radical anions form a bond-over-ring face-to-face dimer through overlap of the singly occupied LUMO, leaving the complex essentially diamagnetic. On the other hand, in (**21**)($TCNQF_4$), a strong antiferromagnetic interaction is believed to associate the $TCNQF_4^{-\bullet}$ and the cation radical azetidino moieties, leaving only the paramagnetic contribution of the nitronyl moiety in **21**.

The two examples described above illustrate that the definition of co-crystals become difficult when electron transfer arises, if one wants to distinguish them from salts. We can mention for example the prototypical (TTF)(TCNQ) charge transfer salt, characterized in the solid state by segregated stacks of TTF and TCNQ molecules. A non-fractional electron transfer takes place between both stacks ($\rho \approx 0.69$), leading to a partial band filling and metallic conductivity.[42] Another borderline situation is provided by the (TTF)(chloranil) neutral co-crystal. It forms stacks of alternating neutral molecules but exhibits at low temperature a neutral-to-ionic transition with electron transfer to give the ($TTF^{+\bullet}$)(chloranil$^{-\bullet}$) ferroelectric structure.[43,44] Situations that can eventually be described as co-crystal salts are those where a given molecule and its radical cation (alt. anion) can be unambiguously identified in the crystal structure.

7.4.2 Mixed-Valence Anion Radical Co-crystal Salts

In contrast to simple 1:1 anion radical salts such as (K^+,$TCNQ^{-\bullet}$),[45] it is also possible that the additional charges on TCNQ are non-equivalently trapped to give a mixed-valence system, where neutral $TCNQ^0$ and radical anion $TCNQ^{-\bullet}$ can be distinguished. The following section will cover a few representative examples of such co-crystal salts containing mixed-valence TCNQ molecules. While salts of TCNQ with a 1:1 or 1:2 ratio of donor and acceptor tend to show uniform charge ($\rho = -1$ or $\rho = -0.5$ respectively) on TCNQ molecules, in 2:3 charge transfer compounds, two negative charges can be either delocalized or trapped over the three TCNQ molecules, as illustrated in Figure 7.10 with the charge-trapped mixed valence system $(Mor)_2(TCNQ)_3$ (Mor^+ = morpholinium).[46,47] The charge of each crystallographically independent TCNQ molecule can be estimated from the intramolecular bond lengths,[48] with one centrosymmetric neutral $TCNQ^0$ ($\rho = 0$) and two non-centrosymmetric anionic $TCNQ^{-\bullet}$ ($\rho = -1$) alternating along the stack as a ($TCNQ^0$/$TCNQ^{-\bullet}$/$TCNQ^{-\bullet}$) triad. Only the anionic $TCNQ^{-\bullet}$ molecules feature an additional hydrogen bond with a NH group of Mor^+. They are associated into dimers through π–π overlap of their SOMOs, with an energy separation between a ground singlet and an excited triplet state of 0.31 eV. As a consequence of this strong charge localization, the electric conductivity is rather poor, with $\sigma_{RT} < 10^{-3}$ S cm^{-1}.

The introduction of different cations in 2:3 TCNQ charge transfer compounds can also favor the co-crystallization of an isolated neutral TCNQ and a $(TCNQ)_2^{2-}$ dimer over the formation of a triad. Such systems are found in $(Me_3NH)_2(TCNQ)_3$,[49] or in $(QND)_2(TCNQ)_3$ (QND = quinuclidinium).[50]

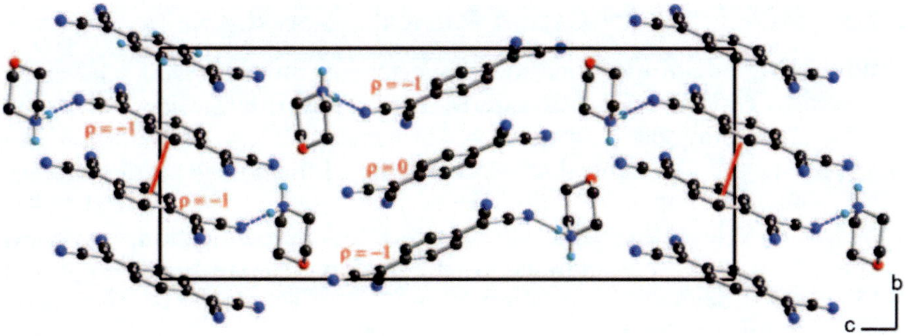

Figure 7.10 Projection view along a of the unit cell of $(Mor)_2(TCNQ)_3$. The thick red lines indicate the overlap interaction between charged ($\rho = -1$) $TCNQ^{-\bullet}$ species. Hydrogen atoms (C–H) were omitted for clarity.

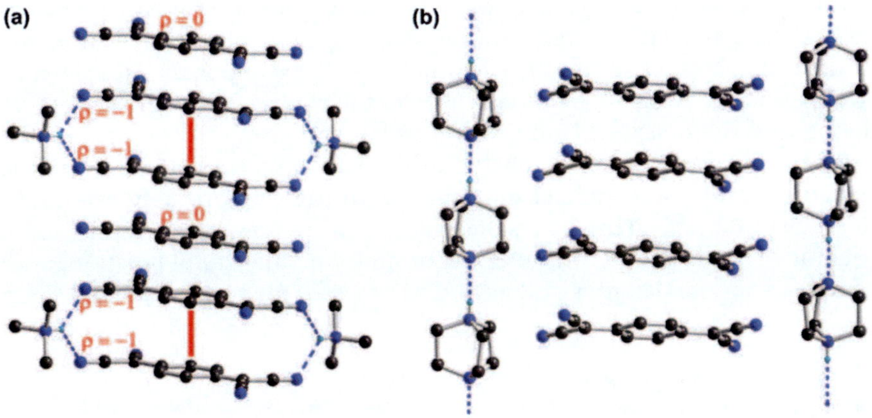

Figure 7.11 Details of the structures of 2 : 3 TCNQ salts, with (a) $(Me_3NH)_2(TCNQ)_3$ and (b) $(HDABCO)_2(TCNQ)_3$. The blue dotted lines indicate the N–H···N hydrogen bonds and the thick red lines the overlap interaction between charged ($\rho = -1$) $TCNQ^{-\bullet}$ species. Hydrogen atoms (C–H) were omitted for clarity.

As shown in Figure 7.11(a) for $(Me_3NH)_2(TCNQ)_3$, only the two negatively charged $TCNQ^{-\bullet}$ are hydrogen bonded to the trimethylammonium cation. These structures contrast with the analogous $(HDABCO)_2(TCNQ)_3$ (DABCO = 1,4-diazabicyclo[2,2,2]octane) salt (Figure 7.11(b)).[51] Here, the only difference between $HDABCO^+$ and QND^+ is the presence of the nitrogen atom instead of the carbon atom. This difference favors the formation of a hydrogen bond between $HDABCO^+$ molecules making a 1D network with themselves, rather than forming a hydrogen bond between the cation and the TCNQ molecules. Consequently, the charge on the TCNQ molecule in $(HDABCO)_2(TCNQ)_3$ is not pinned and is almost uniform along the column, which contributes to at least three orders of magnitude reduction in the electric resistivity.

7.4.3 Mixed-valence Cation Radical Co-crystal Salts

Similar charge disproportionation in electron donor co-crystal salts can be observed in cation radical salts derived from tetrathiafulvalene (TTF) molecules. The simplest example is provided by a 3:2 salt, namely $(TTF)_3(BF_4)_2$.[52,53] The central C=C double bond in TTFs is particularly sensitive to its oxidation state:[54] it increases from 1.34–1.35 Å in neutral TTFs to 1.39–1.40 Å in the cation radical and up to 1.43 Å in the dication.[55] As shown in Figure 7.12(a), the analysis of the intramolecular bond lengths in $(TTF)_3(BF_4)_2$ indeed shows an alternation ...$(TTF^0)(TTF^{+\bullet})(TTF^{+\bullet})$..., with a strong pairing of the $TTF^{+\bullet}$ species into dimer dications. These features explain the low conductivity ($\sigma_{powder} = 2\times10^{-5}$ S cm^{-1}) of this apparently non-stoichiometric TTF salt. A different situation is found in a salt formulated as $(EDT\text{-}TTF\text{-}CONMe_2)_2Br$.[56,57] Despite a 2:1 stoichiometry usually favorable to a uniform charge distribution, this salt already exhibits at room temperature a full charge disproportionation, with alternation of neutral and cationic molecules along both the stacking *a* and transverse *b* directions in the orthorhombic, non-centrosymmetric space group *P*2*nn*. Note that the strongest C–H···O hydrogen bond involves the oxidized TTF derivative as HB donor, and the neutral TTF derivative as HB acceptor.

Finally, we would like to conclude this paragraph on co-crystal salts associating neutral and radical molecules with the extraordinary example of $(EDO\text{-}TTF)_2PF_6$.[58,59] This 2:1 salt crystallizes at room temperature in the triclinic $P\bar{1}$ space group, with one donor molecule in general position, and a central C=C bond length at 1.366(7) Å, as expected for a partial $\rho = +\frac{1}{2}$ charge

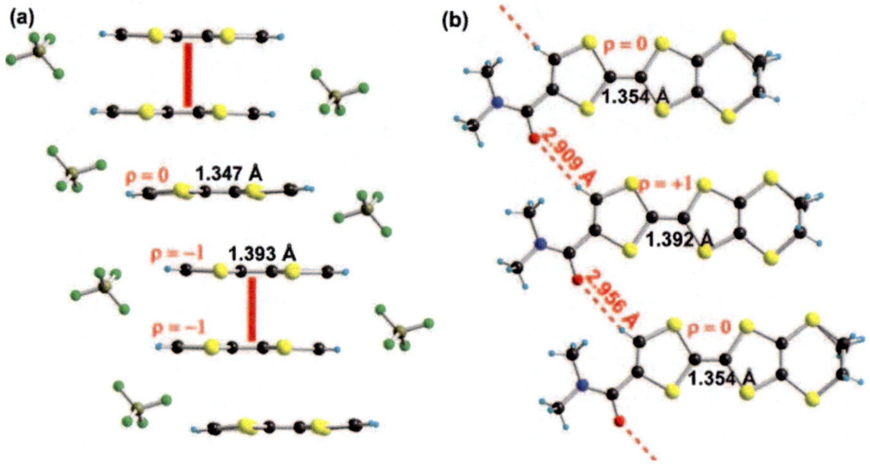

Figure 7.12 (a) Side view of a stack in $(TTF)_3(BF_4)_2$, with indication of the central C=C bond lengths. The thick red line indicates the overlap interaction between charged ($\rho = +1$) $TTF^{+\bullet}$ species. (b) Detail of the structure of the charge-ordered $(EDT\text{-}TTF\text{-}CONMe_2)_2Br$ salt with indication of the central C=C bond lengths. The red dotted lines indicate the C–H···O hydrogen bonds (see text).

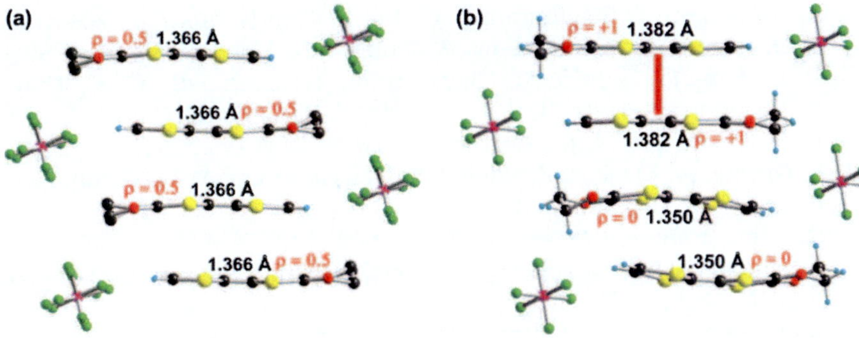

Figure 7.13 Detail of the evolution of the crystal structure of $(EDO\text{-}TTF)_2PF_6$ between (a) room temperature and (b) 100 K, with indication of the central C=C bond lengths.

(Figure 7.13(a)). At 279 K, it exhibits a metal–insulator thermal transition. The nearly uniform stacks giving rise to a quasi-one-dimensional Fermi surface at room temperature transform indeed into a strongly tetramerized form involving a drastic charge ordering phenomenon, with an alternation of two neutral and two oxidized EDO-TTF molecules (Figure 7.13(b)). These charge order phenomena are currently the subject of strong interest, both in one- and in two-dimensional conductors.[60,61]

7.5 Conclusion and Perspectives

We have shown along these lines that the formation of co-crystals from neutral radical species is actually a powerful tool to modify the overlap interactions between paramagnetic species, affording in some (rare) instances novel magnetic structures not observed in the pure radical species. Altogether the strategy, however, tends to push apart radicals from each other and to drastically decrease their magnetic interactions. The extension to co-crystal salts based on radical cations or radical anions is limited to those examples where neutral and radical ions of the same molecule can be identified in the crystal structure. These situations, also refereed as charge ordering, are associated with strong charge localization, with associated low conductivities.

References

1. *Guidance for Industry: Regulatory Classification of Pharmaceutical Co-crystals*, U.S. FDA, Silver Spring, MD, 2011.
2. S. Aitipamula, R. Banerjee, A. K. Bansal, K. Biradha, M. L. Cheney, A. R. Choudhury, G. R. Desiraju, A. G. Dikundwar, R. Dubey, N. Duggirala, P. P. Ghogale, S. Ghosh, P. Kumar Goswami, N. R. Goud, R. R. K. R. Jetti, P. Karpinski, P. Kaushik, D. Kumar, V. Kumar, B. Moulton, A. Mukherjee, G. Mukherjee, A. S. Myerson, V. Puri,

A. Ramanan, T. Rajamannar, C. M. Reddy, N. Rodriguez-Hornedo, R. D. Rogers, T. N. Guru Row, P. Sanphui, N. Shan, G. Shete, A. Singh, C. C. Sun, J. A. Swift, R. Thaimattam, T. S. Thakur, R. K. Thaper, S. P. Thomas, S. Tothadi, V. R. Vangala, N. Variankaval, P. Vishweshwar, D. R. Weyna and M. J. Zaworotko, *Cryst. Growth Des.*, 2012, **12**, 2147.
3. E. Grothe, H. Meekes, E. Vlieg, J. H. ter Horst and R. de Gelder, *Cryst. Growth Des.*, 2016, **16**, 3237.
4. O. Kahn, *Molecular Magnetism*, VCH Publishers, New York, 1993.
5. J. Lagona, P. Mukhopadhyay, S. Chakrabarti and L. Isaacs, *Angew. Chem., Int. Ed.*, 2005, **44**, 4844.
6. J. W. Lee, S. Samal, N. Selvapalam, H.-J. Kim and K. Kim, *Acc. Chem. Res.*, 2003, **36**, 621.
7. E. Mileo, E. Mezzina, F. Grepioni, G. F. Pedulli and M. Lucarini, *Chem. – Eur. J.*, 2009, **15**, 7859.
8. D. Bardelang, K. Banaszak, H. Karoui, A. Rockenbauer, M. Waite, K. Udachin, J. A. Ripmeester, C. I. Ratcliffe, O. Ouari and P. Tordo, *J. Am. Chem. Soc.*, 2009, **131**, 5402.
9. E. V. Peresypkina, V. P. Fedin, V. T. Maurel, A. Grand, P. Rey and K. E. Vostrikova, *Chem. – Eur. J.*, 2010, **16**, 12481.
10. D. N. Polovyanenko, *E. G.* Bagryanskaya, A. Schnegg, K. Moöbius, A. W. Coleman, G. S. Ananchenko, K. A. Udachine and J. A. Ripmeester, *Phys. Chem. Chem. Phys.*, 2008, **10**, 5299.
11. G. S. Ananchenko, K. A. Udachin, A. W. Coleman, D. N. Polovyanenko, E. G. Bagryanskaya and J. A. Ripmeester, *Chem. Commun.*, 2008, 223.
12. H. Kobayashi, T. Ueda, K. Miyakubo, J. Toyoda, T. Eguchi and A. Tanic, *J. Mater. Chem.*, 2005, **15**, 872.
13. A. C. Soegiarto, W. Yan, A. D. Kent and M. D. Ward, *J. Mater. Chem.*, 2011, **21**, 2204.
14. D. Bardelang, M. Giorgi, C. Pardanaud, V. Hornebecq, E. Rizzato, P. Tordo and O. Ouari, *Chem. Commun.*, 2013, **49**, 3519.
15. C. R. Patrick and G. S. Proser, *Nature*, 1960, **187**, 1021.
16. G. W. Coates, A. R. Dumm, L. M. Henling, D. A. Dougherty and R. H. Grubbs, *Angew. Chem., Int. Ed.*, 1997, **36**, 248.
17. S. W. Watt, C. Dai, A. J. Scott, J. M. Burke, R. L. Thomas, J. C. Collings, C. Viney, W. Clegg and T. B. Marder, *Angew. Chem., Int. Ed.*, 2004, **43**, 3061.
18. H. Akpinar, J. T. Mague and P. M. Lahti, *CrystEngComm*, 2013, **15**, 831.
19. H. Akpinar, J. T. Mague, M. A. Novak, J. R. Friedman and P. M. Lahti, *CrystEngComm*, 2012, **14**, 1515.
20. H. Sakurai, A. Izuoka and T. Sugawara, *J. Am. Chem. Soc.*, 2000, **122**, 9723.
21. A. Iruoka, M. Fukada, R. Kumai, M. Itakura, S. Hikami and T. Sugawara, *J. Am. Chem. Soc.*, 1994, **116**, 2609.
22. C. Allen, D. A. Haynes, C. M. Paska and J. M. Rawson, *CrystEngComm*, 2009, **11**, 2048.
23. C. R. Patrick and G. S. Prosser, *Nature*, 1960, **187**, 1021.

24. G. W. Coates, A. R. Dunn, L. M. Henling, J. W. Ziller, E. B. Lobkovsky and R. H. Grubbs, *J. Am. Chem. Soc.*, 1998, **120**, 3641.
25. T. Otsuka, T. Okuno, K. Awaga and T. Inabe, *J. Mater. Chem.*, 1998, **8**, 1157.
26. M. Vérot, N. Bréfuel, J. Pécaut, C. Train and V. Robert, *Chem. – Asian J.*, 2012, **7**, 380.
27. T. Akita, Y. Mazaki and K. Kobayashi, *J. Chem. Soc., Chem. Commun.*, 1995, 1861.
28. Y. Pontillon, T. Akita, A. Grand, K. Kobayashi, E. Lelievre-Berna, J. Pécaut, E. Ressouche and J. Schweizer, *J. Am. Chem. Soc.*, 1999, **121**, 10126.
29. Y. Mazaki, K. Awaga and K. Kobayashi, *J. Chem. Soc., Chem. Commun.*, 1992, 1661.
30. H.-X. Chen, Z.-S. Lib and B.-W. Sun, *Acta Crystallogr., Sect. E*, 2008, **E64**, o1197.
31. T. Otsuka, T. Okuno, M. Ohkawa, T. Inabe and K. Awaga, *Mol. Cryst. Liq. Cryst.*, 1997, **306**, 285.
32. K. Hayakawa, D. Shiomi, T. Ise, K. Sato and T. Takui, *J. Phys. Chem. B*, 2005, **109**, 9195.
33. G. Cavallo, P. Metrangolo, R. Milani, T. Pilati, A. Priimagi, G. Resnati and G. Terraneo, *Chem. Rev.*, 2016, **116**, 2478.
34. M. Fourmigué and J. Lieffrig, *Top. Curr. Chem.*, 2015, **359**, 91.
35. X. Pang and W. J. Jin, *New* J. Chem., 2015, **39**, 5477.
36. G. Minguez Espallargas, A. Recuenco, F. M. Romero, L. Brammer and S. Libri, *CrystEngComm*, 2012, **14**, 6381.
37. G. R. Hanson, P. Jensen, J. McMurtrie, L. Rintoul and A. S. Micallef, *Chem. – Eur. J.*, 2009, **15**, 4156.
38. X. Pang, X. R. Zhao, H. Wang, H.-L. Sun and W. J. Jin, *Cryst. Growth Des.*, 2013, **13**, 3739.
39. K. Boubekeur, J.-L. Syssa-Magalé, P. Palvadeau and B. Schöllhorn, *Tetrahedron Lett.*, 2006, **47**, 1249.
40. S. Nakatsuji, A. Takai, K. Nishikawa, Y. Morimoto, N. Yasuoka, K. Suzuki, T. Enoki and H. Anzai, *Chem. Commun.*, 1997, 275.
41. S. Nakatsuji, A. Takai, K. Nishikawa, Y. Morimoto, N. Yasuoka, K. Suzuki, T. Enoki and H. Anzai, *J. Mater. Chem.*, 1999, **9**, 1747.
42. D. Jérome, Chem. Rev., 2004, **104**, 5565, and references therein.
43. E. Collet, M.-H. Lemée-Cailleau, M. Buron-Le Cointe, H. Cailleau, M. Wulff, T. Luty, S. Y. Koshihara, M. Meyer, L. Toupet, P. Rabiller and S. Techert, *Science*, 2003, **300**, 612.
44. P. Garcia, S. Dahaoui, C. Katan, M. Souhassou and C. Lecomte, *Faraday Disc.*, 2007, **135**, 217.
45. M. Konno, T. Ishii and Y. Saito, *Acta Crystallogr.*, 1977, **B33**, 763.
46. C. J. Fritchie and P. Arthur, *Acta Crystallogr.*, 1966, **21**, 139.
47. T. Sundaresan and S. C. Wallwork, *Acta Crystallogr.*, 1972, **B28**, 491.
48. T. J. Kistenmacher, T. J. Emge, A. N. Bloch and D. O. Cowan, *Acta Crystallogr.*, 1982, **B38**, 1193.

49. H. Kobayashi, T. Danno and Y. Saito, *Acta Crystallogr.*, 1973, **B29**, 2693.
50. A. D. Bandrauk, K. Ishii, K. D. Truong and M. Aubin, *J. Phys. Chem.*, 1985, **89**, 1478.
51. T. Akutagawa, S. Takeda, T. Hasegawa and T. Nakamura, *J. Am. Chem. Soc.*, 2004, **126**, 291.
52. F. Wudl, *J. Am. Chem. Soc.*, 1975, **97**, 1962.
53. J.-P. Legros, M. Bousseau, L. Valade and P. Cassoux, *Mol. Cryst. Liq. Cryst.*, 1983, **100**, 181.
54. P. Guionneau, C. J. Kepert, G. Bravic, D. Chasseau, M. R. Truter, M. Kurmoo and P. Day, *Synth. Met.*, 1997, **86**, 1973.
55. F. Gao, F.-F. Zhu, X.-Y. Wang, Y. Xu, X.-P. Wang and J.-L. Zuo, *Inorg. Chem.*, 2014, **53**, 5321.
56. K. Heuzé, M. Fourmigué, P. Batail, C. Coulon, R. Clérac, E. Canadell, P. Auban-Senzier, S. Ravy and D. Jérome, *Adv. Mater.*, 2003, **15**, 1251.
57. L. Zorina, S. Simonov, C. Mézière, E. Canadell, S. Suh, S. E. Brown, P. Foury-Leylekian, P. Fertey, J.-P. Pouget and P. Batail, *J. Mater. Chem.*, 2009, **19**, 6980.
58. S. Aoyagi, K. Kato, A. Ota, H. Yamochi, G. Saito, H. Suematsu, M. Sakata and M. Takaka, *Angew. Chem., Int. Ed.*, 2004, **43**, 3670.
59. T. Murata, X. Shao, Y. Nakano, H. Yamochi, M. Uruichi, K. Yakushi, G. Saito and K. Tanaka, *Chem. Mater.*, 2010, **22**, 3121.
60. H. Seo, C. Hotta and H. Fukuyama, *Chem. Rev.*, 2004, **104**, 5005.
61. O. Jeannin, E. Canadell, P. Auban-Senzier and M. Fourmigué, *Chem. Commun.*, 2016, **52**, 308.

CHAPTER 8

Hydrogen-bonded Semiconductor Co-crystals

KRISTIN M. HUTCHINS[a] AND LEONARD R. MACGILLIVRAY*[b]

[a] Department of Chemistry & Biochemistry, Texas Tech University, Lubbock, TX 79409-1061, USA; [b] Department of Chemistry, University of Iowa, Iowa City, IA 52242-1294, USA
*Email: len-macgillivray@uiowa.edu

8.1 Introduction

The field of organic semiconductors is a rapidly expanding area of research with efforts to design materials with appreciably high conductivities as a replacement for inorganic-based systems. Organic materials are of interest[1,2] owing to lower costs, inherent flexibilities, and low processing temperatures, and are being investigated for making low-cost flexible electronics (*e.g.* RFID tags, displays, electronic paper).[3–5] While a variety of organic oligomers and polymers are studied as semiconductor materials,[6,7] molecular-based architectures endowed with π-conjugation (*e.g.* pentacene, sexithiophene) have received the most attention. Indeed, computational and structural studies of π-rich organic semiconductors show that face-to-face, or co-facial, stacking of π-surfaces in the solid state can lead to relatively high charge mobilities owing to increased orbital–orbital overlap of neighboring molecules.[8–10] Face-to-face π-stacking of molecules in solids, however, is often unfavorable, especially in the case of electron-rich aromatics.[11] Consequently, many π-rich organic semiconductor molecules are dominated by herringbone or edge-to-face crystal packing, which is not an ideal geometry to achieve maximum performance of organic semiconductor materials owing to poor orbital overlap. Although some materials that pack in herringbone motifs can be conductive, maximum conductivity is often not reached.[12] It is, thus, of great importance and value to establish control of π-orbital overlap within organic semiconductor solids and related materials.

In this context, the field of crystal engineering involves the understanding of intermolecular interactions (*e.g.* π-stacking) in the context of crystal packing and the utilization of such understanding to design solids with desired physical and chemical properties.[13] A major aim of crystal engineering is to establish reliable connections between molecular and supramolecular structure by investigating intermolecular forces that direct and define crystal packing.[14] Noncovalent forces are typically utilized in crystal engineering to dictate packings as finite assemblies (*i.e.* zero-dimensional or 0D) and/or extended (*i.e.* 1D, 2D, or 3D) networks. The areas of molecular recognition and self-assembly can, in this regard, provide fertile grounds for exploring how molecular and supramolecular structures can be developed and utilized to influence properties of organic solid-state materials.

It is with these ideas in mind that we describe here work performed in our laboratory during the past decade that employs principles of crystal engineering to control packings and properties of organic semiconductor molecules and materials.[15–19] Specifically, we show how two-component solids in the form of co-crystals[20–23] can be used to direct packings of π-rich organic semiconductor molecules in the form of thiophenes and acenes to achieve face-to-face π-stacked geometries. Hydrogen bonding between the components that make up the co-crystals is a central theme, whereby small ditopic molecules are employed to enforce π-stacking within finite and extended structures. We also show how the co-crystal approach provides an avenue for supporting intermolecular covalent-bond forming reactions in the solid state in the form [2 + 2] photodimerizations of olefins decorated with semiconductor functional groups.[17,18] We begin by providing a discussion on the utility of co-crystals and then outline our work to achieve face-to-face π–π stacking. Recent and related efforts to develop a metal-organic approach[19] to affect charge conductivity in the solid state will also be addressed.

8.2 Supramolecular Construction and Co-crystals

In recent years, chemists have developed and applied a synthetic approach to form solids based on co-crystals to influence and control properties of organic solid-state materials. Co-crystals are multicomponent solids with molecular species assembled in combination to afford properties different than those exhibited by the individual solid components.[20–23] A co-crystal consists of a target molecule crystallized with at least one second molecule, or co-crystal former (CCF), that is used in a deliberate way to influence specific properties of the target (*e.g.* solubility). The CCF is generally designed to interact with the target *via* attractive intermolecular forces (*e.g.* hydrogen bonds) that are employed to unite the components in the solid. The use of a CCF to construct a solid and influence properties is important because a main purpose of the CCF is to compete with deleterious effects of crystal packing and, in doing so, aid the construction of functional solids based on the target molecule (Scheme 8.1).

8.3 Co-crystals for Crystal Engineering Semiconductors

Over the last decade, we have employed a co-crystal approach to generate assemblies of molecules in the organic solid state that are designed to organize semiconductor molecules into face-to-face π-stacked geometries.[15] CCFs are used to assemble and preorganize a semiconductor molecule (*e.g.* acene, thiophene) in a face-to-face geometry with plane-to-plane separations on the order of 4 Å. It is well established that the extent of face-to-face π-overlap of semiconductor molecules in the solid state plays a significant role in facilitating efficient movement of charge (Scheme 8.2).[8–10] Moreover, we expected that being able to enforce face-to-face stacking of semiconductors using a CCF may grant access to solid-state materials that exhibit relatively high charge mobilities so as to be useful for applications in areas related to flexible organic electronics (*e.g.*, smart cards, electronic paper).[24,25]

8.3.1 Benefits of the Co-crystal Approach

Aromatic molecules (*e.g.* benzene) commonly exhibit edge-to-face, as opposed to face-to-face, stacking in the solid state.[26,27] Consequently, there

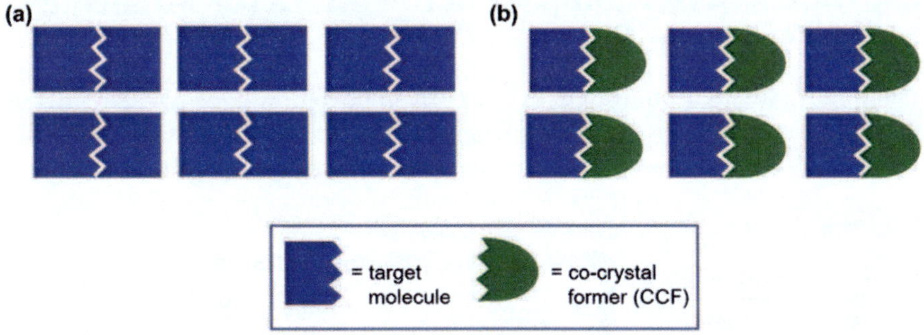

Scheme 8.1 Representation of: (a) a single-component solid and (b) a co-crystal.

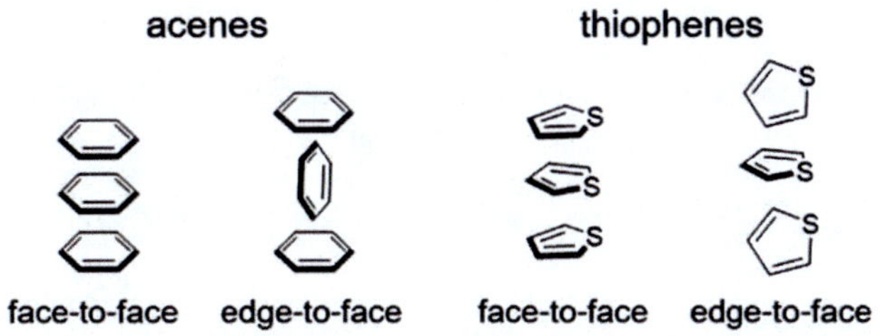

Scheme 8.2 Face-to-face and edge-to-face stackings for semiconductor molecules.

have been a variety of efforts to develop methods that guide the assembly of aromatics into face-to-face geometries in order to control properties of organic solids (*e.g.* mobility, reactivity, optical). Many of the approaches have relied on using covalently attached substituents to dictate and govern packing *via* attractive (*e.g.* hydrogen bonds, halogen–halogen interactions, charge-transfer forces),[28–33] as well as repulsive (*e.g.* steric),[34–36] noncovalent forces. The substituent approach, however, has invariably encountered difficulties in competing with the subtle structural demands of dense crystal packing, which is the driving force for crystallization.[37,38] The difficulties are largely a result of the fact that the substituents, while providing a capacity to govern self-assembly *via* attractive forces, will introduce a change in shape to a substrate molecule that can adversely affect packing. In this context, we hypothesized that a second molecule in the form of a CCF with a structure designed to assemble and preorganize aromatics into face-to-face geometries could provide an added measure of control to confront crystal packing to reliably achieve stacking of semiconductors in solids. The CCF would, in effect, provide a means to decouple effects of long-range packing from a desired crystal geometry. Given that our group had reported uses of CCFs based on resorcinol (res) to direct [2 + 2] photodimerizations of stacked aromatics in the solid state, we hypothesized that the co-crystal approach based on res may be exploited to not only to direct stacking but also achieve photoreactions of semiconductor molecules (Scheme 8.3).[39,40]

8.4 Finite Assemblies of Semiconductor Molecules in Co-crystals

Specifically, we demonstrated in an early report how face-to-face stacking of organic semiconductor molecules can be achieved using CCFs in two-component co-crystals.[39,41] Similar to our work on cycloaddition reactions,

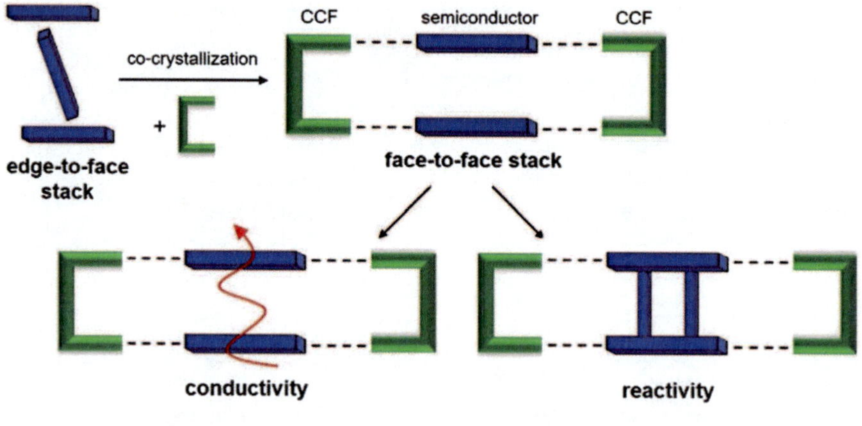

Scheme 8.3 Utilization of a CCF to stack aromatics for conductivity and solid-state reactivity.

the method involved co-crystallizing an appropriately functionalized semiconductor molecule with a res-based CCF that enforces face-to-face π-stacking *via* hydrogen bonds. More specifically, we employed res-based CCFs to enforce co-facial stacking of acenes and thiophenes functionalized with pyridyl-ethynyl groups; namely, 9,10-bis(4-pyridylethynyl)anthracene (9,10-DEA) and 2,5-bis(4-pyridylethynyl)thiophene (DPT). In the original report, we showed that 5-iodo-res (5-I-res) and 5-methyl-res (5-Me-res) can be used to enforce π–π stacking of an anthracene and thiophene as substrates in co-crystals of 2(5-I-res)·2(9,10-DEA) and 2(5-Me-res)·2(DPT), respectively (Figure 8.1).[15] The acene and thiophene moieties were directed into face-to-face geometries with separation distances on the order of 3.45 Å. The hydrogen-bonded complexes further self-assembled into 1D extended face-to-face arrangements, which supported the generation of extended π-stacked structures. The acene and thiophene were also demonstrated to pack in the solid state as pure forms such that they lack face-to-face stacking in the absence of a CCF.

8.4.1 Polymorphism of Pure Thiophene Substrates

Organic semiconductor molecules are often composed of multiple aromatic rings (*e.g.* oligothiophenes), which can be conformationally flexible. Changes in conformation at the molecular level are known to impact electrical transport properties of solid-state materials.[42,43] Thus, understanding both the packing structure of the organic semiconductor substrate and geometrical conformation of the rings (*e.g.* thiophenes) in single crystals and

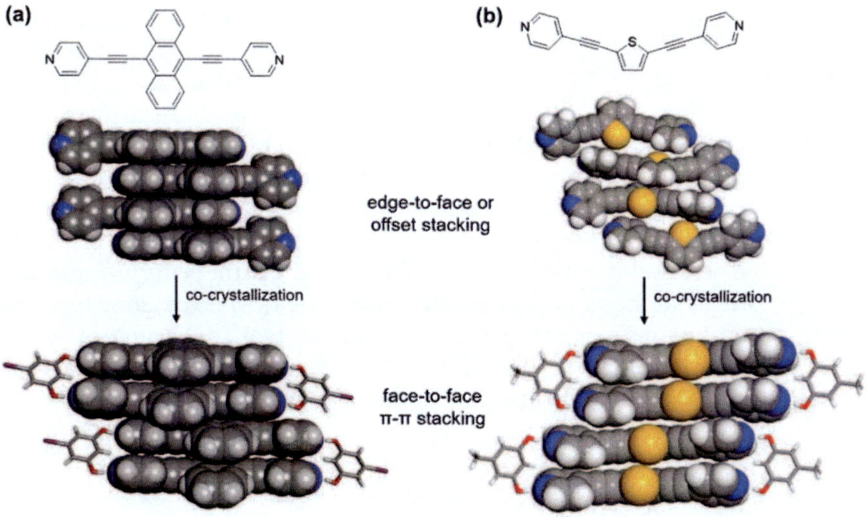

Figure 8.1 Chemical and X-ray structures: (a) single-component 9,10-DEA and co-crystal with 5-I-res and (b) single-component DPT and co-crystal with 5-Me-res. Anthracene and thiophene shown in space filling.

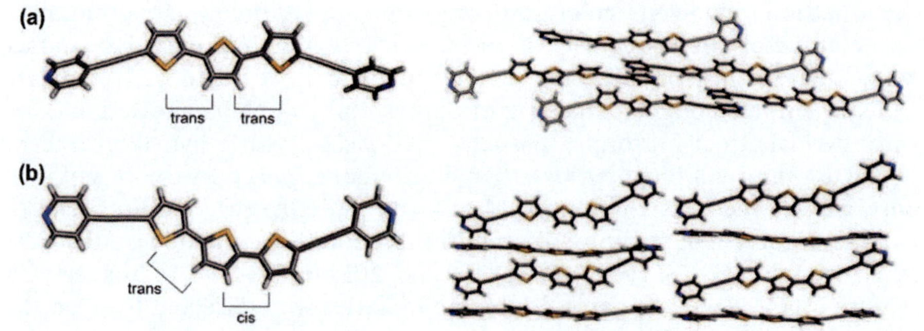

Figure 8.2 X-ray structures of DPPT showing conformations and extended packings: (a) achiral form (*trans–trans*) and (b) chiral form (*cis–trans*).

thin films is important for developing high-performance electrical devices.[43–45]

In the course of our studies to develop co-crystals based on higher oligothiophenes, we discovered the ability of 2,5″-bis(4-pyridylethynyl)-5,2′,5′,2″-terthienyl (DPTT) as a pure form to exhibit conformational polymorphism. In particular, crystallization of DPTT from toluene was determined to afford achiral crystals wherein DPTT adopts the common *trans–trans* conformation (Figure 8.2(a)).[16] In the achiral form, DPTT is packed in a planar, slipped-stack arrangement with neighboring thiophene rings engaged in minimal π-stacking. In contrast, crystallization of DPTT from benzene affords chiral crystals wherein DPTT adopts a rare *cis–trans* conformation (Figure 8.2(b)).[16] In the chiral form, the thiophene rings of DPTT are stacked in a herringbone motif dominated by C–H···π forces. The phases were determined to readily interconvert upon recrystallization from the appropriate solvent.

The conformation of DPTT in each polymorph was also confirmed using solid-state IR spectroscopy. Specifically, the achiral and chiral forms displayed two and three unique C–H bands, respectively. The larger number of bands for the latter was ascribed to the lower symmetry of the conformation of the molecule.[16] Using IR spectroscopy to assign conformations of oligothiophenes in solid media is a general challenge owing to inherent difficulties to isolate solids that exhibit different and distinct conformations of a specific oligothiophene molecule. The use of the IR spectroscopy to assign conformation enabled an improved understanding of the structures of the thiophene-based solids,[46,47] which can be important in cases where conformation may influence device performance.

8.4.2 Related Approaches for Hydrogen-bonded Semiconductors

Related strategies involving charge-assisted hydrogen bonds were subsequently reported for the construction of conductive organic solids. Thus,

(a) (b)

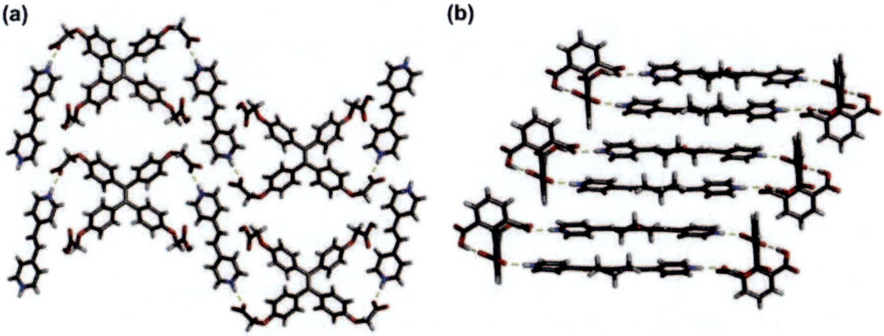

Figure 8.3 X-ray structures of two-component crystals sustained by charge-assisted hydrogen bonds: (a) tetraphenyl ethylene tetracarboxylate and 4,4′-bispyridinium ethylene and (b) 2′,6′-dicarboxybiphenyl-2,6-dicarboxylate and 2,6-bis(pyridinium-4-ylmethylidene) cyclohexanone.

Pigge and co-workers reported on the generation of two-component crystals composed of a tetracarboxylic acid and a series of bipyridine molecules that were assembled *via* carboxylic acid-pyridine charge-assisted hydrogen bonds (Figure 8.3(a)).[48] In particular, tetraphenyl ethylene tetracarboxylic acid when reacted with systematically modified bipyridines afforded solids that exhibited changes in color, band gap, and conductivity. Likewise, a semiconductor solid based on a tetraacid and bipyridine was reported by Biradha and co-workers. In the report, reaction of 2,2′,6,6′-tetracarboxybiphenyl and 2,6-bis(4-pyridyl methylidene) cyclohexanone generated a solid sustained by carboxylate-pyridinium forces (Figure 8.3(b)). The solid material was shown to exhibit both reversible photochromism and conductivity.[49]

8.5 Solid-state Reactivity of Thiophene Substrates in Co-crystals

The tendency of thiophenes to adopt edge-to-face packings in the solid state cannot only limit conduction in semiconductor,[50] thin-film transistor,[51] light-emitting diode,[52] and photovoltaic cell materials,[53] but prohibit such molecules from undergoing [2 + 2] photocycloaddition reactions in the solid state. For photodimerization to occur in a solid, carbon–carbon (C=C) bonds of adjacent stacked olefins are expected to lie parallel and separated less than 4.2 Å.[54]

From our work on generating semiconductor co-crystals, we turned to apply the co-crystal approach to achieve intermolecular [2 + 2] photodimerizations of thiophenes in the solid state. Intermolecular [2 + 2] photodimerizations of thiophenes in solids, as well as solution, had been extremely rare,[55,56] with regiocontrol of the cycloaddition being addressed in limited cases (Figure 8.4(a)). Very few styrylthiophenes had been shown to react to form cyclobutanes. The reactions were shown to often proceed in low yields (*i.e.* <30%),[57–61] with high-yielding photodimerizations being rare.[58,62]

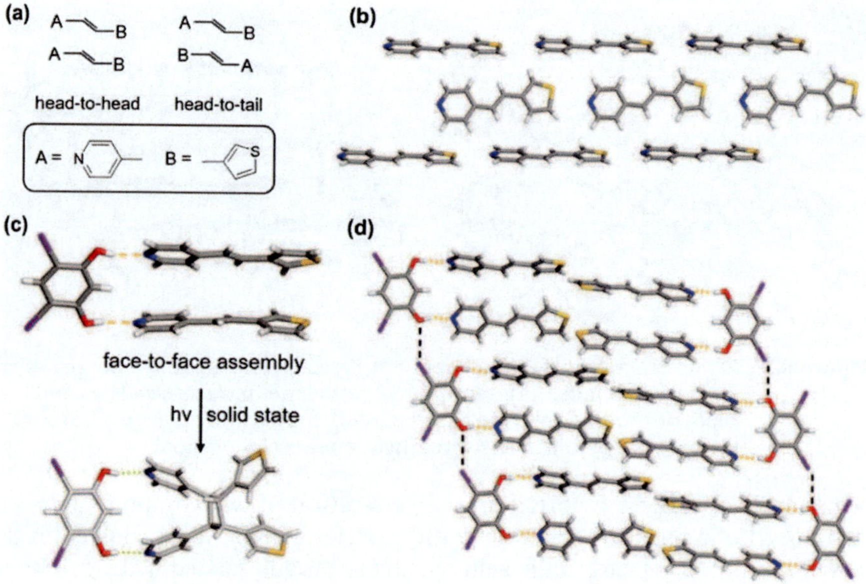

Figure 8.4 Assembly of thiophenes: (a) head-to-head and head-to-tail arranged decorated olefins, (b) X-ray structure of β-PTE, (c) X-ray structures of (4,6-diI-res)·2(β-PTE) co-crystal and photoproduct (4,6-diI-res)·(1,2-PTCB) and (d) extended packing of (4,6-diI-res)·2(β-PTE).

8.5.1 Head-to-head Photodimerization

To achieve an intermolecular [2 + 2] photodimerization of a thiophene, we studied CCFs based on res to enforce face-to-face π-stackings of the β-substituted thiophene *trans*-1-(4-pyridyl)-2-(3-thienyl)ethylene (β-PTE). As a pure form, we showed that β-PTE adopts edge-to-face π-stacking, which renders the molecule photostable (Figure 8.4(b)).[17] Moreover, we demonstrated that 4,6-diiodo-res (4,6-diI-res) can be used to achieve face-to-face stacking of β-PTE in the co-crystal (4,6-diI-res)·2(β-PTE) (Figure 8.4(c)).[17] The olefin was arranged in a head-to-head geometry and separated by 4.15 and 4.30 Å (Figure 8.4(a)). Upon UV irradiation, regiocontrolled photodimerization occurred in up to 90% yield to generate the corresponding head-to-head photodimer *rctt*-1,2-bis(4-pyridyl)-3,4-bis(3-thienyl) cyclobutane (1,2-PTCB). Halogen bonds based on I···O forces[63] in (4,6-diI-res)·2(β-PTE) were also demonstrated to support the assembly of β-PTE into an extended π-stacked structure (Figure 8.4(d)). When 4,6-dichloro-res (4,6-diCl-res) was used as a CCF in (4,6-diCl-res)·2(β-PTE), an edge-to-face arrangement of β-PTE resulted, which was ascribed to the presence of weaker Cl···O forces.[17] Thus, without secondary interactions to help guide crystal packing, the propensity of the thiophene to stack in an edge-to-face geometry dominated.

8.5.2 Head-to-tail Photodimerization

While a CCF based on res was successfully used to generate the head-to-head photodimer 1,2-PTCB, a CCF based on a dicarboxylic acid enabled the formation of the head-to-tail product *rctt*-1,3-bis(4-pyridyl)-2,4-bis(3-thienyl)cyclobutane (1,3-PTCB). Specifically, co-crystallization of succinic acid (SA) with β-PTE afforded the co-crystal (SA)·2(β-PTE) wherein the components formed a three-component assembly sustained by COOH···N hydrogen bonds (Figure 8.5(a)).[18] The assemblies packed such that β-PTE was arranged in a head-to-tail and face-to-face π-stacked geometry, with the C=C bonds being separated at approximately 4.0 Å (Figure 8.5(b)). Upon UV irradiation, photodimerization proceeded in up to 86% yield and afforded the head-to-tail photodimer 1,3-PTCB (Figure 8.5(c)).

8.6 Metal-organic Approach

We have discussed purely organic π-rich molecules that have been explored as potential building blocks for semiconductor materials, but integration of metal atoms into the crystal lattice may also provide an avenue for increasing electrical conductivity. Furthermore, coordination bonds provide directionality, which can be utilized to achieve face-to-face stacking in aromatics, as well as enhancement of conductivity.[64–66] Incorporation of metals into organic materials while achieving favorable conductivity still lies at a nascent stage.[67,68] In this context, transition metal-organic complexes with relatively high conductivity or mobility have been reported for silver-,[67] nickel-,[69] and zinc-based[70] metal-organic complexes. For example, early work of Chen and co-workers reported a silver complex that comprises Ag(I) ions coordinated to two phenanthroline-based ligands that stack in a face-to-face geometry to afford a solid with a conductivity of 14 S cm^{-1}.[67]

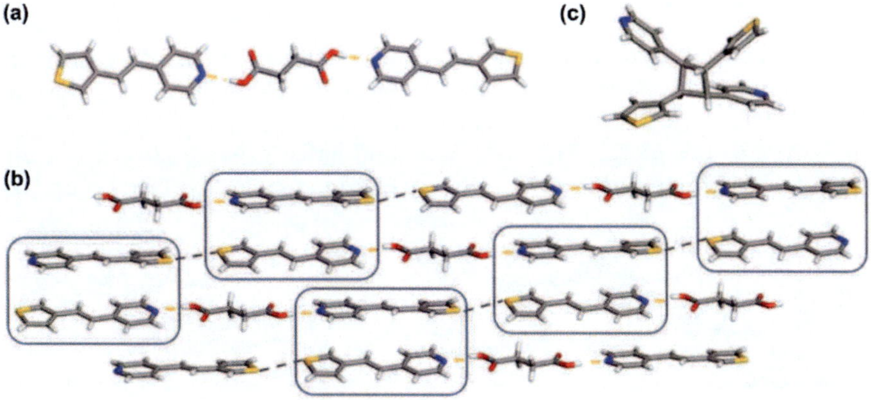

Figure 8.5 X-ray structures: (a) three-component assembly of (SA)·2(β-PTE), (b) extended packing with β-PTE in a head-to-tail arrangement, and (c) photoproduct 1,3-PTCB.

8.6.1 Electrical Conductivity Before and After [2+2] Photodimerization

Our group had previously reported the use of Ag-N(pyridyl) bonds in combination with Ag···Ag forces to achieve face-to-face arrangements of olefins to facilitate [2+2] photodimerizations in the solid state.[71] From the work, we sought to pair our interests in organic semiconductors and solid-state reactivity to explore the possible impact that a photodimerization may have on electrical conductivity of a metal-organic semiconductor solid.

Specifically, we utilized a Ag(I) salt to assemble and preorganize *trans*-1-(4-pyridyl)-2-phenylethylene (4-stilbz) into a face-to-face stacked π-geometry in $[Ag_2(4\text{-stilbz})_4][CF_3SO_3]_2$.[19] The olefins in the complex were shown to be separated by 3.90 Å. Upon UV irradiation, a cycloaddition of the olefins was determined to proceed regiospecifically and in quantitative yield to produce 1,2-bis(4-pyridyl)-3,4-bis(phenyl)cyclobutane (4-pyr-ph-cb) in the 1D coordination polymer $[Ag_2((4\text{-pyr-ph-cb})_2][CF_3SO_3]_2$ (Figure 8.6). The solid-state reaction was also shown to proceed in a rare single-crystal-to-single-crystal (SCSC) transformation[72] to generate the polymer. Materials that undergo SCSC reactions are attractive for applications in photoactivated molecular switches and actuators,[73–75] as well as 3D data storage.[76] Importantly, before the photodimerization, the complex $[Ag_2(4\text{-stilbz})_4][CF_3SO_3]_2$ was determined to exhibit an electrical conductivity of 20.8 ± 1.3 S cm^{-1}. After the SCSC photoreaction, the electrical conductivity was determined to increase to 37.0 ± 4.1 S cm^{-1}. The increase in conductivity was supported by density of states calculations.[19] Resistivity measurements were also performed on a single crystal during the course of the photodimerization, which demonstrated the increase in conductivity.

8.6.2 Electrical Conductivity in Related Metal-organic Materials

Dincă and co-workers, in related work, had reported a series of isostructural metal-organic frameworks based on a tetrathiafulvalene (TTF)-tetrabenzoate ligand wherein the S···S distance between TTF cores changed depending on the nature of the metal. The electrical conductivities were shown to range from 3.95×10^{-6} to 2.86×10^{-4} S cm^{-1} (Figure 8.7).[70,77,78] The group also

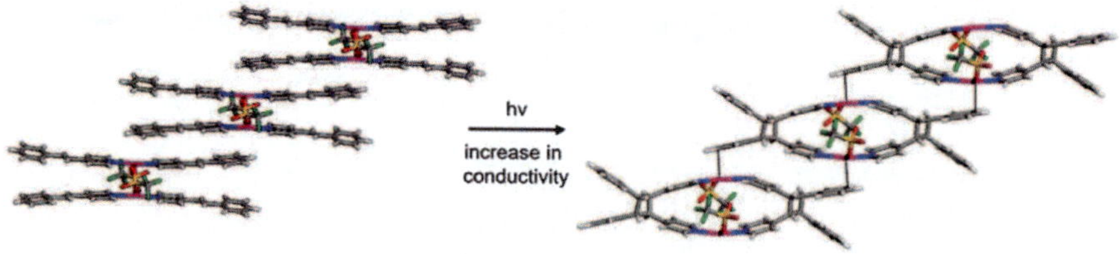

Figure 8.6 X-ray structures of disilver complex $[Ag_2(4\text{-stilbz})_4][CF_3SO_3]_2$ and photoproduct $[Ag_2((4\text{-pyr-ph-cb})_2][CF_3SO_3]_2$.

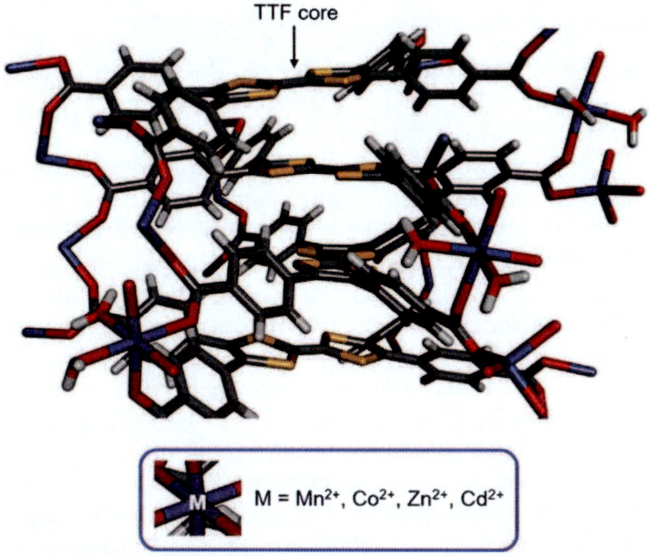

Figure 8.7 X-ray structure of a TTF-tetrabenzoate metal-organic framework with variable metal.

reported a nickel complex that exhibited a graphene-like structure when coordinated to 2,3,6,7,10,11-hexaiminotriphenylenesemiquinonate. The solid was determined to exhibit a conductivity of 40 S cm^{-1} as a thin film.[69]

8.7 Summary and Outlook

In this chapter, we have shown that finite assemblies of molecules can be developed in the organic solid state to direct face-to-face π-stacking of organic semiconductor molecules. The assemblies are achieved using a co-crystal approach[20–23] wherein hydrogen-bond-donor CCFs support the assembly of the molecular substrates *via* hydrogen bonds.[15,39,41] The geometries of the assemblies effectively decouple typical packing properties of the substrates from the effects of long-range packing, so that the solid state may be employed to construct supramolecular targets and enforce face-to-face arrangements of the semiconductor molecules that undergo photodimerizations. In addition to expanding the co-crystal method, we are focusing on how more complex assembly processes, which could encompass networks and higher order co-crystals (*e.g.* ternary solids), can be developed to confront problems associated with supramolecular synthesis as related to the crystal engineering of bulk electrical and reactivity properties.

Acknowledgements

LRM gratefully acknowledges the National Science Foundation (DMR-1408834) and the University of Iowa for supporting this work.

References

1. J. Mei, Y. Diao, A. L. Appleton, L. Fang and Z. Bao, Integrated Materials Design of Organic Semiconductors for Field-Effect Transistors, *J. Am. Chem. Soc.*, 2013, **135**, 6724–6746.
2. S. E. Root, S. Savagatrup, A. D. Printz, D. Rodriquez and D. J. Lipomi, Mechanical Properties of Organic Semiconductors for Stretchable, Highly Flexible, and Mechanically Robust Electronics, *Chem. Rev.*, 2017, **117**, 6467–6499.
3. H. L. Dong, C. L. Wang and W. P. Hu, High performance organic semiconductors for field-effect transistors, *Chem. Commun.*, 2010, **46**, 5211–5222.
4. H. N. Tsao and K. Mullen, Improving polymer transistor performance via morphology control, *Chem. Soc. Rev.*, 2010, **39**, 2372–2386.
5. Y. Yamashita, Organic semiconductors for organic field-effect transistors, *Sci. Technol. Adv. Mater.*, 2009, **10**, 1–9.
6. H. Tsuji and E. Nakamura, Design and Functions of Semiconducting Fused Polycyclic Furans for Optoelectronic Applications, *Acc. Chem. Res.*, 2017, **50**, 396–406.
7. M. Jeffries-El, B. M. Kobilka and B. J. Hale, Optimizing the Performance of Conjugated Polymers in Organic Photovoltaic Cells by Traversing Group 16, *Macromolecules*, 2014, **47**, 7253–7271.
8. J. Cornil, D. Beljonne, J. P. Calbert and J. L. Bredas, Interchain interactions in organic pi-conjugated materials: Impact on electronic structure, optical response, and charge transport, *Adv. Mater.*, 2001, **13**, 1053–1067.
9. M. D. Curtis, J. Cao and J. W. Kampf, Solid-state packing of conjugated oligomers: from pi-stacks to the herringbone structure, *J. Am. Chem. Soc.*, 2004, **126**, 4318–4328.
10. L. J. Wang, G. J. Nan, X. D. Yang, Q. Peng, Q. K. Li and Z. G. Shuai, Computational methods for design of organic materials with high charge mobility, *Chem. Soc. Rev.*, 2010, **39**, 423–434.
11. C. R. Martinez and B. L. Iverson, Rethinking the term "pi-stacking", *Chem. Sci.*, 2012, **3**, 2191–2201.
12. J. Cornil, J. P. Calbert and J. L. Brédas, Electronic structure of the pentacene single crystal: Relation to transport properties, *J. Am. Chem. Soc.*, 2001, **123**, 1250–1251.
13. G. R. Desiraju, *Crystal Engineering: The Design of Organic Solids*. Elsevier: Amsterdam, 1989.
14. G. R. Desiraju, Supramolecular Synthons in Crystal Engineering – a New Organic-Synthesis, *Angew. Chem., Int. Ed.*, 1995, **34**, 2311–2327.
15. A. N. Sokolov, T. Friščić and L. R. MacGillivray, Enforced face-to-face stacking of organic semiconductor building blocks within hydrogen-bonded molecular cocrystals, *J. Am. Chem. Soc.*, 2006, **128**, 2806–2807.
16. A. N. Sokolov, J. C. Sumrak and L. R. MacGillivray, Conformational polymorphism facilitates assignment of trans and cis-conformers of an

alpha-substituted oligothiophene via IR spectroscopy, *Chem. Commun.*, 2010, **46**, 82–84.

17. K. M. Hutchins, J. C. Sumrak and L. R. MacGillivray, Resorcinol-Templated Head-to-Head Photodimerization of a Thiophene in the Solid State and Unusual Edge-to-Face Stacking in a Discrete Hydrogen-Bonded Assembly, *Org. Lett.*, 2014, **16**, 1052–1055.
18. K. M. Hutchins, J. C. Sumrak, D. C. Swenson and L. R. MacGillivray, Head-to-tail photodimerization of a thiophene in a co-crystal and a rare adipic acid dimer in the presence of a heterosynthon, *CrystEngComm*, 2014, **16**, 5762–5764.
19. K. M. Hutchins, T. P. Rupasinghe, L. R. Ditzler, D. C. Swenson, J. R. G. Sander, J. Baltrusaitis, A. V. Tivanski and L. R. MacGillivray, Nanocrystals of a Metal-Organic Complex Exhibit Remarkably High Conductivity that Increases in a Single-Crystal-to-Single-Crystal Transformation, *J. Am. Chem. Soc.*, 2014, **136**, 6778–6781.
20. C. B. Aakeröy, J. Desper, M. Fasulo, I. Hussain, B. Levin and N. Schultheiss, Ten years of co-crystal synthesis; the good, the bad, and the ugly, *CrystEngComm*, 2008, **10**, 1816–1821.
21. C. B. Aakeröy and D. J. Salmon, Building co-crystals with molecular sense and supramolecular sensibility, *CrystEngComm*, 2005, **7**, 439–448.
22. M. J. Zaworotko, Molecules to crystals, crystals to molecules ... and back again?, *Cryst. Growth Des.*, 2007, **7**, 4–9.
23. D. Braga, F. Grepioni and L. Maini, The growing world of crystal forms, *Chem. Commun.*, 2010, **46**, 6232–6242.
24. H. Sirringhaus, Materials and Applications for Solution-Processed Organic Field-Effect Transistors, *Proc. IEEE*, 2009, **97**, 1570–1579.
25. C. Reese, M. Roberts, M. M. Ling and Z. N. Bao, Organic thin film transistors, *Mater. Today*, 2004, **7**, 20–27.
26. R. Azumi, M. Goto, K. Honda and M. Matsumoto, Conformation and packing of odd-numbered alpha-oligothiophenes in single crystals, *Bull. Chem. Soc. Jpn.*, 2003, **76**, 1561–1567.
27. D. Holmes, S. Kumaraswamy, A. J. Matzger and K. P. C. Vollhardt, On the nature of nonplanarity in the [N]phenylenes, *Chem. – Eur. J.*, 1999, **5**, 3399–3412.
28. G. W. Coates, A. R. Dunn, L. M. Henling, J. W. Ziller, E. B. Lobkovsky and R. H. Grubbs, Phenyl-perfluorophenyl stacking interactions: Topochemical[2 + 2] photodimerization and photopolymerization of olefinic compounds, *J. Am. Chem. Soc.*, 1998, **120**, 3641–3649.
29. C. V. K. Sharma, K. Panneerselvam, L. Shimoni, H. Katz, H. L. Carrell and G. R. Desiraju, 3-(3′,5′-Dinitrophenyl)-4-(2′,5′-Dimethoxyphenyl) Cyclobutane-1,2-Dicarboxylic Acid – Engineered Topochemical Synthesis and Molecular and Supramolecular Properties, *Chem. Mater.*, 1994, **6**, 1282–1292.
30. J. A. R. P. Sarma and G. R. Desiraju, The Role of Cl···Cl and C–H···O Interactions in the Crystal Engineering of 4 Å Short-Axis Structures, *Acc. Chem. Res.*, 1986, **19**, 222–228.

31. H. Moon, R. Zeis, E. J. Borkent, C. Besnard, A. J. Lovinger, T. Siegrist, C. Kloc and Z. N. Bao, Synthesis, crystal structure, and transistor performance of tetracene derivatives, *J. Am. Chem. Soc.*, 2004, **126**, 15322–15323.
32. M. Halik, H. Klauk, U. Zschieschang, G. Schmid, S. Ponomarenko, S. Kirchmeyer and W. Weber, Relationship between molecular structure and electrical performance of oligothiophene organic thin film transistors, *Adv. Mater.*, 2003, **15**, 917–922.
33. M. Mushrush, A. Facchetti, M. Lefenfeld, H. E. Katz and T. J. Marks, Easily processable phenylene-thiophene-based organic field-effect transistors and solution-fabricated nonvolatile transistor memory elements, *J. Am. Chem. Soc.*, 2003, **125**, 9414–9423.
34. J. E. Anthony, Functionalized acenes and heteroacenes for organic electronics, *Chem. Rev.*, 2006, **106**, 5028–5048.
35. J. E. Anthony, J. S. Brooks, D. L. Eaton and S. R. Parkin, Functionalized pentacene: Improved electronic properties from control of solid-state order, *J. Am. Chem. Soc.*, 2001, **123**, 9482–9483.
36. C. D. Sheraw, T. N. Jackson, D. L. Eaton and J. E. Anthony, Functionalized pentacene active layer organic thin-film transistors, *Adv. Mater.*, 2003, **15**, 2009–2011.
37. V. A. Russell and M. D. Ward, Molecular crystals with dimensionally controlled hydrogen-bonded nanostructures, *Chem. Mater.*, 1996, **8**, 1654–1666.
38. K. T. Holman, A. M. Pivovar, J. A. Swift and M. D. Ward, Metric engineering of soft molecular host frameworks, *Acc. Chem. Res.*, 2001, **34**, 107–118.
39. L. R. MacGillivray, G. S. Papaefstathiou, T. Friščić, T. D. Hamilton, D.-K. Bučar, Q. Chu, D. B. Varshney and I. G. Georgiev, Supramolecular control of reactivity in the solid state: From templates to ladderanes to metal-organic frameworks, *Acc. Chem. Res.*, 2008, **41**, 280–291.
40. L. R. MacGillivray, J. L. Reid and J. A. Ripmeester, Supramolecular control of reactivity in the solid state using linear molecular templates, *J. Am. Chem. Soc.*, 2000, **122**, 7817–7818.
41. L. R. MacGillivray and J. L. Atwood, Structural classification and general principles for the design of spherical molecular hosts, *Angew. Chem., Int. Ed.*, 1999, **38**, 1019–1034.
42. M. A. Summers, P. R. Kemper, J. E. Bushnell, M. R. Robinson, G. C. Bazan, M. T. Bowers and S. K. Buratto, Conformation and luminescence of isolated molecular semiconductor molecules, *J. Am. Chem. Soc.*, 2003, **125**, 5199–5203.
43. V. Podzorov, V. M. Pudalov and M. E. Gershenson, Field-effect transistors on rubrene single crystals with parylene gate insulator, *Appl. Phys. Lett.*, 2003, **82**, 1739–1741.
44. S. C. B. Mannsfeld, A. Virkar, C. Reese, M. F. Toney and Z. N. Bao, Precise Structure of Pentacene Monolayers on Amorphous Silicon Oxide and Relation to Charge Transport, *Adv. Mater.*, 2009, **21**, 2294–2298.

45. Z. Bao, J. Locklin, *Organic Field-Effect Transistors*. CRC Press: Boca Raton, FL, 2007.
46. P. Lang, R. Hajlaoui, F. Garnier, B. Desbat, T. Buffeteau, G. Horowitz and A. Yassar, IR Spectroscopy Evidence for a Substrate-Dependent Organization of Sexithiophene Thin Films Vacuum Evaporated onto SiH/Si and SiO2/Si, *J. Phys. Chem.*, 1995, **99**, 5492–5499.
47. C. Ehrendorfer and A. Karpfen, Theoretical Vibrational Spectra of a Bipolaronic Defect in Oligothiophenes and in Polythiophene: A Combined Semiempirical and ab Initio Study, *J. Phys. Chem.*, 1995, **99**, 5341–5353.
48. P. P. Kapadia, L. R. Ditzler, J. Baltrusaitis, D. C. Swenson, A. V. Tivanski and F. C. Pigge, Semiconducting Organic Assemblies Prepared from Tetraphenylethylene Tetracarboxylic Acid and Bis(pyridine)s via Charge-Assisted Hydrogen Bonding, *J. Am. Chem. Soc.*, 2011, **133**, 8490–8493.
49. S. Roy, S. P. Mondal, S. K. Ray and K. Biradha, A Photoswitchable and Photoluminescent Organic Semiconductor Based On Cation-pi and Carboxylate-Pyridinium Interactions: A Supramolecular Approach, *Angew. Chem., Int. Ed.*, 2012, **51**, 12012–12015.
50. A. R. Murphy and J. M. J. Fréchet, Organic semiconducting oligomers for use in thin film transistors, *Chem. Rev.*, 2007, **107**, 1066–1096.
51. H. Meng, J. Zheng, A. J. Lovinger, B.-C. Wang, P. G. Van Patten and Z. Bao, Oligofluorene–Thiophene Derivatives as High-Performance Semiconductors for Organic Thin Film Transistors, *Chem. Mater.*, 2003, **15**, 1778–1787.
52. J. H. Burroughes, D. D. C. Bradley, A. R. Brown, R. N. Marks, K. Mackay, R. H. Friend, P. L. Burn and A. B. Holmes, Light-Emitting-Diodes Based on Conjugated Polymers, *Nature*, 1990, **347**, 539–541.
53. J. G. Xue, S. Uchida, B. P. Rand and S. R. Forrest, Asymmetric tandem organic photovoltaic cells with hybrid planar-mixed molecular heterojunctions, *Appl. Phys. Lett.*, 2004, **85**, 5757–5759.
54. G. M. J. Schmidt, Photodimerization in the solid state, *Pure Appl. Chem.*, 1971, **27**, 647–678.
55. V. Baret, A. Gandini and E. Rousset, Photodimerization of heteroarylene-vinylenes, *J. Photochem. Photobiol., A*, 1997, **103**, 169–175.
56. A. Usta, A. Yasar, N. Yilmaz, C. Gulec, N. Yayh, S. A. Karaoglu and N. Yayh, Synthesis, configuration, and antimicrobial properties of novel substituted and cyclized '2′,3″-thiazachalcones', *Helv. Chim. Acta*, 2007, **90**, 1482–1490.
57. H. Wynberg, M. B. Groen and R. M. Kellogg, Photodimerization of Some Thiophene Analogs of Chalcone, *J. Org. Chem.*, 1970, **35**, 2828–2830.
58. A. Elgavi, B. S. Green and G. M. J. Schmidt, Reactions in Chiral Crystals – Optically-Active Heterophotodimer Formation from Chiral Single-Crystals, *J. Am. Chem. Soc.*, 1973, **95**, 2058–2059.
59. B. S. Green and L. Heller, Solution and Solid-State Photodimerization of Some Styrylthiophenes, *J. Org. Chem.*, 1974, **39**, 196–201.

60. Y. Ito, H. Hosomi and S. Ohba, Compelled orientational control of the solid-state photodimerization of trans-cinnamamides: Dicarboxylic acid as a non-covalent linker, *Tetrahedron*, 2000, **56**, 6833–6844.
61. D. Vuk, Z. Marinic, K. Molcanov, B. Kojic-Prodic and M. Sindler-Kulyk, Photochemical transformation of beta,beta′-dithienyl substituted o-divinylbenzenes leading to 1,2-dihydronaphthalenes or fused pentacyclic compounds: first evidence of electrocyclization process via 2,3-dihydronaphthalene intermediates, *Tetrahedron*, 2012, **68**, 6873–6880.
62. M. E. Amato, G. Musumarra, G. Scarlata, D. Lamba and R. Spagna, X-Ray Structure Determination of Cyclobutane Photodimers from (Z)-Alpha, Beta-Diarylacrylonitriles, *J. Crystallogr. Spectrosc. Res.*, 1989, **19**, 791–808.
63. Y. V. Nelyubina, M. Y. Antipin and K. A. Lyssenko, Extremely short halogen bond: the nature and energy of iodine-oxygen interactions in crystalline iodic acid, *Mendeleev Commun.*, 2011, **21**, 250–252.
64. M. Munakata, G. L. Ning, Y. Suenaga, T. Kuroda-Sowa, M. Maekawa and T. Ohta, A one-dimensional metallocyclophane with columnar aromatic stacking: The silver(I) eta(2)-coordination complex of 1,2-benztriphenylene, *Angew. Chem., Int. Ed.*, 2000, **39**, 4555–4557.
65. M. Munakata, L. P. Wu, T. Kuroda-Sowa, M. Maekawa, Y. Suenaga, G. L. Ning and T. Kojima, Supramolecular silver(I) complexes with highly strained polycyclic aromatic compounds, *J. Am. Chem. Soc.*, 1998, **120**, 8610–8618.
66. Y. S. Yao, W. T. Shen, B. Nohra, C. Lescop and R. Reau, Coordination-Driven Hierarchical Organization of pi-Conjugated Systems: From Molecular to Supramolecular pi-Stacked Assemblies, *Chem. - Eur. J.*, 2010, **16**, 7143–7163.
67. S. L. Zheng, J. P. Zhang, W. T. Wong and X. M. Chen, A novel, highly electrical conducting, single-component molecular material: [Ag_2(ophen)$_2$] (Hophen = 1H-[1,10]phenanthrolin-2-one), *J. Am. Chem. Soc.*, 2003, **125**, 6882–6883.
68. R. D. Schmitt, R. M. Wing and A. H. Maki, Donor-Acceptor Complexes of Inorganic Pi Acceptor, Bis-Cis-(1,2-Perfluoromethylethene-1,2-Dithiolato) Nickel, *J. Am. Chem. Soc.*, 1969, **91**, 4394–4401.
69. D. Sheberla, L. Sun, M. A. Blood-Forsythe, S. Er, C. R. Wade, C. K. Brozek, A. Aspuru-Guzik and M. Dinca, High Electrical Conductivity in Ni-$_3$(2,3,6,7,10,11-hexaiminotriphenylene)$_2$, a Semiconducting Metal-Organic Graphene Analogue, *J. Am. Chem. Soc.*, 2014, **136**, 8859–8862.
70. T. C. Narayan, T. Miyakai, S. Seki and M. Dincă, High Charge Mobility in a Tetrathiafulvalene-Based Microporous Metal-Organic Framework, *J. Am. Chem. Soc.*, 2012, **134**, 12932–12935.
71. I. G. Georgiev, D.-K. Bučar and L. R. MacGillivray, Stereospecific and quantitative photodimerisation of terminal olefins in the solid state, *Chem. Commun.*, 2010, **46**, 4956–4958.
72. Q. L. Chu, D. C. Swenson and L. R. MacGillivray, A single-crystal-to-single-crystal transformation mediated by argentophilic forces converts

a finite metal complex into an infinite coordination network, *Angew. Chem., Int. Ed.*, 2005, **44**, 3569–3572.
73. T. Kim, M. K. Al-Muhanna, S. D. Al-Suwaidan, R. O. Al-Kaysi and C. J. Bardeen, Photoinduced Curling of Organic Molecular Crystal Nanowires, *Angew. Chem., Int. Ed.*, 2013, 52.
74. S. X. Lu, Y. Liu, N. Shao and B. Panchapakesan, Nanotube micro-opto-mechanical systems, *Nanotechnology*, 2007, **18**, 1–7.
75. I. Willner and B. Willner, Layered molecular optoelectronic assemblies, *J. Mater. Chem.*, 1998, **8**, 2543–2556.
76. F. Li, J. P. Zhuang, G. Y. Jiang, H. H. Tang, A. D. Xia, L. Jiang, Y. L. Song, Y. L. Li and D. B. Zhu, A rewritable optical data storage material system by [2 + 2] photocycloreversion-photocycloaddition, *Chem. Mater.*, 2008, **20**, 1194–1196.
77. S. S. Park, E. R. Hontz, L. Sun, C. H. Hendon, A. Walsh, T. Van Voorhis and M. Dincă, Cation-Dependent Intrinsic Electrical Conductivity in Isostructural Tetrathiafulvalene-Based Microporous Metal–Organic Frameworks, *J. Am. Chem. Soc.*, 2015, **137**, 1774–1777.
78. L. Sun, S. S. Park, D. Sheberla and M. Dincă, Measuring and Reporting Electrical Conductivity in Metal–Organic Frameworks: Cd_2(TTFTB) as a Case Study, *J. Am. Chem. Soc.*, 2016, **138**, 14772–14782.

CHAPTER 9

Co-crystallization as a Versatile Tool in Separations Technology

ABHIJEET S. SINHA* AND CHRISTER B. AAKERÖY

Department of Chemistry, Kansas State University, Manhattan, Kansas 66506, USA
*Email: sinha@ksu.edu

9.1 Introduction

A significant challenge in any chemical synthesis not only involves the separation and purification of the target molecule, but also of the intermediates to be used for subsequent steps during synthesis. Various processes such as crystallization, liquid–liquid extraction and chromatography are used in separations.[1] Although crystallization is the least costly of these methods,[2] in some cases impurities/side-products become substituted into the crystalline lattice of the compound being purified, thus reducing the selectivity and efficiency of the separation process. A separation process that can either selectively isolate the target molecule or remove the impurity without the involvement of further separation processes would be of tremendous use to academia and industries alike.

Over the past decade, co-crystallization has been used for improving the physical or chemical properties of crystalline materials.[3] Particularly in the pharmaceutical industry, co-crystallization of an active pharmaceutical ingredient (API) with a co-former is a common technology for product formulation and for enhancement of solubility, stability, dissolution rate, and bioavailability of the API.[4] Co-crystallization has also been used as an effective tool to modify the performance of energetic materials.[5] In addition, co-crystallization also finds use in the improvement of properties of semiconductors,[6] magnetic materials,[7] and agrochemicals.[8]

Another important application of co-crystallization is in the area of product isolation and separation.[9,10] The solubility of co-crystals is usually distinct from that of the pure components.[11] Also, the solubility can be tuned by the choice of the co-former (co-crystal former) involved.[12] Modifying the solubility of co-crystals can facilitate crystallization at decreased concentrations where pure components do not crystallize, and upon regeneration both the target molecule and the co-former can be liberated in the absence of any undesired impurities. Also, depending on functional group compatibility and synthon considerations, there are multiple available co-formers for co-crystallization. The differing solubility coupled with the myriad of available options for co-formers make co-crystallization a very interesting prospect as a separation technology.

The use of co-crystallization as a separation technique was first envisaged by Vostokov *et al.* in 1976[13] and 1977.[14] Their first study involved the isolation of *m*-xylylenediamine from mixtures containing both *m*- and *p*-xylylenediamine by co-crystallization with naphthalene.[13] A difference in their solid–liquid phase distribution coefficients resulted in the successful isolation of the *m*-isomer from the *p*-isomer. A follow-up study by the same authors showed the successful separation of 4-diaminodiphenylmethane from its 2,2- and 2,4-isomers by co-crystallization with benzil.[14] Despite these early efforts, the idea of using co-crystallization as a separation technique did not really take off for nearly two decades. Finally, in 1992 Lovrien and co-workers recognized that instead of crystallizing or precipitating amino acids and proteins as homogenous products, it was often easier to co-precipitate or co-crystallize these biomolecules with suitable ligands.[15] The authors noted that organic ionic ligands with large apolar groups would bind to the solute or compound to be isolated and the resulting complexes would precipitate as co-crystals.

The first systematic study that involved a careful consideration of not only solubility differences between co-formers, but also the interplay between hydrogen- and halogen-bond synthons,[16] was carried out by Resnati and co-workers in 2000.[17] The authors reasoned that when the molecular recognition pattern controlling self-assembly is based on either hydrogen- or halogen-bonding, then under appropriate conditions either the halogen or the hydrogen bond can dominate the self-assembly process, leading to the formation of supramolecular architectures involving one of the synthons. They first co-crystallized a series of aromatic perfluorinated iodo-compounds and aromatic diols with 1,2-bis(4-pyridyl)-ethane and noted that each of the co-formers yielded 1-D infinite chains with bipyridyl molecules. Furthermore, they examined the competition between these two intermolecular forces of interaction by combining equimolar amounts of both the halogen- and hydrogen-bond donor with the pyridyl-based acceptor in a single pot (Figure 9.1). It was observed that the halogen-bond donor exclusively formed co-crystals with the pyridyl-based acceptor and precipitated from solution, whereas the hydrogen-bond donor remained in solution as confirmed by micro-analysis, thin-layer chromatography, and NMR spectroscopy.[17]

Furthermore, they also showed that this competition/selectivity in driving self-assembly processes exists not only for aromatic systems but also holds true for aliphatic systems (Figure 9.2).[17]

Although this seminal study was not aimed at exploring the use of co-crystallization as a separation technique, it did provide important indications that a careful consideration of not only the solubility differences in co-crystals, but also of the interplay between the different intermolecular forces of interactions involved in the self-assembly of supramolecular architectures can aid us in designing systems that can be used for chemical

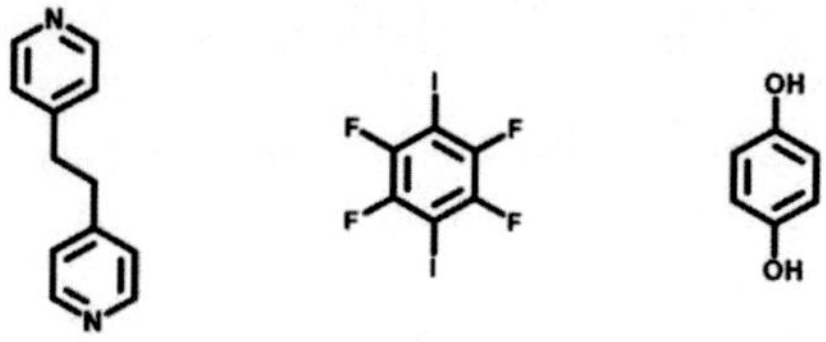

1a 2a 4a

1a + 2a ⟶ ····1a····2a····1a····2a···· 3a

1a + 4a ⟶ ····1a····4a····1a····4a···· 5a

1a + 2a + 4a ⟶ ····1a····2a····1a····2a···· 3a

Figure 9.1 Competition between aromatic halogen- and hydrogen-bond donors in the self-assembly of supramolecular architectures.[17]

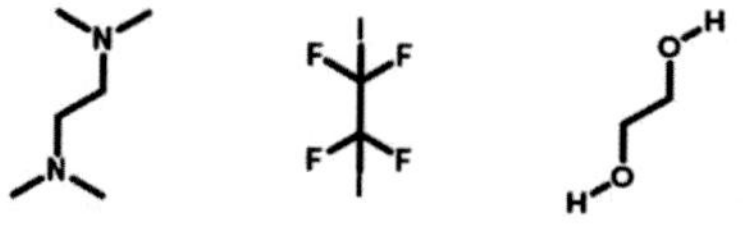

1b 2b 4b

1b + 2b ⟶ ····1b····2b····1b····2b···· 3b m.p. = 105 °C

1b + 4b ⟶ ····1b····4b····1b····4b···· 5b liquid at r.t.

1b + 2b + 4b ⟶ ····1b····2b····1b····2b···· 3b

Figure 9.2 Competition between aliphatic halogen- and hydrogen-bond donors in the self-assembly of supramolecular architectures.[17]

separation. Thus, the aim of this chapter is to provide a perspective on research in the nascent field of chemical separations *via* co-crystallization. We will endeavor to highlight some of the key studies in this field, and also identify the remaining challenges. We have divided our chapter into sections based on not only the thermodynamics behind chemical separations *via* co-crystallization, but also on specific applications such as industrial separations and chiral resolution. This will enable readers to gain a clearer understanding of both the theoretical and experimental factors behind designing a separation method using co-crystallization.

9.2 Thermodynamics of Separation *via* Co-crystallization

Any reliable and optimal crystallization process in solution ideally requires the construction of appropriate phase diagrams.[18] This also applies to resolution/separation of enantiomers/isomers using co-crystallization as the separation technique. Construction of phase diagrams not only aids in the choice of an appropriate co-former, but also helps to understand the nature of solvent required for regenerating the pure components.[19] In short, a better understanding of phase diagrams is very helpful in the pursuit of optimal conditions required for a co-crystallization based separation process.

Although crystallization can be used to isolate the product, quite often the product solubility is higher than the limiting product concentration,[20] which in turn prevents crystallization from taking place. Thus, it is necessary to decrease the product solubility below the limiting product concentration. This can be achieved by converting the pure component crystals of the product to a salt or a co-crystal, where the choice of an appropriate counterion or co-former aids in tuning the solubility of the product. Co-crystallization involves the use of neutral co-formers and regeneration of the co-crystal provides both the product and the co-former,[21] thus avoiding the production of waste salt. Ter Horst and co-workers in 2010 demonstrated the use of co-crystallization as a separation technique for isolating the product in fermentations using the model compound cinnamic acid (CA) (Figure 9.3).[22] Their strategy involved three steps: (a) co-crystallize CA with 3-nitrobenzamide (NBA) to form the reported 1 : 1 CA-NBA co-crystal; (b) determine the solubility decrease for the

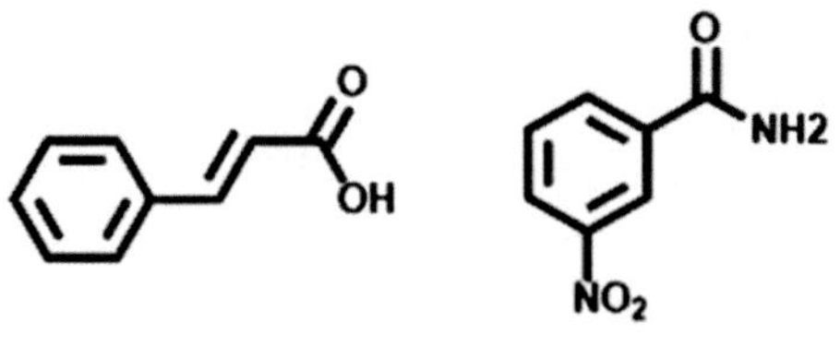

Figure 9.3 Molecular structure of cinnamic acid (CA, left) and 3-nitrobenzamide (NBA, right).[22]

co-crystal compared to the pure component crystals of CA; and (c) replicate the results in a simulated fermentation experiment.

CA forms a stable co-crystal with NBA in both ethanol and water, as is evidenced by the identical powder X-ray diffraction pattern of the solids obtained from both solvents. A ternary phase diagram for this co-crystal in water was constructed to examine its stability as well as to determine the apparent solubility of CA in water (Figure 9.4).[22]

The lines indicated by X_{CA}* and X_{NBA}* represent the solubility lines of pure CA and NBA, respectively. At 30 °C, the experimentally determined solubility of CA and NBA in water is 0.069 mmol mol^{-1} and 0.357 mmol mol^{-1}, repectively.[22] The solubility line of the co-crystal is depicted by $(X_{CA}{\cdot}X_{NBA})^*$. Low pure component solubility in water results in differences between the ideal and the actual phase diagrams of CA-NBA co-crystals. In region L, the solution is in the single stable phase, whereas in regions L+CA, L+NBA, and

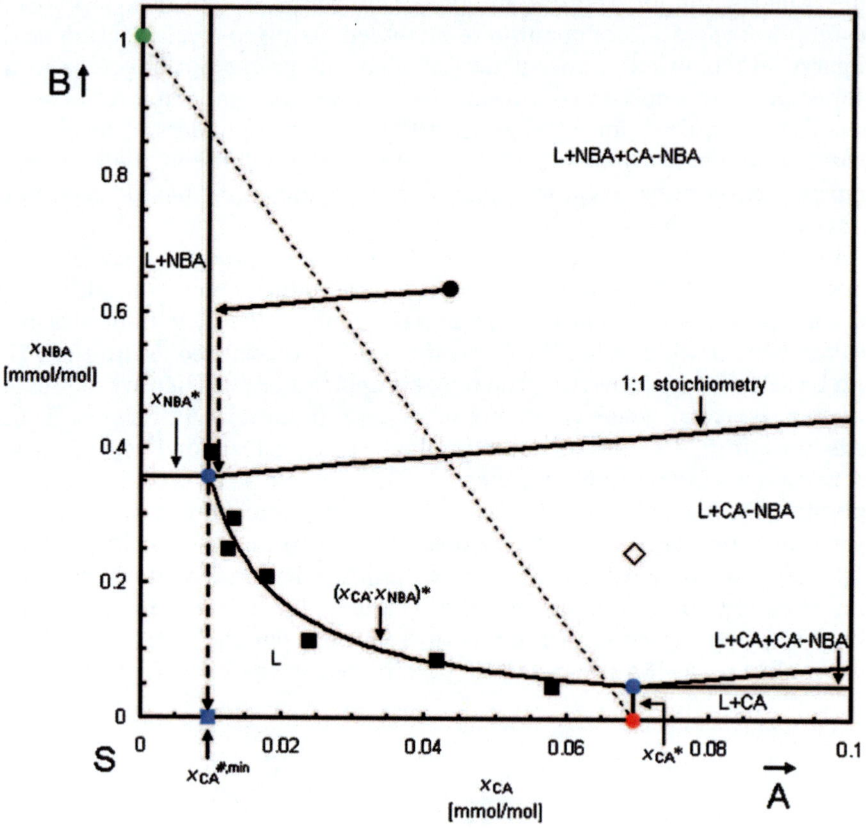

Figure 9.4 The ternary phase diagram at 30 °C of cinnamic acid (CA) and 3-nitrobenzamide (NBA) in water.[22]
Reprinted with permission from J. Urbanus, C. P. M. Roelands, D. Verdoes, P. J. Jansens and J. H. ter Horst, *Cryst. Growth Des.*, 2010, **10**, 1171–1179, Copyright 2010 American Chemical Society.

L+CA-NBA, an equilibrium exists between the solution and the CA, NBA, or CA-NBA crystals, respectively. A solution composition eventually will equilibrate to a position on the CA (X_{CA}*), NBA (X_{NBA}*), or CA-NBA ($X_{CA}{\cdot}X_{NBA}$)* line. Lastly, in regions L+CA+CA-NBA and L+NBA+CA-NBA, the solution equilibrates to the three-phase equilibrium point (blue dots in the phase diagram). At these points, where the co-crystal and the pure component solubility lines intercept, the clear solution L, the co-crystal CA-NBA, and one of the pure component crystalline phases (CA or NBA) are in equilibrium. The black dot also indicates this composition in region L+NBA+CA-NBA, which equilibrates to the three-phase equilibrium point with the solution, co-crystal CA-NBA and pure component NBA.[22] The dashed arrow connecting the black dot with the corresponding three-phase equilibrium point is split into two parts to indicate that the co-crystals are formed in a 1 : 1 stoichiometry, simultaneously where NBA crystals are being formed (vertical dashed arrow). The phase diagram not only provides critical information about the operating window for co-crystallization of CA and NBA at 30 °C in water, but also tells us whether and in which region different crystalline materials will form together with the corresponding solution composition.

Upon co-crystallizing near the three-phase equilibrium point both CA and NBA are removed from solution along with a decrease in the solution mole fractions.[22] Thus, to maintain the mole fractions at the requisite levels near the three-phase equilibrium point, both CA and NBA need to be added to the solution. In the case of fermentations involving cinnamic acid, CA would be provided by the biocatalyst, whereas NBA would have to be added. Alternatively, a suspension of NBA crystals could act as a NBA buffer and maintain the requisite concentration of NBA near the three-phase equilibrium point during the product removal process. The minimum concentration for co-crystallization of CA is determined by the position of the three-phase equilibrium point, which in turn depends on the solubility of pure co-former. Phase diagrams of CA-NBA co-crystals show a seven-fold decrease in the apparent solubility of CA at solution compositions near the three-phase equilibrium point.[22] Also, they assist in establishing the optimal conditions for designing an efficient process for the isolation of CA by co-crystallization.

Many APIs are chiral and are mostly synthesized as racemic mixtures with an additional step included for chiral resolution during the synthesis/scale-up process. Leyssens and co-workers in 2012 carried out the chiral resolution of an API *via* co-crystallization based on a combined knowledge of functional group compatibility and ternary phase diagrams.[23] Usually, an enantiopure API only forms a co-crystal with one of the two enantiomers of a chiral co-former and, unlike chiral salts, a diastereomeric pair cannot be obtained. (*RS*)-2-(2-oxopyrrolidin-1-yl)butanamide was selected as the model system because the *S*-enantiomer is biologically active[24] and no viable salts of this compound are known (Figure 9.5).[23] Also, the only known method of resolving these enantiomers is by chiral chromatography.[25]

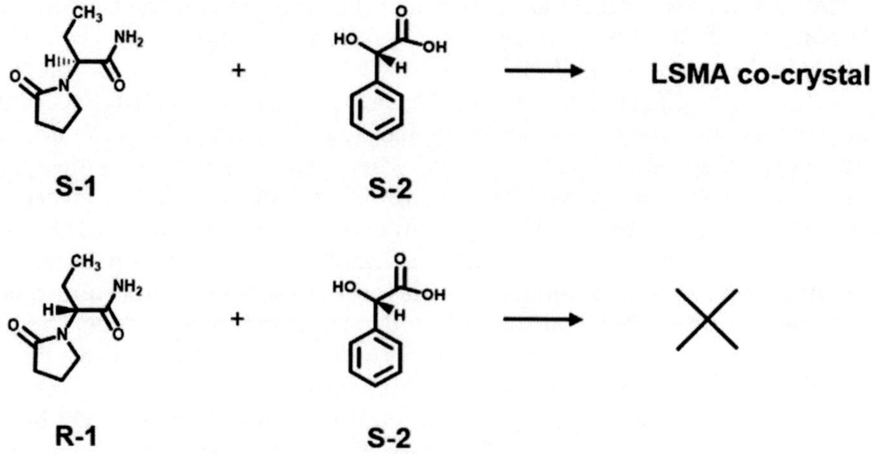

Figure 9.5 Levetiracetam (S-1) co-crystallizes with *S*-mandelic acid (S-2). R-1 does not form a co-crystal with *S*-mandelic acid (S-2).[23]

The authors had in a previous study already demonstrated that only the *S*-isomer of the API co-crystallizes with either *S*-mandelic acid or *S*-tartaric acid, whereas it does not form a co-crystal with *R*-mandelic acid or *R*-tartaric acid.[26] This enantiospecific behavior of a target molecule shows that a racemic mixture of this molecule can be resolved through co-crystallization in solution. Leyssens and co-workers co-crystallized the *S*-isomer of the API with *S*-mandelic acid in acetonitrile, and constructed the experimental ternary phase diagrams through screening of different total compositions in acetonitrile at a given temperature (Figure 9.6).[23] Supersaturated solutions of different compositions were obtained by dissolving all solids at higher temperatures, where the mixtures were seeded with all possible solid states (RS-1 form I, S-1, S-2, and the LSMA co-crystal) and kept at a specified temperature. After two weeks, it was assumed that the system had attained equilibrium and the liquid phase was analyzed by chiral and achiral high-performance liquid chromatography to determine both the solution concentration and enantiomeric excess. Finally, the solid phase was analyzed by powder X-ray diffraction to distinguish the different solid forms.[23]

The zones that are observed in the ternary phase diagrams depend both on the amount of solvent and the temperature. In this case, zone III, in which only the *S*-isomer is stable, is not observed. Consequently, the zones in which the co-crystal and the racemate are stable become adjacent and merge into a zone where their mixture is stable (zone II + IV).[23] Ideally, for the successful separation of the target molecule from the racemic mixture using co-crystallization, two conditions must be met: (a) the chosen crystallization conditions (nature and amount of solvent, and temperature) must allow co-crystal formation; and (b) zone II or alternatively zone I + II must cross the racemic composition line (red dotted line). As seen in the phase diagram, both conditions are met for this system. Furthermore, to test the

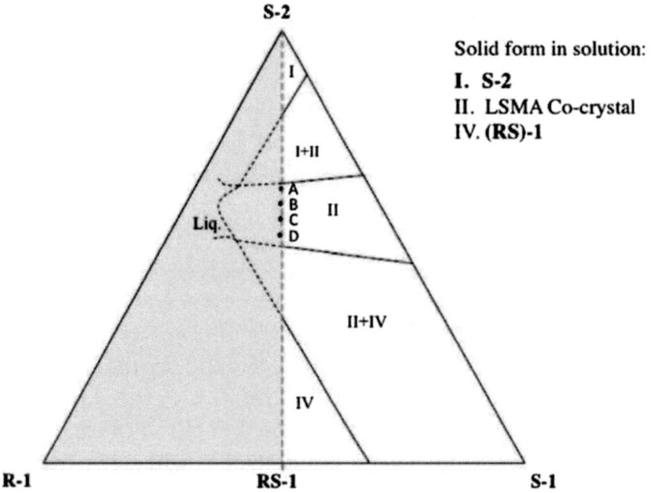

Figure 9.6 Schematic ternary phase diagram of the R-1/S-1/S-2 system in acetonitrile, based on experimental results at −10 °C.[23]
Reprinted with permission from G. Springuel and T. Leyssens, Cryst. Growth Des., 2012, 12, 3374–3378, Copyright 2012, American Chemical Society.

efficiency of their resolution technique, four initial racemic mixtures were selected along the racemic composition line, *i.e.* A, B, C, and D. At 9 °C, enantiospecific co-crystallization was observed at only one condition (point C), where the enantiomerically resolved co-crystal was the only stable form in suspension (zone II). The remaining solution was enriched in R-1 ($ee = 7.92\%$), and 14.67% of S-1 was recovered in the solid phase. The remaining three conditions (A, B, and D) did not show any crystallization. Enantiospecific co-crystallization of the LSMA co-crystal (zone II) was observed for all conditions at 3 °C and −10 °C. Lowering the crystallization temperature led to a more efficient resolution, as confirmed by the results where the *ee* in the remaining solution increased to 27.80% and 53.40%, respectively, at 3 °C and −10 °C.[23] Lower temperature conditions resulted in the selective recovery of 69.60% of the initial amount of S-1 in the solid phase.

Others have also examined ternary[27] and quaternary[28] phase diagrams as tools in chiral resolution *via* co-crystallization. These studies show that an understanding of the thermodynamic behavior of co-crystal systems along with functional group compatibility is critical for designing approaches to efficient enantiospecific resolution of chiral APIs.

9.3 Industrial Separation and Co-crystallization

Separation and purification processes are employed in many industries to isolate intermediates and target molecules. Of these, the pharmaceutical and food industries manufacture strictly regulated products of high quality.

Consequently, it is essential for them to develop efficient and economic separation processes. The key features of such a process are high selectivity (purity), capacity (maximum applicable amount of material), and recovery (yield). Co-crystallization, being a relatively cheaper and 'greener' method when compared to other commonly used separation techniques such as liquid–liquid extraction and chromatography, can be used as an alternative route to industrial chemical separations.

Typically, most studies examine the separation of the target molecules from the impurities *via* co-crystallization. However, Myerson and co-workers first developed a strategy to separate an impurity from the intermediate by selective co-crystal formation.[29] The premise of their strategy was to decrease the solubility of the impurity through co-crystal formation, thus allowing for the removal of the impurity while retaining the target molecule in solution. The success of this protocol relied on three critical conditions: (a) selection of co-formers that would exclusively bind with the impurity; (b) co-crystals of the impurity with decreased solubility; and (c) ideal conditions for the separation process, such as co-former concentration and solvent system. The authors selected ibuprofen (IBU) and ketoprofen (KETO), both well-known structurally similar APIs, as the impurity and target, respectively (Figure 9.7). Both molecules have the carboxylic acid group available for hydrogen-bonding. An extensive search of the Cambridge Structural Database (CSD)[30] to identify an appropriate co-former led to 4,4′-bipyridine (BIPY) as the ideal co-former since BIPY forms a co-crystal with IBU,[31] while there were no reported co-crystals of BIPY with KETO (possibly due to the steric effect of the methyl group in KETO).

Selective 2 : 1 IBU : BIPY co-crystal formation was observed in competitive slurry experiments involving 2 : 2 : 1 KETO : IBU : BIPY ratios, thus highlighting the preference of the co-former to bind only to the impurity.[29] Furthermore, it was observed that the solubility of ibuprofen in the IBU : BIPY co-crystal decreased by a factor of eight in ethyl acetate, when compared to ibuprofen itself. Although, this was a significant decrease in solubility, it still did not meet the threshold for the separation standard. Consequently, the solvent system was changed to a 1 : 1 water/ethanol mixture in which the solubility of IBU decreased nine-fold from 54.5 mg g^{-1} to 5.9 mg g^{-1}. Choice of the solvent system was further confirmed by separation experiments, which showed a 60% decrease in concentration of the impurity in 1 : 1 water/ethanol mixture (Table 9.1).

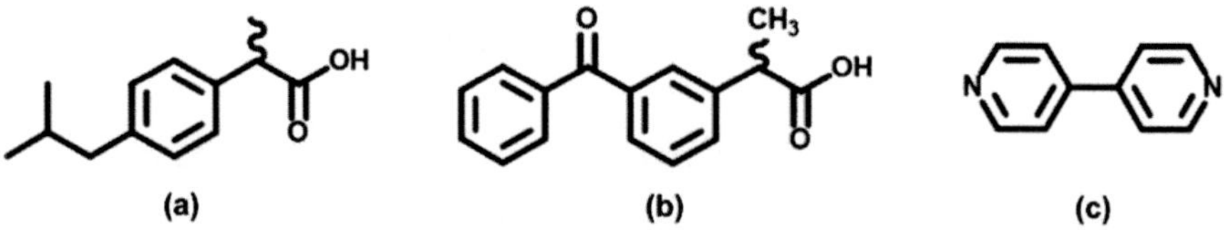

Figure 9.7 Structure of: (a) ibuprofen (impurity); (b) ketoprofen (target); and (c) 4,4′-bipyridine (co-former).[29]

Table 9.1 Compositions for the separation experiments.[29 a]

Solvents	Initial [KETO] (mg g^{-1})	Initial [IBU] (mg g^{-1})	Final [IBU] (mg g^{-1})	Final [BIPY] (mg g^{-1})
Ethyl acetate	183.8	149.1	103.9	11.3
1 : 1 water/ ethanol	84.6	68.6	28.0	4.2

[a]Reproduced from ref. 29 with permission from The Royal Society of Chemistry.

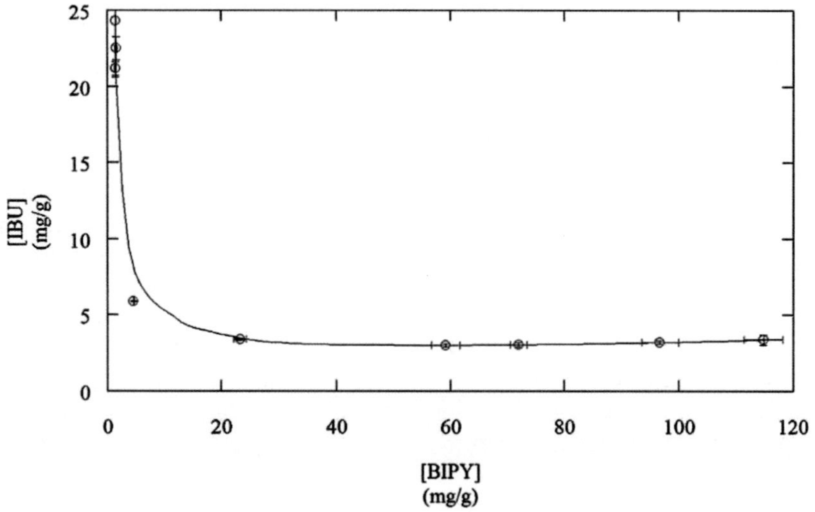

Figure 9.8 Phase solubility diagram of IBU in the presence of BIPY at 20 °C and 1 atm. Reproduced from Ref. 29 with permission from The Royal Society of Chemistry.

Lastly, the authors determined the ideal stoichiometric co-former : impurity ratio for maximum removal of the impurity from solution by constructing phase solubility diagrams (Figure 9.8).[32] It is interesting that the stoichiometry that resulted in the lowest impurity concentration was 1 : 2 IBU : BIPY, which did not match the stoichiometry of the co-crystal (2 : 1 IBU : BIPY).[29]

Using model compounds, the authors not only demonstrated the successful removal of an impurity from solution using co-crystallization, but also established a workflow for designing such experiments, *i.e.* choice of the co-former, decreased solubility of the co-crystal, ideal co-former concentration, and the solvent system of choice for maximum separation.

Another study by Myerson and co-workers in 2013 focused on purifying structurally similar compounds by preventing impurities from substituting into the crystal lattice of target molecules.[33] Work by Higuchi in the 1950s on caffeine complexes in solution showed that complexation can affect solid properties.[34] Consequently, the authors employed a similar strategy of choosing an ideal co-former that would selectively hydrogen bond with the impurity in the target molecule/impurity mixture, and assumed that co-formers that can form co-crystals with impurities will likely form complexes with the impurity in solution. These impurity complexes will in turn no longer fit in the crystal lattice of the target molecule due to steric effects. The model systems selected were benzamide (BAM) and cinnamamide (CAM) with benzoic acid (BA) and cinnamic acid (CA) as the impurities, respectively (Figure 9.9). Based on functional group compatibility and an extensive search of the CSD,[30] the selected co-formers for complexation with the impurities were isonicotinamide (INA), 2-amino-4,6-dimethylpyrimidine (DMP), and dimethylglyoxime (DMG).[33]

Phase solubility diagrams for the most successful combinations, *i.e.* BA-DMG and CA-DMP for which maximum reduction in the impurity concentration was observed, were used to determine the ideal stoichiometry and the binding constants of the complexes formed in solution.[33] It should be noted that the complexation in the separation experiments was not measured by spectroscopic techniques. Hence, binding constants were the best indicators of complexation in solution, with the authors only considering the interaction between the impurity and the co-former. In the case of the BAM/BA system, although the maximum binding constant was observed for the BA-DMP complex, it was the BA-DMG complex that showed the highest removal of impurity from solution. Whereas, in the case of the CAM/CA system, the highest binding energy for the CA-DMP complex coincided with maximum removal of impurity. This highlights that solution binding constants alone are not enough to predict the ideal co-former for impurity removal. Nevertheless, this study shows a pathway to the purification of structurally similar compounds by the addition of co-formers that have the potential to complex with the impurities in solution.

Myerson and co-workers demonstrated an application of the strategy for the isolation of impurities from target molecules by purifying amoxicillin trihydrate (AMCT), a major β-lactam antibiotic widely used against a broad spectrum of bacteria.[35] The traditional large-scale synthesis of AMCT is

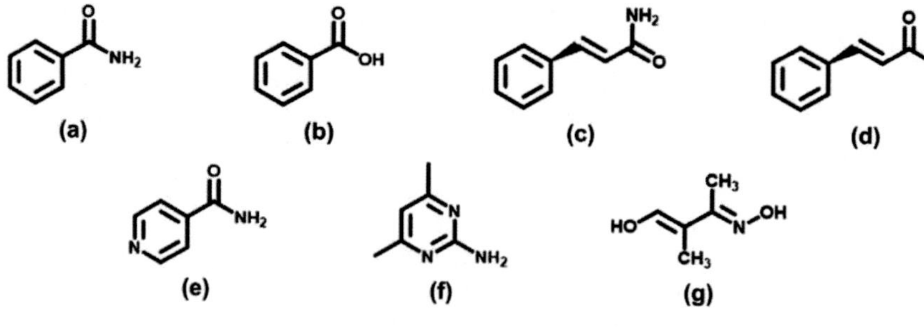

Figure 9.9 Structures of the model systems: benzamide (a) and its impurity benzoic acid (b); cinnamamide (c) and its impurity cinnamic acid (d); and the selected co-formers for complexation (e)–(g).[33]

based on the Dane salt route with yields higher than 90%. However, the need for low temperatures, protection/deprotection/activation steps, and several reagents and solvents combined with significant waste generation makes this synthetic route less desirable (Figure 9.10).[36]

A 'greener' enzymatic route to AMCT synthesis can be completed in fewer steps in aqueous solution at neutral pH and ambient conditions;[36] however, this involves a hydrolysis reaction, which leads to the formation of side-products of which 4-hydroxyphenylglycine (4HPG) is the primary impurity (Figure 9.11).[36,37] Despite all efforts, the formation of 4HPG is inevitable during this enzymatic route. Several studies have been conducted to improve the yield of AMCT, but none have offered a cheap and sustainable alternative.[36–38]

An extensive search of the CSD[30] for co-formers that would form heteromeric hydrogen bonds with 4HPG led to a shortlist of forty-seven compounds.[35] Solid-state grinding of both the impurity and the target molecule with each of the initially identified co-formers, and analysis of the ground materials, allowed for the shortening of the list to eleven compounds, which selectively formed hydrogen bonds with the impurity and not the target molecule (Figure 9.12). Several factors such as molecular weight of the co-former, functional groups on the co-former, and whether the co-former had a phenyl group or not, were examined for their effect on co-crystal formation with the impurity. It was observed that co-formers with a molecular weight below 150 $g\,mol^{-1}$ had a higher chance of forming co-crystals with 4HPG, possibly due to steric effects. It was also observed that co-formers that can engage with both hydrogen-bonding functional groups on 4HPG, *i.e.* carboxylic acid and amine groups, were more likely to selectively form co-crystals

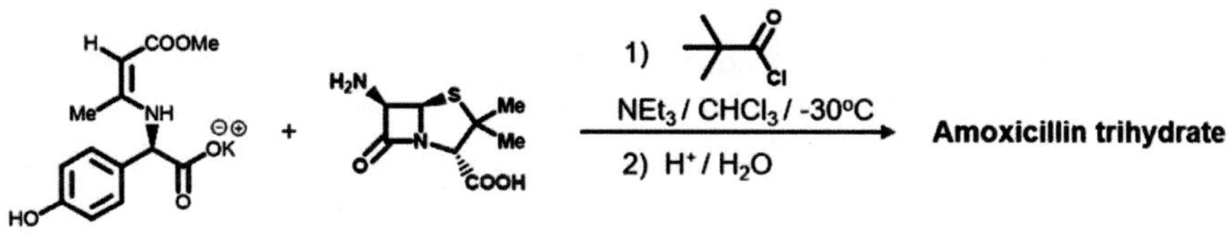

Figure 9.10 Traditional large-scale synthesis of amoxicillin trihydrate (AMCT).[36]

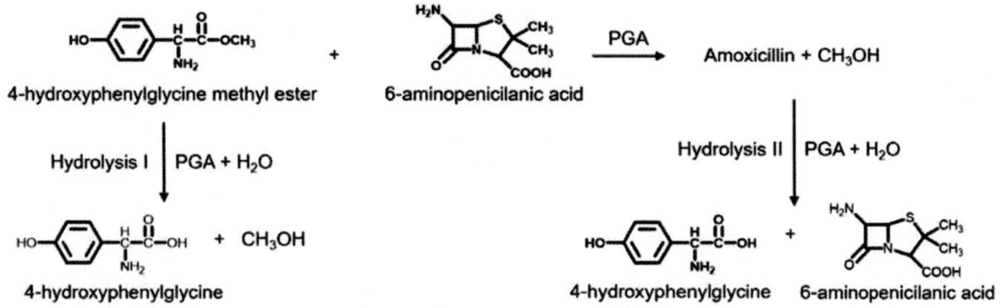

Figure 9.11 Enzymatic synthesis of AMCT using PGA as the catalyst.[35]

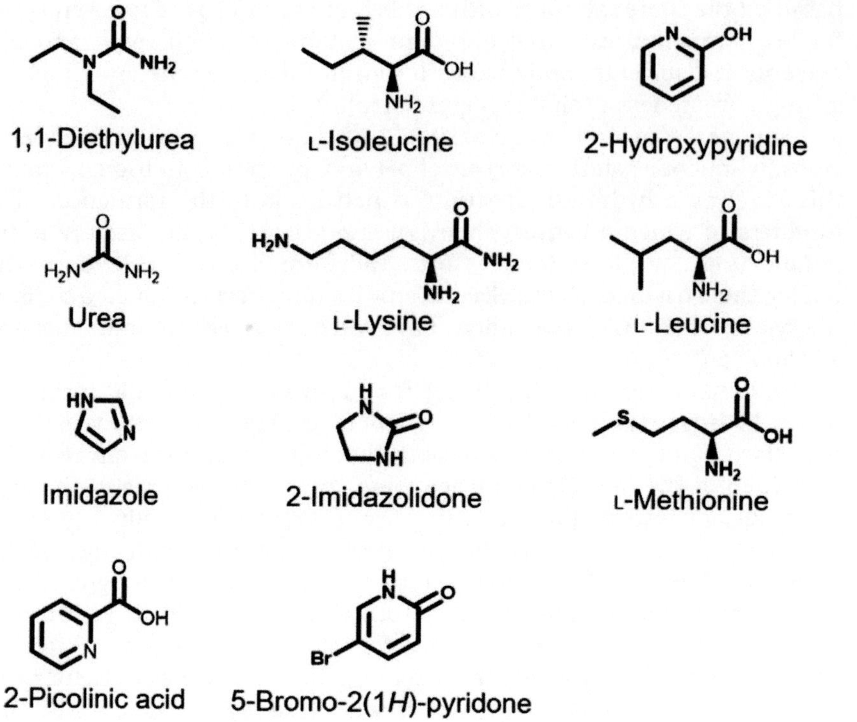

Figure 9.12 Co-formers that selectively form co-crystals with 4HPG.[35]

with 4HPG. Lastly, there was no clear correlation between the presence of phenyl rings on the co-formers and co-crystal formation.[35]

Of the eleven shortlisted compounds, only four co-formers (2-picolinic acid, L-lysine, L-leucine, and L-isoleucine) displayed maximum enhancement in the purity of AMCT.[35] This is because these four co-formers have functional groups that can bind to both functional groups on the impurity to form two stable cyclic hydrogen bonds. Thus, an enhancement in the purity of AMCT was a result of increased complexation in solution of the impurity with the co-formers. The purity was much higher than without the addition of co-former [0.12(±0.01) wt% from 0.98(±0.01) wt%] as well as significantly higher than that obtained from successive crystallizations from fresh solvent. This study not only paves the way for co-crystallization to be used as a viable method in chemical separations, but also sheds light on some of the important factors (steric effect (molecular weight) and functional groups) that need to be considered when designing separation experiments.

A second example of practical applications of co-crystallization as a separation technique is the purification of rhein, an active ingredient in traditional Chinese medicine, from its inherent impurities by co-crystallizing it with arginine as the co-former.[39] The authors managed to purify the active ingredient from the crude extract in excess of 99.9%. Lee and co-workers

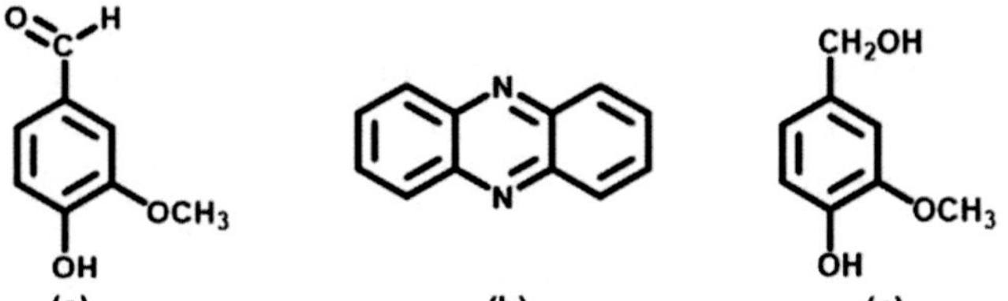

Figure 9.13 Structure of vanillin (a), phenazine (b), and vanillyl alcohol (c).[40]

isolated vanillin (extensively used in the food industry) from vanillyl alcohol (major impurity in vanillin production) by selectively co-crystallizing vanillin with phenazine as the co-former (Figure 9.13).[40] Their strategic approach initially involved the selection of appropriate solvents that would not only promote selective vanillin-phenazine co-crystal[41] formation, but also minimize the incorporation of the impurity in the crystal lattice and reduce the environmental burden. Finally, they explored the scale-up and solubility effects of the process. They performed their co-crystallization experiments in toluene (solvent of choice) and later successfully regenerated the two components in acetone as form-α phenazine and vanillin solution. They even demonstrated the scale-up of this purification process using both continuous and batch modes of crystallization.[40]

In the above examples, the choice of co-former was based on the selectivity of co-crystal formation between the target molecule/impurity and the co-formers. Such an approach yields only qualitative data and is tedious when there are few known co-crystals of the target molecule and requires more work to rank the effectiveness of the co-formers. Furthermore, this approach does not account for the solution behavior of the co-formers, or of solute-solvent interactions. Hence, a quantitative approach that can predict the successful complexing agents without the need for extensive screening of potential compounds would be extremely valuable. Myerson and co-workers developed a quantitative approach to complexing agent selection and tested their method using model crystallizations with cinnamamide (CAM) and benzamide (BAM) as the targets, and cinnamic acid (CA) and benzoic acid (BA) as the impurities, respectively (Figure 9.9(a)–(d).[42] The selected acids when used as additives not only affect the crystal growth and habit of the corresponding amides, but also form strong hydrogen bonds with them.[43]

Quantitative thermodynamic data for the interactions between the impurity and the selected complexing agents were determined by isothermal titration calorimetry (ITC), a highly sensitive technique for examining solution association.[44] A correlation between the association energy and the improvement in purification of both CAM and BAM was observed, which further showed that the choice of an optimal complexing agent in a rationally selected crystallization solvent can lead to over 96% enhancement in purity for the target molecules, which is higher than the purification achieved by successive crystallization experiments.[42] Furthermore, they also demonstrated the applicability of their quantitative complexing strategy by purifying fenofibrate in excess of 94% from its major impurity fenofibric acid by complexing with 1,3-di-*o*-tolylguanidine (Figure 9.14). Thus, thermodynamic data and infrared spectroscopy, coupled with theoretical density functional theory (DFT) calculations were used to not only determine the strength of target···co-former solution association, but also to develop a strategy for the potential purification of organic crystals using a combination of both experimental and theoretical techniques.

An expansion of the above quantitative solution-phase chemical complexation technique of separations was the isolation of nitro-phenol isomers, which highlighted the utility of solution-phase chemical complexation in purifying structural isomers without affecting product crystal form or decreasing yield.[45] Furthermore, this was the first implementation of this technique on systems that did not contain carboxylic acid functionalities. Targeting the differences in the relative proximity of the functional groups present within molecules that are structural isomers, it was shown that lattice-impurity incorporation of 3-nitrophenol into 4-nitrophenol could be prevented using 3-aminobenzoic acid as the complexing agent (Figure 9.15).[45] The outcome was purity enhancements of more than 80% (mass) with respect to non-complexed controls with no reduction in yield. A nanofiltration membrane was used to experimentally verify the formation of complexes in solution.

Myerson and co-workers also demonstrated the inhibition of nucleation of benzoic acid in water through chemical complexation with 1,3-di-*o*-tolylguanidine.[46] They monitored the shifts in the nucleation rates upon addition of the complexing agent to characterize the inhibition effect, and showed that complexation led to significant nucleation inhibition. This work

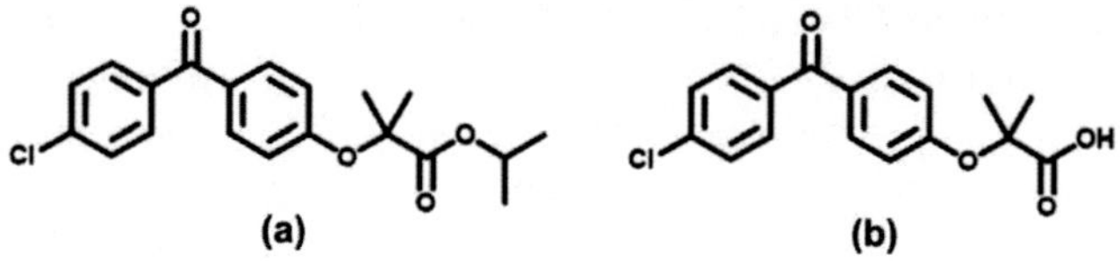

Figure 9.14 Structure of the API fenofibrate (a) and its major impurity finofibric acid (b).[42]

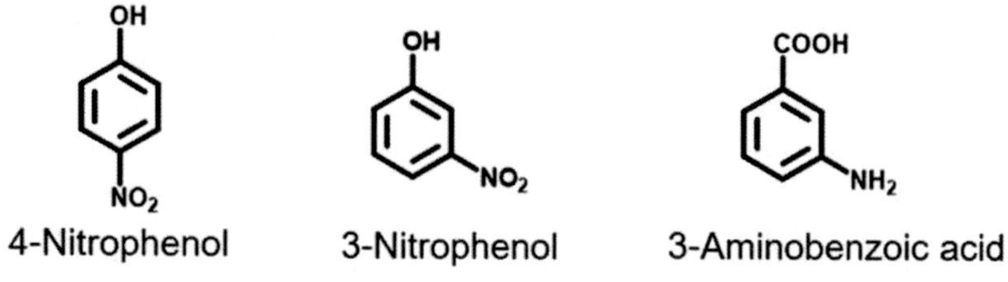

Figure 9.15 Structures of 4-nitrophenol (target), 3-nitrophenol (impurity), and 3-aminobenzoic acid (co-former).[45]

highlights the use of tailored intermolecular interactions in controlling the onset of nucleation. Furthermore, this shows that small-molecule additives can be used as potential precipitation inhibitors, thus providing molecular tools for crystallization control in industrial separations. Recently, Lee and co-workers developed new strategies for direct co-crystal assembly where they synthesized the 1 : 1 acetaminophen-caffeine co-crystal.[47] In their study they also highlighted an efficient isomeric separation of acetaminophen, an analgesic, by co-crystallizing with theophylline.

These studies show that a judicious choice of the co-former (based on CSD data) and solvent system, coupled with a reduction in the solubility of the co-crystal can lead to the isolation of the target molecule from its impurity. Also, more recent efforts have highlighted that an understanding of the strength of association in solution phase using both experimental and theoretical techniques can go a long way in designing co-crystallization chemical separation experiments that can be used as a viable alternative in industrial separations.

9.4 Chiral Resolution *via* Co-crystallization

Molecular recognition[48] has long been applied in classical resolution of acids or bases by formation of salts with distinctly different solubilities. This is the basis of 'Dutch resolution',[49] in which families of closely related resolving agents are used for separating racemic mixtures. Common techniques for chiral resolution of enantiomers are diastereomeric salt formation[50] and chiral chromatography.[51] Diastereomeric salt formation is a less expensive process and involves the addition of an enantiopure resolving agent that converts the racemic compound into its two distinct diastereomeric salts. These salts have their own unique physical properties and can be separated by crystallization. In the case of compounds that cannot readily be converted to their corresponding salts, chiral resolution using expensive chiral chromatography is the only viable option.[52]

Enantiospecific co-crystallization is an innovative chiral resolution technique based on principles of molecular recognition.[48] Its advantages are not limited to being a cheaper and 'greener' alternative to traditional methods of chiral resolution. This method can be extended to resolve molecules that are not prone to salt formation. Also, considering the vast library of available co-formers coupled with the expanded knowledge of functional-group compatibility, it is possible to design a system-specific co-crystallization method for the chiral resolution of compounds. Although most of the work in the field of chiral resolution by co-crystallization has been done in the past decade, the first successful attempt was by Caira *et al.* in 1996, when they resolved racemic lactam 4-amino-*p*-chlorobutyric acid into its enantiomers by co-crystallizing the racemate with the common resolving agent (2*R*,3*R*)-(+)-tartaric acid (Figure 9.16).[53] The resolving agent selectively formed 1 : 2 co-crystals with the *R*-isomer of lactam. Extensive hydrogen bonding between the secondary amide moiety on the lactam and the acid

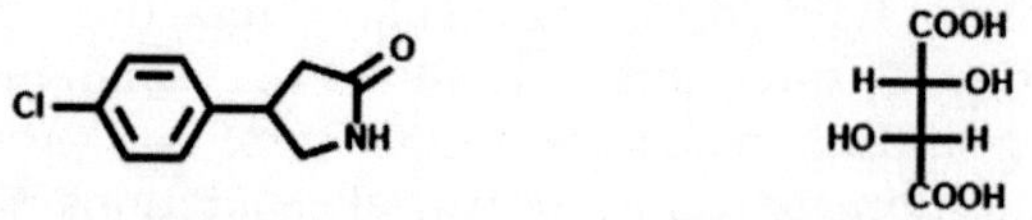

4-Amino-*p*-chlorobutyric acid lactam (2R,3R)-(+)-Tartaric acid

Figure 9.16 Structure of racemic lactam (left) and resolving agent (right).[53]

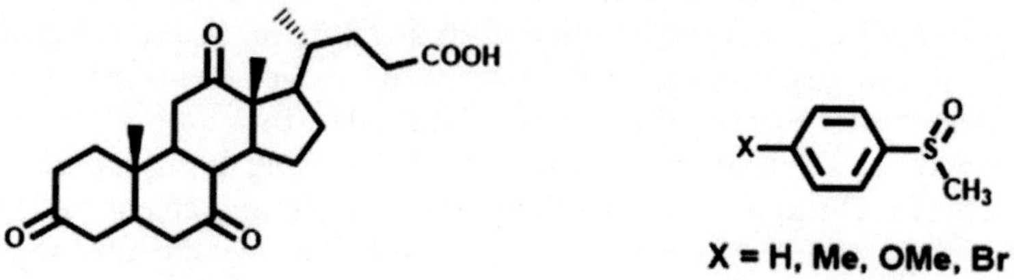

Dehydrocholic acid (host) Arylmethylsulfoxides (guest)

Figure 9.17 Structure of enantiopure 'host' (left) and chiral arylmethylsulfoxides (right).[56]

moiety on the resolving agent resulted in enantiopure co-crystals, where enantioselectivity did not directly involve the chiral center of the racemic lactam molecule.

Other early examples of resolution of racemic compounds by co-crystallization involved enantioselective inclusion of 'small' chiral guest molecules in the crystal lattice of large and readily available naturally occurring enantiopure host compounds such as steroids.[54] This method of lattice inclusion of chiral compounds was used by Fantin and co-workers to separate chiral arylmethylsulfoxides by selective precipitation of highly enriched 1 : 1 co-crystals of the sulfoxides with dehydrocholic acid as the host (Figure 9.17).[55] They obtained enantiopurification of the *R*-isomer in all but one case at the preparative scale with enantiomeric excess as high as 99%. The selectivity was attributed to small, but non-negligible, differences in the geometry and energetics between the inclusion modes of the two enantiomers within the same chiral 'host'. Furthermore, they also characterized these co-crystals by single-crystal X-ray diffraction.[56]

Blackmond and co-workers in 2007 achieved enantioenrichment in chiral amino acid solutions of eutectic compositions of aqueous mixtures of L and D amino acids by co-crystallization with achiral dicarboxylic acids.[57] In several cases, these mixtures of achiral co-former and chiral amino acids yielded new eutectic compositions of 98% *ee* or higher. Although these seminal studies laid the groundwork for the development of co-crystallization as a viable technique in chiral resolution, most of these early efforts did not focus on functional group compatibility between the target racemate and the resolving agent.

In this regard, a more systematic study by Tanaka and co-workers on the facile separation of *cis* and *trans* isomers of cyclohexanediamines by co-crystallization with readily available low molecular weight diols is intriguing (Figure 9.18).[58] Although the authors do not mention this, their choice of hydroxy-based co-formers for co-crystallization with amines is understandable, as alcohols are known to form stable hydrogen-bonded $R_4^4(8)$[59] tetramers with amines.[60] As expected, the co-crystals characterized by single-crystal X-ray diffraction in this study displayed the $R_4^4(8)$ tetrameric hydrogen-bonding motif (Figure 9.19). Also, a complete fulfillment of all hydrogen-bond donors and acceptors was observed, leading to a close-packed 3-D supramolecular architecture. These extensive hydrogen-bonding networks dominate the packing of the preferred co-crystals, thus potentially increasing the lattice energy and decreasing the solubility of these preferred co-crystals.[58] Six out of seven of these co-crystals showed a significant enrichment of one cyclohexanediamine enantiomer over the other, thus highlighting a simple but effective method for knowledge-based isomeric enrichment.

Aher and co-workers demonstrated an important application of enantiomeric resolution *via* co-crystallization to synthetic organic chemistry by

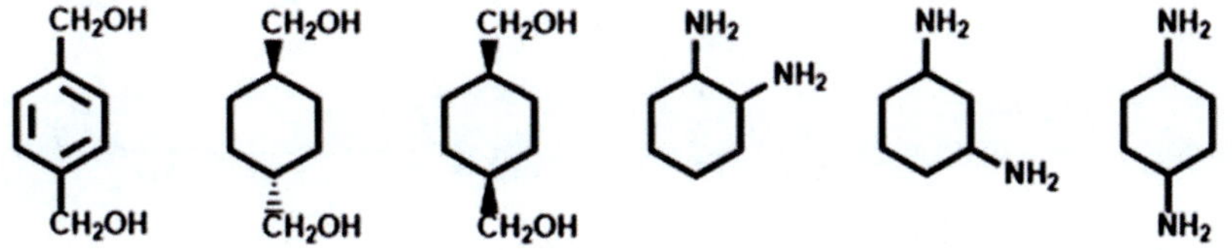

1 2 3 4 5 6

Figure 9.18 Low molecular weight diols (1-3) and isomers of cyclohexanediamines (4-6).[58]

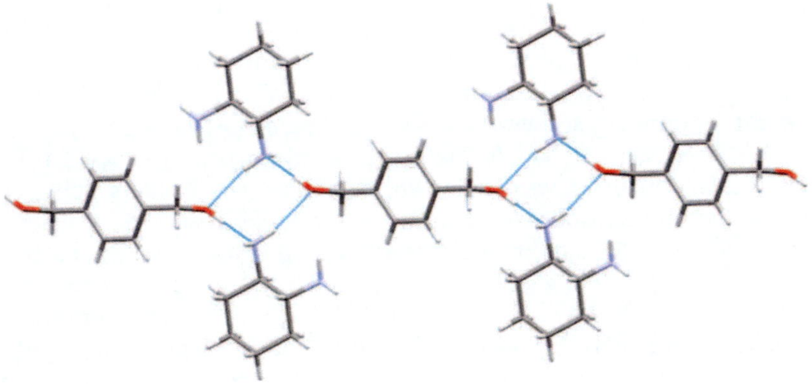

Figure 9.19 Observed $R_4^4(8)$ hydrogen-bonding motifs in co-crystals of diols with amines.[58]

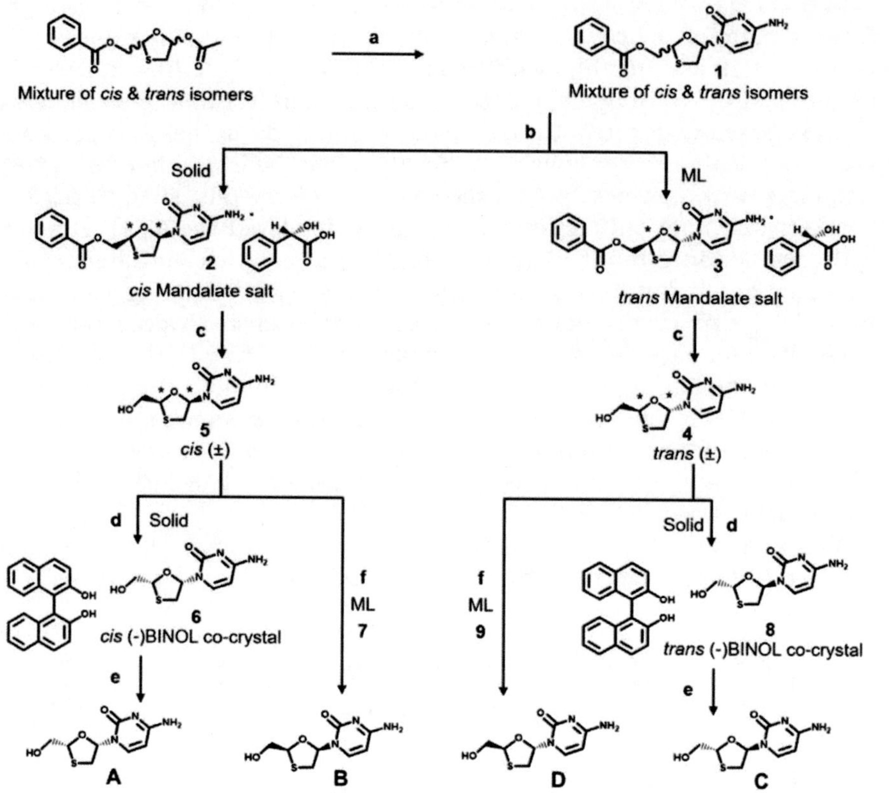

Figure 9.20 Resolution of racemic lamivudine: (a) cytosine, HMDS, TMCS; (b) *S*-(+)-mandelic acid, ethyl acetate, 70 °C to room temperature; (c) 25% methanolic ammonia, ethanol; (d) (S)-BINOL, methanol, 65 °C; (e) ethyl acetate, water, aqueous HCl, NaOH, 25 °C; (f) ethyl acetate, water.[61]

carrying out the synthesis of (-)-[2*R*,5*S*]-4-amino-1-[2-(hydroxymethyl)-1,3-oxathiolan-5-yl]-2(1H)-pyrimidin-2-one (lamivudine) (Figure 9.20).[61] Both isomers of lamivudine inhibit the human immunodeficiency virus 1 and 2 *in vitro*,[62] but the *cis*-(-)-enantiomer is considerably less cytotoxic than the other three optical isomers.[63] Also, resolution of racemic lamivudine through diastereomeric salt formation has largely been unsuccessful with several acids.[64] Thus, it was important to resolve the isomers using an alternative technique, and racemic lamivudine was resolved by co-crystallization with (*S*)-BINOL, a known versatile chiral ligand used in asymmetric synthesis.[65]

Only the *cis*-(−) and *trans*-(−) optical isomers formed 1 : 1 co-crystals with (*S*)-BINOL, leaving behind *cis*-(+) and *trans*-(+) optical isomers in solution.[61] The resulting co-crystals (70–75% yield) were treated with dilute hydrochloric acid to regenerate the two components, and the remaining isomers in solution were isolated from the mother liquor by washing with ethyl acetate to remove excess (*S*)-BINOL. All four isomers (**A–D**) were separated and characterized in an enantiomeric excess of more than 99%, with the recovery and recycling of the resolving agent also established during the process. Interestingly, the authors noted that it is important to first resolve the *cis*-(±) and *trans*-(±) isomers through (*S*)-(+)-mandelate salt formation before the co-crystallization step to further resolve the isolated (±)-isomers in high *ee*'s.

An important step in chiral resolution by co-crystallization is screening for co-formers that will form stable co-crystals with the racemic compound. Differences in solubility of the API and the potential co-formers can lead to precipitation of the two components as separate phases during solution crystallization.[66] As a result, it is possible to miss potential co-crystal combinations during solution based co-crystal screens. Also, a comprehensive co-crystal screen involves crystallization from multiple solvents, which makes the whole process tedious and time consuming. Jones and co-workers suggested an alternative route to chiral co-crystal screening using mechanochemical milling,[67] either in the absence/presence (liquid-assisted grinding)[68] of a minimum amount of solvent. Screening by mechanochemical milling avoids the difficulties caused by differences in solubility of co-formers and is a more efficient process.[67] Their goal was to investigate the possibility of converting a racemic compound into a pair of diastereomeric co-crystals by milling with an enantiopure co-former, where the model system included racemic malic acid and enantiopure tartaric acid.[69] During the initial screen with enantiopure compounds, they were able to replicate the already reported form I[70] of the L-malic acid : L-tartaric acid co-crystal (milling in polar solvent) as well as synthesize two new forms, *i.e.* form II (milling in non-polar solvent) and form III (heating form I and form II to above 150 °C). They also synthesized the L-malic acid : D-tartaric acid co-crystal by milling in the presence of polar/non-polar solvents.[69] The co-crystals L-malic acid : L-tartaric acid and L-malic acid : D-tartaric acid are diastereomeric pairs having different crystal structures and properties such as solubility and thermal behavior. To test their hypothesis of co-crystal screening of chiral compounds by mechanochemical means, the authors milled racemic malic acid with a single enantiomer of tartaric acid (Figure 9.21).

Grinding equimolar amounts of DL-malic acid with L-tartaric acid resulted in a mixture of L-malic acid : L-tartaric acid and L-malic acid : D-tartaric acid co-crystals (characterized by PXRD) with no residual starting materials.[69] This suggested that the two diastereomeric co-crystals could be separated by washing the ground mixture with a solvent, which will completely dissolve the more soluble co-crystal. A similar screen between racemic tartaric acid and enantiopure malic acid did not result in chiral resolution of the enantiomers (possibly due to higher thermodynamic stability of racemic tartaric acid). This was the first example of successful resolution of racemic malic acid into its diastereomeric co-crystals by mechanochemical milling, and thus highlights the potential of this approach to complement the traditional solution-based methods in the co-crystal screen of chiral compounds.

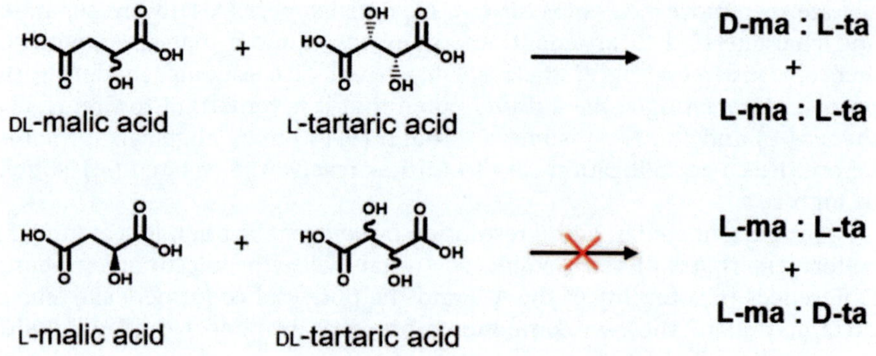

Figure 9.21 Co-crystal screening of chiral compounds by mechanochemical means.[69]

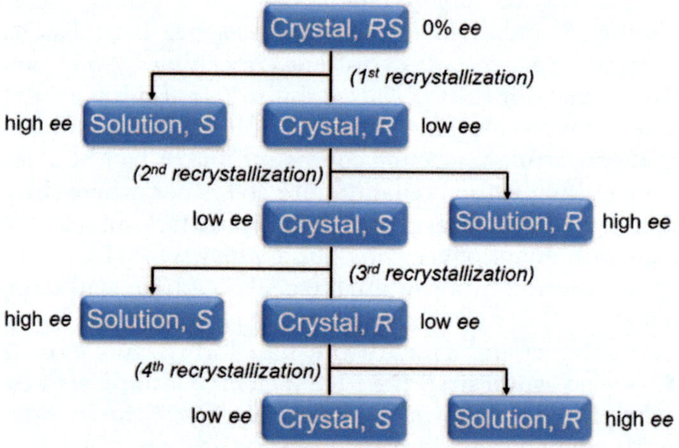

Scheme 9.1 Schematic of PE in the case of substantial enrichment of the S enantiomer in solution after the first recrystallization of a racemic crystal.[74]

A long-held belief has been that resolution of racemic compounds by simple crystallization is not possible without the presence of an external chiral resolving agent. However, Tamura and co-workers discovered an unusual phenomenon of spontaneous enantiomeric resolution of racemic crystals called preferential enrichment (PE).[71] PE is a unique chiral-symmetry breaking technique of spontaneous enantiomeric resolution caused by polymorphic transition of an incipient metastable polymorphic form into a thermodynamically stable form during crystallization from a supersaturated solution of specific racemic mixed crystals (Scheme 9.1).[72] Many examples of resolution of racemic compounds by PE have been demonstrated and a mechanism has also been proposed.[73]

There are five important requirements for the occurrence of PE: (a) sufficient solubility difference (pure enantiomer > racemate); (b) occurrence of polymorphic transition; (c) formation of unique crystal structures involving heterochiral cyclic RS dimer chains as well as homochiral 1D R chains and S chains after the polymorphic transition; (d) selective redissolution of the excess enantiomer from the transformed fragile crystals; and (e) eventual deposition of stable non-racemic mixed crystals capable of 'memorizing' the event of chiral symmetry breaking.[75] Polymorphism in co-crystals is common and co-crystals have the potential to undergo polymorphic transitions during crystallization, thus making them ideal candidates for PE. A case study of PE in co-crystallization was conducted by Tamura and co-workers on DL-arginine and its co-crystals with fumaric acid (**1**), malonic acid (**2**) and oxalic acid (**3**) (Figure 9.22).[74]

All these co-crystals either partially (**2** and **3**) or fully (**1**) had the structural attributes for PE, with head-to-tail cyclic heterochiral DL dimer assembly of arginine involving two cyclic hydrogen bonds [graph set: $R_2^2(8)$] linking the carboxylate and guanidyl groups. PE experiments were carried out on all three co-crystals in H_2O/EtOH (1/1 v/v) solvent mixture under various conditions and observed that only co-crystal **1** showed an excellent PE phenomenon (Scheme 9.2).[74]

Recrystallization of crystals of non-racemic **1** (L-rich, 5.0% *ee*), obtained by recrystallization of **1**, and subsequently deposited crystals (2.7–5.1% *ee*) from an eight-fold supersaturated H_2O/EtOH (1/1 v/v) solution at 5 °C produced alternating enrichment of the two enantiomers up to 92.7% *ee* in the mother liquors and a slight enrichment (<5.1% *ee*) of the opposite enantiomer in the deposited co-crystals.[74] Upon collecting mother liquors of the same handedness, an efficient separation of the two enantiomers (>90% *ee*) was achieved with a good recovery yield (>30%) after four consecutive recrystallizations. Powder X-ray diffraction and differential scanning calorimetry data showed that the deposited crystals were monophasic. Also, no time-dependent change in the obtained crystalline product was detected, leading to the conclusion that in this case chiral symmetry breaking, *i.e.* the PE phenomenon, is a consequence of direct nucleation (phase transition) from the molecular clusters present in the supersaturated solution to give stable mixed crystalline phase and not because of polymorphic transitions.[74] This study was the first example of enantiomeric resolution of DL-arginine, an essential amino acid,[76] in the absence of an external chiral source. Tamura and co-workers also successfully conducted similar preferential enrichment studies on the co-crystals of DL-phenylalanine with fumaric

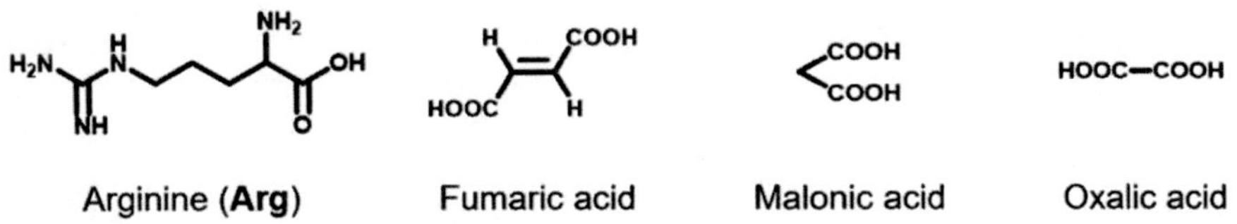

Figure 9.22 Structure of arginine and of the co-formers used by Tamura and co-workers.[74]

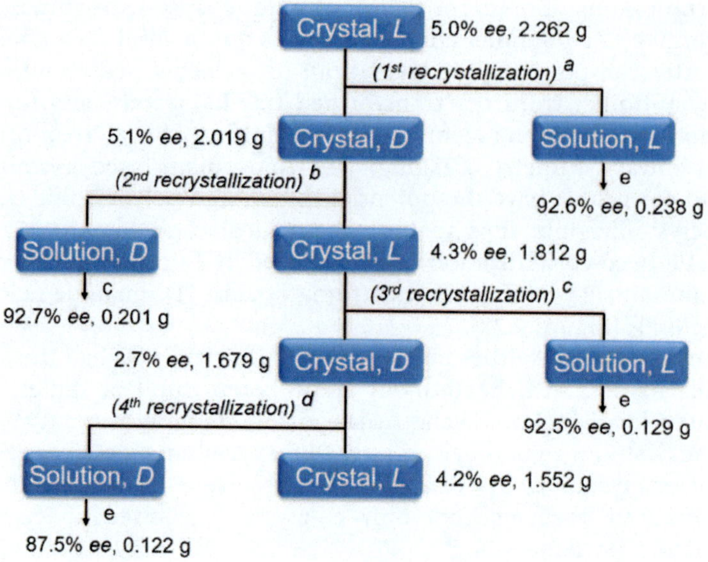

Scheme 9.2 PE of co-crystal **1** by using an eightfold supersaturated solution in H_2O/EtOH (1/1 v/v). Conditions: (a) H_2O/EtOH (8.8 ml) at 5 °C for 6 d; (b) H_2O/EtOH (7.9 ml) at 5 °C for 4 d; (c) H_2O/EtOH (7.1 ml) at 5 °C for 5 d; (d) H_2O/EtOH (6.6 ml) at 5 °C for 4 d; (e) removal of the solvent by evaporation.[74]

acid,[77] and DL-leucine with oxalic acid,[78] to achieve enantiomeric resolution with high *ee* values in the absence of an external chiral source for both amino acids.

Enantioselective co-crystal screening of chiral APIs can be expensive and time-consuming, especially during early stages of drug research when only a limited amount of the API is available. Springuel *et al.* showed that molecules possessing similar chemical structures and functional groups will likely form co-crystals with the same co-formers.[26] Following a similar approach, Leyssens and co-workers investigated the propensity of co-crystal formation between enantiopure APIs and different co-formers, when these co-formers are already known to form co-crystals with the corresponding racemic compounds.[79] The authors selected levetiracetam (LEVI), a nootropic drug used as an anticonvulsant in the treatment of epilepsy,[80] and its corresponding racemic compound etiracetam (ETI) as the target molecules for their co-crystal screen with 152 non-chiral co-formers having at least one alcohol, aldehyde, amide, amine, or carboxylic acid functional group (Figure 9.23).[79] The co-crystals were prepared by solvent-drop grinding and characterized by powder and single-crystal X-ray diffraction techniques.

A comparable success rate of about 10% (18/152 and 14/152) with both the racemic and enantiopure compounds, respectively, was observed.[79] Interestingly, 13 of the 14 co-formers that formed co-crystals with the

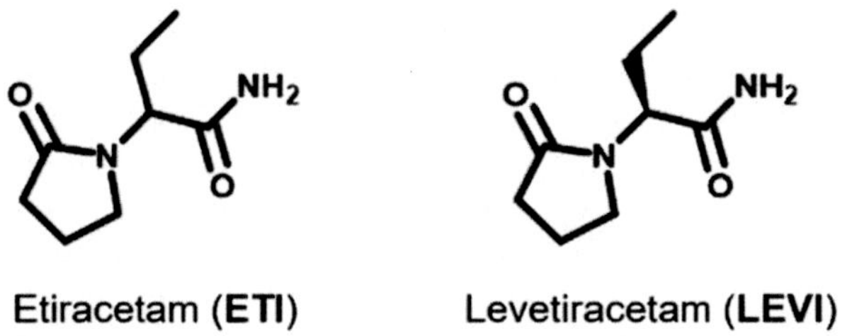

Figure 9.23 Chemical structures of etiracetam (left) and levetiracetam (right).[79]

enantiopure compound were common to those that gave a positive hit with the racemic compound. If the co-former list for the enantiopure screen is reduced to only the 18 compounds that were successful during the racemic screen, then the success rate jumps to 72% (13/18). This highlights the tendency of both the enantiopure compound (LEVI), and its corresponding racemic compound (ETI), to form co-crystals with identical non-chiral partners, which in turn shows that co-crystal screens of enantiopure compounds can be performed more efficiently if the co-crystal screen of the racemic compound has already been carried out. Also, all co-crystals showed a dense hydrogen-bonding network with each of the potential hydrogen-bond donors and acceptors involved in hydrogen bonding. This is contrary to the hydrogen bonding in the structures of pure LEVI and ETI, which partly explains the propensity of these compounds for co-crystallization.[79] Furthermore, formation of 3D supramolecular architectures is observed in all co-crystals with layers held together by hydrophobic interactions. This shows that although hydrogen bonds are the primary driving force towards co-crystal formation of chiral APIs, other forces such as steric hindrance and hydrophobic interactions (π–π and van der Waals) also need to be considered.[81] This two-step method of first performing an extended co-crystal screen on the racemic compound, followed by a focused screen with the enantiopure API using co-formers that gave positive hits in the first screen, represents an inexpensive and time-saving alternative to traditional solution-based enantioselective co-crystal screens of chiral APIs during early drug research.

Although relatively rare, when compared to diastereomeric salt formation, it is possible to obtain diastereomeric co-crystals of chiral compounds.[82] In such cases, it is important to develop strategies based on differences in the physical and physicochemical properties of the diastereomeric co-crystal pair to not only separate them, but also to regenerate the enantiomers. In this context, Höpfl and co-workers investigated the co-crystallization of praziquantel (PZQ),[83] an essential drug used in the treatment of gastrointestinal parasites,[84] with L-malic acid (L-MA). Praziquantel is marketed as the racemate despite *R*-PZQ being the biologically active component.[85] Also, the API suffers from poor solubility under physiological conditions (0.40 mg ml^{-1} in water at 25 °C).[86] Other efforts to separate the enantiomers have largely been unsuccessful,[87] thus making the idea of enantiomeric resolution of PZQ *via* co-crystallization an interesting prospect. Liquid-assisted grinding of *RS*-PZQ with L-MA led to the formation of the diastereomeric co-crystal pair (*R*-PZQ : L-MA and *S*-PZQ : L-MA). Analysis of the hydrogen- bond patterns as well as of melting point differences between the two co-crystals indicated that *R*-PZQ : L-MA co-crystal is more stable. Also, solubility experiments on the individual co-crystals indicated significant solubility differences in some solvents. Using the above information, the authors designed a two-step strategy to achieve chiral resolution of racemic PZQ *via* fractional crystallization from acetone or ethyl acetate in the presence of L-MA, followed by a phase-decomposition of the *R*-PZQ : L-MA co-crystal by treating with water (Scheme 9.3).[83] Also, since the *R*-PZQ : L-MA co-crystal exhibits a better solubility profile than *R*-**PZQ**·0.5H_2O and *RS*-PZQ, the novel co-crystal can be used as an alternative drug form to the racemate.

Crystal structure prediction (CSP) is a computational approach for predicting the crystal structure of single and multi-component crystals and for understanding the role played by different intermolecular interactions in the solid-state.[88] Consequently, there has been much research into CSP in the past few decades, and the goal is the development of a reliable computational method for predicting how a molecule will crystallize in the solid-state, with only the chemical diagram and the crystallization conditions known.[89] The main methodology used in CSP is global lattice energy minimization, wherein structure searching methods are used to generate the different possible ways of packing a molecule into a crystal, which are then ranked on the basis of their calculated energies.[90] CSP can also be used for chiral co-crystal screening, where co-crystal structures and structures for diastereomeric partners that cannot be determined experimentally can be visualized by CSP, which in turn assists in the study of the structural elements and intermolecular interactions present in chiral co-crystals.[82] Habgood applied CSP to three reported chiral systems, *i.e.* co-crystals of levetiracetam with mandelic acid and tartaric acid, and co-crystal of tartaric acid with malic acid, to assess the ability of CSP to correctly predict the experimental structures and crystal energy landscapes of chiral molecules.[82] Also, this study was aimed at obtaining detailed information on the intermolecular interactions responsible for the observed enantioselectivity during co-crystallization.

In this case, an examination of both the experimentally determined and the CSP predicted hypothetical crystal structures strongly suggested that the differences in stability between co-crystals of different enantiomers could be rationalized by the presence of the highly stabilizing eight-membered double hydrogen-bonded ring [R_2^2] motif.[82] A significant energy gap in co-crystals of tartaric acid with malic acid or levetiracetam is a consequence of large changes in other interaction motifs, which in turn are a result of exchange of one enantiomer for another while maintaining the $R_2^2(8)$ motif. The presence of an alternate highly stabilizing motif in the levetiracetam–mandelic acid co-crystal led to a smaller energy gap despite the exchange of enantiomers.

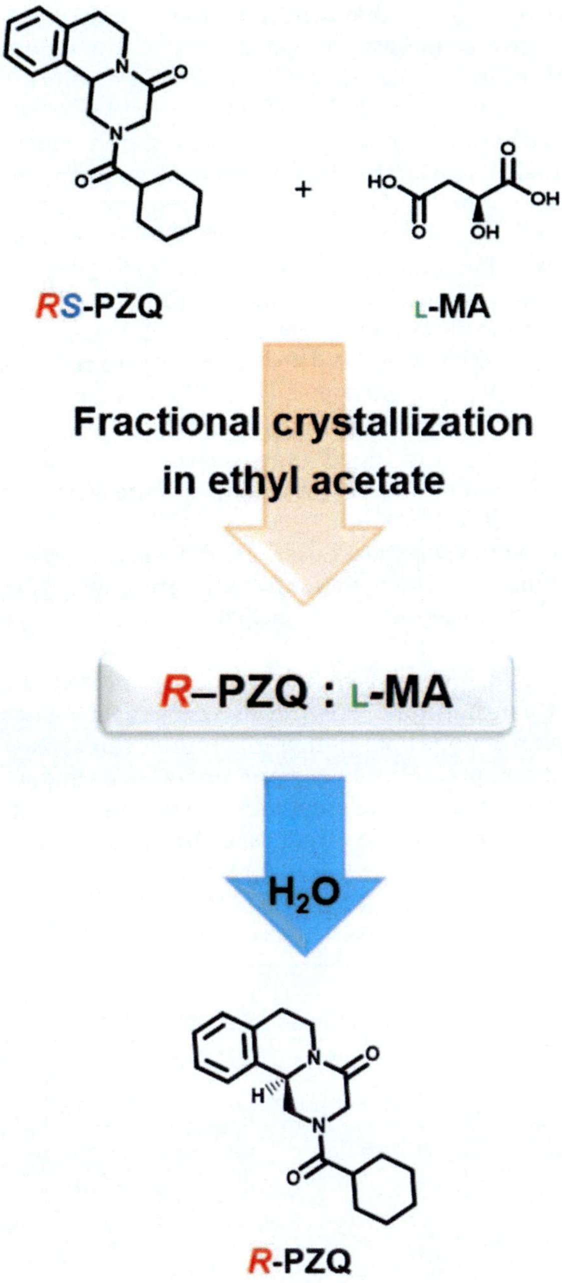

Scheme 9.3 Racemic resolution of *RS*-praziquantel (*RS*-PZQ) *via* fractional crystallization in the presence of L-malic acid followed by treatment with water.[83]

This shows that systems that can potentially rearrange the secondary interactions to give a moderately stable co-crystal containing the primary ring motif with either enantiomer will produce a diastereomeric co-crystal pair, whereas systems which lack such options will lead to enantiospecific co-crystals. These results are critical in highlighting some guidelines that need to be considered during the selection of co-formers that can act as effective resolving agents: (a) need for strong, orientationally restrictive interaction motif; (b) avoiding competing motifs; and (c) presence of multiple weaker secondary motifs that can be disrupted by the exchange of one enantiomer for another within the primary motif.[82]

Leyssens and co-workers combined experimental and theoretical methods (CSP) to screen a series of chiral APIs to investigate the propensity for two optically active molecules to co-crystallize in an enantiospecific manner.[91] Using CSP, they calculated the lattice energy landscapes for the different solid-state forms and concluded that the calculated structure would be formed only if the lattice energy was lower than the sum of its component's lattice energies (Figure 9.24).

It is known that co-crystals rarely exhibit high stabilization energies, with the stabilization free enthalpy of formation hardly surpassing 10 kJ mol^{-1}.[82,92] Consequently, lattice energies of co-crystals are smaller (around −200 kJ mol^{-1}),[91] when compared to that of most salts (−600 kJ mol^{-1}),[93] which also explains the low success rate in typical co-crystal screens. Furthermore, this means that small changes in structure, such as a switch from one enantiomer to another, can reduce the chances of co-crystal formation due to stronger steric interactions and loss of stabilizing secondary interactions.[91] These secondary interactions are of the same order of magnitude as the total co-crystal stabilization energy and their loss

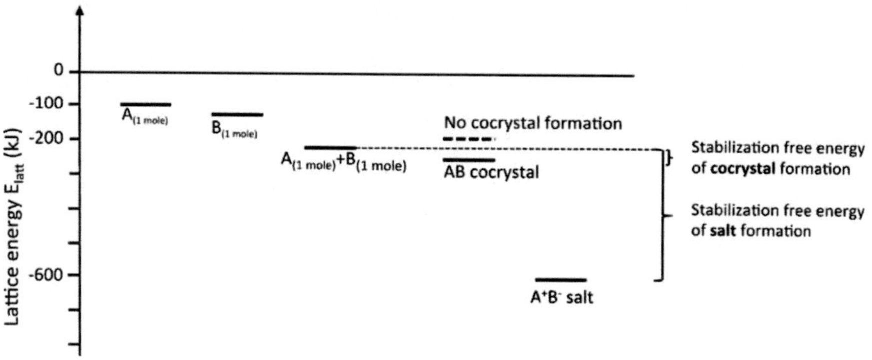

Figure 9.24 Illustration of lattice energies of isolated components (about −100 kJ mol^{-1}), the sum of components' lattice energy, lattice energy of co-crystal formation (about −200 kJ mol^{-1}), and lattice energy of salt formation (about −600 kJ mol^{-1}).[91]
Reprinted with permission from G. Springuel, K. Robeyns, B. Norberg, J. Wouters and T. Leyssens, *Cryst. Growth Des.*, 2014, **14**, 3996–4004, Copyright 2014 American Chemical Society.

explains why one enantiomer can form a co-crystal, whereas the other enantiomer cannot, even though they both possess the same primary hydrogen-bonding motifs. It also explains why almost 85% of the time co-crystals of chiral compounds display enantiospecific behavior, which is an advantage when compared to diastereomeric salt screens.

9.5 Summary

Co-crystallization can be an effective method for modifying the physicochemical properties of compounds. Thus, it is understandable that co-crystallization has attracted tremendous interest as a technique for chemical separations and chiral resolution. However, a wide range of factors needs to be carefully considered when designing co-crystallization based separation processes. First, it is important to understand the thermodynamic behavior of co-crystal systems by constructing ternary and quaternary phase diagrams, as this can allow us to identify ideal solvent system and stoichiometry, and optimum crystallization conditions. Second, a combination of thermodynamics and judicious choice of co-former (based on CSD data), coupled with a reduction in the solubility of the co-crystal can potentially prevent the inclusion of impurities in the crystal lattice of the target molecule. Other factors such as steric effects (molecular weight) and functional groups also need to be considered. Furthermore, recent studies have highlighted that a quantitative understanding of the strength of association in solution-phase using both experimental (phase solubility diagrams) and theoretical techniques can directly identify co-crystallization-based chemical separation experiments that can be used as a viable alternative to conventional methods. Third, co-crystallization, in conjunction with methods such as mechanochemical milling and preferential enrichment, can also be successfully used to resolve racemic compounds into their diastereomers in high enantiomeric excess. In the case of enantiopure APIs that are available in limited quantities in early drug research, a two-step method of first performing an extended co-crystal screen on the racemic compound, followed by a focused screen with the enantiopure API using co-formers that gave positive hits in the first screen represents an inexpensive and time-saving alternative to traditional solution-based co-crystal screens. Lastly, CSP has been used to establish guidelines that need to be considered during the selection of co-formers that can act as effective resolving agents. A successful protocol requires (a) a strong and orientationally restrictive interaction motif; (b) the absence of potentially competing motifs; and (c) the presence of multiple weaker secondary motifs that can be disrupted by the exchange of one enantiomer for another within the primary motif. This means that small changes in molecular structure, such as a switch from one enantiomer to another, can reduce the propensity for co-crystal formation due to stronger steric interactions and loss of stabilizing secondary interactions, which explains why one enantiomer can form a co-crystal while the other cannot.

References

1. K. Valko, *Separation Methods in Drug Synthesis and Purification*, Elsevier Science, Amsterdam, 1st edn, 2000, vol. 1.
2. B. Y. Shekunov and P. York, *J. Cryst. Growth*, 2000, **211**, 122–136.
3. S. Cherukuvada, R. Kaur and T. N. Guru Row, *CrystEngComm*, 2016, **18**, 8528–8555.
4. N. K. Duggirala, M. L. Perry, Ö. Almarsson and M. J. Zaworotko, *Chem. Commun.*, 2016, **52**, 640–655.
5. J. Zhang and J. M. Shreeve, *CrystEngComm*, 2016, **18**, 6124–6133.
6. (a) A. N. Sokolov, T. Friščić and L. R. MacGillivray, *J. Am. Chem. Soc.*, 2006, **128**, 2806–2807; (b) D. Braga, F. Grepioni and L. Maini, *Chem. Commun.*, 2010, **46**, 6232–6242; (c) K. M. Hutchins, J. C. Sumrak, D. C. Swenson and L. R. MacGillivray, *CrystEngComm*, 2014, **16**, 5762–5764.
7. (a) J. W. Lee, S. Samal, N. Selvapalam, H.-J. Kim and K. Kim, *Acc. Chem. Res.*, 2003, **36**, 621–630; (b) E. Mileo, E. Mezzina, F. Grepioni, G. F. Pedulli and M. Lucarini, *Chem. - Eur. J.*, 2009, **15**, 7859–7862; (c) E. V. Peresypkina, V. P. Fedin, V. Maurel, A. Grand, P. Rey and K. E. Vostrikova, *Chem. Eur. J.*, 2010, **16**, 12481–12487.
8. E. Nauha and M. Nissinen, *J. Mol. Struct.*, 2011, **1006**, 566–569.
9. X. Pang and W. J. Jin, *Top. Curr. Chem.*, 2014, **359**, 115–146.
10. (a) S. Jiang, R. S. Patil, C. L. Barnes and J. L. Atwood, *Cryst. Growth Des.*, 2017, **17**, 2919–2922; (b) S. Jiang, R. S. Patil, C. L. Barnes and J. L. Atwood, *Cryst. Growth Des.*, 2017, **17**, 4060–4063.
11. (a) R. A. Chiarella, R. J. Davey and M. L. Peterson, *Cryst. Growth Des.*, 2007, **7**, 1223–1226; (b) J. H. ter Horst, M. A. Deij and P. W. Cains, *Cryst. Growth Des.*, 2009, **9**, 1531–1537.
12. (a) C. B. Aakeröy, S. Forbes and J. Desper, *J. Am. Chem. Soc.*, 2009, **131**, 17048–17049; (b) P. Sanphui and L. Rajput, *Acta Crystallogr., Sect. B: Struct. Sci., Cryst. Eng. Mater.*, 2014, **70**, 81–90.
13. L. G. Pakhomov, V. A. Mitrofanova and I. A. Vostokov, *Fiz.-Khim. Metody Anal.*, 1976, **1**, 73–74.
14. L. G. Pakhomov, V. A. Mitrofanova and I. A. Vostokov, *Zh. Obshch. Khim.*, 1977, **47**, 1103–1106.
15. M. J. Conroy and R. E. Lovrien, *J. Cryst. Growth*, 1992, **122**, 213–222.
16. G. R. Desiraju, *Angew. Chem., Int. Ed. Engl.*, 1995, **34**, 2311–2327.
17. E. Corradi, S. V. Meille, M. T. Messina, P. Metrangolo and G. Resnati, *Angew. Chem., Int. Ed.*, 2000, **39**, 1782–1786.
18. (a) J. W. Schroer, C. Wibowo and K. M. Ng, *AIChE J.*, 2001, **47**, 369; (b) L. Codan, M. U. Bäbler and M. Mazzotti, *Cryst. Growth Des.*, 2010, **10**, 4005; (c) V. S. Sistla, H. Lorenz and A. Seidel-Morgenstern, *Cryst. Growth Des.*, 2013, **13**, 2548; (d) L. Codan, M. U. Bäbler and M. Mazzotti, *Org. Process Res. Dev.*, 2012, **16**, 294.
19. (a) K. Chadwick, R. Davey, G. Sadiq, W. Cross and R. Pritchard, *CrystEngComm*, 2009, **11**, 412; (b) K. Guo, G. Sadiq, C. Seaton, R. Davey and Q. Yin, *Cryst. Growth Des.*, 2010, **10**, 268; (c) D. H. Leung, S. Lohani,

R. G. Ball, N. Canfield, Y. Wang, T. Rhodes and A. Bak, *Cryst. Growth Des.*, 2012, **12**, 1254.
20. A. J. J. Straathof, *Biotechnol. Prog.*, 2003, **19**, 755–762.
21. C. B. Aakeröy, M. E. Fasulo and J. Desper, *Mol. Pharmaceutics*, 2007, **4**, 317–322.
22. J. Urbanus, C. P. M. Roelands, D. Verdoes, P. J. Jansens and J. H. ter Horst, *Cryst. Growth Des.*, 2010, **10**, 1171–1179.
23. G. Springuel and T. Leyssens, *Cryst. Growth Des.*, 2012, **12**, 3374–3378.
24. B. S. Sekhon, *ARS Pharm.*, 2009, **50**, 99–117.
25. S. P. Kotkar and A. Sudalai, *Tetradron Lett.*, 2006, **47**, 6813–6815.
26. G. Springuel, B. Norberg, K. Robeyns, J. Wouters and T. Leyssens, *Cryst. Growth Des.*, 2012, **12**, 475–484.
27. P. Billot, P. Hosek and M.-A. Perrin, *Org. Process Res. Dev.*, 2013, **17**, 505–511.
28. G. Springuel, L. Collard and T. Leyssens, *CrystEngComm*, 2013, **15**, 7951–7958.
29. K. H.-Y. Hsi, K. Chadwick, A. Fried, M. Kenny and A. S. Myerson, *CrystEngComm*, 2012, **14**, 2386–2388.
30. C. R. Groom, I. J. Bruno, M. P. Lightfoot and S. C. Ward, *Acta Crystallogr.*, 2016, **B72**, 171–179.
31. R. D. B. Walsh, M. W. Bradner, S. Fleischman, L. A. Morales, B. Moulton, N. Rodríguez-Hornedo and M. J. Zaworotko, *Chem. Commun.*, 2003, **0**, 186–187.
32. S. J. Nehm, B. Rodríguez-Spong and N. Rodríguez-Hornedo, *Cryst. Growth Des.*, 2006, **6**, 592–600.
33. K. H. Hsi, M. Kenny, A. Simi and A. S. Myerson, *Cryst. Growth Des.*, 2013, **13**, 1577–1582.
34. (a) T. Higuchi and D. A. Zuck, *J. Am. Pharm. Assoc.*, 1953, **42**, 132–138; (b) T. Higuchi and D. A. Zuck, *J. Am. Pharm. Assoc.*, 1953, **42**, 138–145; (c) T. Higuchi and J. L. Lach, *J. Am. Pharm. Assoc.*, 1954, **43**, 349–354; (d) T. Higuchi and J. L. Lach, *J. Am. Pharm. Assoc.*, 1954, **43**, 524–527; (e) T. Higuchi and I. H. Pitman, *J. Pharm. Sci.*, 1973, **62**, 55–58.
35. K. H. Y. Hsi, A. J. Concepcion, M. Kenny, A. A. Magzoub and A. S. Myerson, *CrystEngComm*, 2013, **15**, 6776–6781.
36. A. Bruggink, *Synthesis of β-Lactam Antibiotics: Chemistry, Biocatalysis & Process Integration*, Springer, 2001.
37. R. C. Giordano, M. P. A. Ribeiro and R. L. C. Giordano, *Biotechnol. Adv.*, 2006, **24**, 27–41.
38. (a) L. R. B. GonCalves, R. Fernandez-Lafuente, J. M. Guisán and R. L. C. Giordano, *Appl. Biochem. Biotechnol.*, 2000, **84–86**, 931–945; (b) I. Alemzadeh, G. Borghei, L. Vafi and R. Roostaazad, *Sci. Iran., Trans. C*, 2010, **17**, 106–113.
39. Y. Hua, Q. Shan, X. Cong and B. Cai, *Zhonghua Zhongyiyao Xuekan*, 2013, **31**, 727–730.
40. T. Lee, H. R. Chen, H. Y. Lin and H. L. Lee, *Cryst. Growth Des.*, 2012, **12**, 5897–5907.

41. D. Braga, F. Grepioni, L. Maini, P. P. Mazzeo and K. Rubini, *Thermochim. Acta*, 2010, **507–508**, 1–8.
42. C. C. Weber, G. P. F. Wood, A. J. Kunov-Kruse, D. E. Nmagu, B. L. Trout and A. S. Myerson, *Cryst. Growth Des.*, 2014, **14**, 3649–3657.
43. L. Addadi, Z. Berkovitch-Yellin, I. Weissbuch, M. Lahav and L. Leiserowitz, *Mol. Cryst. Liq. Cryst.*, 1983, **96**, 1–17.
44. F. P. Schmidtchen, in *Analytical Methods in Supramolecular Chemistry*, ed. C. A. Schalley, Wiley-VCH: Weinheim, Germany, 2nd edn, 2012, vol. 1, p. 67.
45. C. A. Pons-Siepermann, S. Huang and A. S. Myerson, *CrystEngComm*, 2016, **18**, 7487–7493.
46. C. A. Pons-Siepermann, S. Huang and A. S. Myerson, *Cryst. Growth Des.*, 2017, **17**, 2646–2653.
47. H. L. Lee and T. Lee, *CrystEngComm*, 2015, **17**, 9002–9006.
48. J.-M. Lehn, *Supramolecular Chemistry: Concepts and Perspectives*, VCH, Weinheim, 1995.
49. S. Müller, M. C. Afraz, R. de Gelder, G. J. A. Ariaans, B. Kaptein, Q. B. Broxterman and A. Bruggink, *Eur. J. Org. Chem.*, 2005, **2005**, 1082–1096.
50. (a) H. Lorenz, F. Capla, D. Polenske, M. P. Elsner and A. Seidel-Morgenstern, *J. Univ. Chem. Technol. Metall.*, 2007, **42**, 5–16; (b) K. A. Lyseng-Williamson, *Drugs*, 2011, **71**, 489–514; (c) E. Fogassy, M. Nógrádi, D. Kozma, G. Egri, E. Pálovics and V. Kiss, *Org. Biomol. Chem.*, 2006, **4**, 3011–3030; (d) V. Kumar and S. V. Malhotra, *ACS Symp. Ser.*, 2010, **1038**, 1–12; (e) S. H. Neau, *Water-Insoluble Drug Formulation*, 2000, 405–425; (f) N. R. Srinivas, *Biomed. Chromatogr.*, 2004, **18**, 207–233.
51. (a) B. Chankvetadze, in *Molecular Biology in Medicinal Chemistry, Book Series: Methods and Principles in Medicinal Chemistry*, ed. T. Dingermann, D. Steinhilber and G. Folkers, Wiley-VCH Verlag GmbH & Co. KGaA, New York, 2004, ch. 21, pp. 181–210; (b) J. Wu, P. Liu, Q. Wang, H. Chen, P. Gao, L. Wang and S. Zhang, *Chromatographia*, 2011, **74**, 789–797; (c) T. J. Ward, *Anal. Chem.*, 2002, **74**, 2863–2872.
52. G. Blaschke and B. Chankvetadze, in *New Trends in Synthetic Medicinal Chemistry, Book Series: Methods and Principles in Medicinal Chemistry*, ed. F. Gualtieri, Wiley-VCH Verlag GmbH, New York, 2008, ch. 6, pp. 139–173.
53. M. R. Caira, L. R. Nassimbeni, J. L. Scott and A. F. Wildervanck, *J. Chem. Crystallogr.*, 1996, **26**, 117–122.
54. (a) F. Toda, K. Tanaka and S. Nagamatsu, *Tetrahedron Lett.*, 1984, **25**, 4929–4932; (b) K. Ogura, T. Uchida, M. Noguchi, M. Minoguchi, A. Murata, M. Fujita and K. Ogata, *Tetrahedron Lett.*, 1990, **31**, 3331–3334; (c) M. Akazome, M. Noguchi, O. Tanaka, A. Sumikawa, T. Uchida and K. Ogura, *Tetrahedron*, 1997, **53**, 8315–8322.
55. O. Bortolini, G. Fantin, M. Fogagnolo, A. Medici and P. Pedrini, *Chem. Commun.*, 2000, 365–366.

56. G. Fantin, M. Fogagnolo, O. Bortolini, N. Masciocchi, S. Galli and A. Sironi, *New J. Chem.*, 2004, **28**, 1295–1300.
57. M. Klussmann, T. Izumi, A. J. P. White, A. Armstrong and D. G. Blackmond, *J. Am. Chem. Soc.*, 2007, **129**, 7657–7660.
58. J. L. Scott, S. Hachiken and K. Tanaka, *Cryst. Growth Des.*, 2008, **8**, 2447–2452.
59. (a) M. C. Etter, *Acc. Chem. Res.*, 1990, **23**, 120–126; (b) J. Bernstein, R. E. Davis, L. Shimoni and N.-L. Chang, *Angew. Chem., Int. Ed. Engl.*, 1995, **34**, 1555–1573.
60. O. Ermer and A. Eling, *J. Chem. Soc., Perkin Trans. 2*, 1994, 925–944.
61. B. N. Roy, G. P. Singh, D. Srivastava, H. S. Jadhav, M. B. Saini and U. P. Aher, *Org. Process Res. Dev.*, 2009, **13**, 450–455.
62. (a) B. Belleau, D. Dixit, N. Nguyen-Ba and J. Kraus, Fifth International Conference on AIDS, Montreal, Canada, June, 1989, paper T.C.O.1; (b) B. Belleau and N. Nguyen-Ba, U. S. Pat., 5 047 407, 1991.
63. (a) J. A. Coates, N. Cammack, H. J. Jenkinson, A. J. Jowett, M. I. Jowett, B. A. Pearson, C. R. Penn, P. L. Rouse, K. C. Viner and J. M. Cameron, *Antimicrob. Agents Chemother.*, 1992, **36**, 733–739; (b) C. K. Chu, J. W. Beach, L. S. Jeong, B. G. Choi, F. I. Comer, A. J. Alves and R. F. Schinazi, *J. Org. Chem.*, 1991, **56**, 6503–6505; (c) L. S. Jeong, A. J. Alves, S. W. Carrigan, H. O. Kim, J. W. Beach and C. K. Chu, *Tetrahedron Lett.*, 1992, **33**, 595–598.
64. D. C. Liotta, R. F. Schinazi and W.-B. Choi, *U. S. Pat.*, 6 069 252, 2000.
65. (a) J. M. Brunel, *Chem. Rev.*, 2005, **105**, 857–898; (b) R. Noyori, *Asymmetric Catalysis in Organic Synthesis*, John Wiley & Sons, New York, 1994.
66. R. A. Chiarella, R. J. Davey and M. L. Peterson, *Cryst. Growth Des.*, 2007, **7**, 1223–1226.
67. (a) M. R. Caira, L. R. Nassimbeni and A. F. Wildervanck, *J. Chem. Soc., Perkin Trans. 2*, 1995, 2213–2216; (b) T. Friščić and W. Jones, *Cryst. Growth Des.*, 2009, **9**, 1621–1637.
68. N. Shan, F. Toda and W. Jones, *Chem. Commun.*, 2002, 2372–2373.
69. M. D. Eddleston, M. Arhangelskis, T. Friščić and W. Jones, *Chem. Commun.*, 2012, **48**, 11340–11342.
70. C. B. Aakeröy, T. I. Cooke and M. Nieuwenhuyzen, *Supramol. Chem.*, 1996, **7**, 153–156.
71. (a) T. Ushio, R. Tamura, H. Takahashi and K. Yamamoto, *Angew. Chem., Int. Ed. Engl.*, 1996, **35**, 2372–2374; (b) R. Tamura, H. Takahashi, K. Hirotsu, Y. Nakajima, T. Ushio and F. Toda, *Angew. Chem. Int. Ed.*, 1998, **37**, 2876–2878.
72. (a) R. Tamura, D. Fujimoto, Z. Lepp, K. Misaki, H. Miura, H. Takahashi, T. Ushio, T. Nakai and K. Hirotsu, *J. Am. Chem. Soc.*, 2002, **124**, 13139–13153; (b) D. Fujimoto, R. Tamura, Z. Lepp, H. Takahashi and T. Ushio, *Cryst. Growth Des.*, 2003, **3**, 973–979.
73. (a) S. Iwama, M. Horiguchi, H. Sato, Y. Uchida, H. Takahashi, H. Tsue and H. R. Tamura, *Cryst. Growth Des.*, 2010, **10**, 2668–2675; (b) R. G. Gonnade,

S. Iwama, R. Sugiwake, K. Manoj, H. Takahashi, H. Tsue and R. Tamura, *Chem. Commun.*, 2012, **48**, 2791–2793.
74. S. Iwama, K. Kuyama, Y. Mori, K. Manoj, R. G. Gonnade, K. Suzuki, C. E. Hughes, P. A. Williams, K. D. M. Harris, S. Veesler, H. Takahashi, H. Tsue and R. Tamura, *Chem. - Eur. J.*, 2014, **20**, 10343–10350.
75. (a) R. Tamura, S. Iwama and R. G. Gonnade, *CrystEngComm*, 2011, **13**, 5269–5280; (b) R. Tamura, S. Iwama and H. Takahashi, *Symmetry*, 2010, **2**, 112–135; (c) R. Tamura, H. Takahashi, D. Fujimoto and T. Ushio, *Top. Curr. Chem.*, 2007, **269**, 53–82.
76. (a) E. Courvoisier, P. A. Williams, G. K. Lim, C. E. Hughes and K. D. M. Harris, *Chem. Commun.*, 2012, **48**, 2761–2763; (b) H. Tapiero, G. Mathé, P. Couvreur and K. D. Tew, *Biomed. Pharmacother.*, 2002, **56**, 439–445.
77. R. G. Gonnade, S. Iwama, Y. Mori, H. Takahashi, H. Tsue and R. Tamura, *Cryst. Growth Des.*, 2011, **11**, 607–615.
78. K. Manoj, H. Takahashi, Y. Morita, R. G. Gonnade, S. Iwama, H. Tsue and R. Tamura, *Chirality*, 2015, **27**, 405–410.
79. F. George, N. Tumanov, B. Norberg, K. Robeyns, Y. Filinchuk, J. Wouters and T. Leyssens, *Cryst. Growth Des.*, 2014, **14**, 2880–2892.
80. B. Hurtado, M. J. Koepp, J. W. Sander and P. J. Thompson, *Epilepsy Behav.*, 2006, **8**, 588–592.
81. (a) P. W. Baures, J. R. Rush, A. V. Wiznycia, J. Desper, B. A. Helfrich and A. M. Beatty, *Cryst. Growth Des.*, 2002, **2**, 653–664; (b) N. Takata, K. Shiraki, R. Takano, Y. Hayashi and K. Terada, *Cryst. Growth Des.*, 2008, **8**, 3032–3037.
82. M. Habgood, *Cryst. Growth Des.*, 2013, **13**, 4549–4558.
83. O. Sánchez-Guadarrama, F. Mendoza-Navarro, A. Cedillo-Cruz, H. Jung-Cook, J. I. Arenas-García, A. Delgado-Díaz, D. Herrera-Ruiz, H. Morales-Rojas and H. Höpfl, *Cryst. Growth Des.*, 2016, **16**, 307–3314.
84. D. Cioli and L. Pica-Mattoccia, *Parasitol. Res.*, 2003, **90**, S3–S9.
85. T. Meyer, H. Sekljic, S. Fuchs, H. Bothe, D. Schollmeyer and C. Miculka, *PLoS Neglected Trop. Dis.*, 2009, **3**, e357.
86. P. de La Torre, S. Torrado and S. Torrado, *Chem. Pharm. Bull.*, 1999, **47**, 1629–1633.
87. M. Woelfle, J.-P. Seerden, J. De Gooijer, K. Pouwer, P. Olliaro and M. H. Todd, *PLoS Neglected Trop. Dis.*, 2011, **5**, e1260.
88. G. M. Day, *Crystallogr. Rev.*, 2011, **17**, 3–52.
89. D. A. Bardwell, C. S. Adjiman, Y. A. Arnautova, E. Bartashevich, S. X. M. Boerrigter, D. E. Braun, A. J. Cruz-Cabeza, G. M. Day, R. G. Della Valle, G. R. Desiraju, B. P. van Eijck, J. C. Facelli, M. B. Ferraro, D. Grillo, M. Habgood, D. W. M. Hofmann, F. Hofmann, K. V. Jovan Jose, P. G. Karamertzanis, A. V. Kazantsev, J. Kendrick, L. N. Kuleshova, F. J. J. Leusen, A. V. Maleev, A. J. Misquitta, S. Mohamed, R. J. Needs, M. A. Neumann, D. Nikylov, A. M. Orendt, R. Pal, C. C. Pantelides, C. J. Pickard, L. S. Price, S. L. Price,

H. A. Scheraga, J. van de Streek, T. S. Thakur, S. Tiwari, E. Venuti and I. K. Zhitkov, *Acta Crystallogr., Sect. B.*, 2011, **B67**, 535.

90. A. V. Kazantsev, P. G. Karamertzanis, C. S. Adjiman, C. C. Pantelides, S. L. Price, P. T. A. Galek, G. M. Day and A. J. Cruz-Cabeza, *Int. J. Pharm.*, 2011, **418**, 168.
91. G. Springuel, K. Robeyns, B. Norberg, J. Wouters and T. Leyssens, *Cryst. Growth Des.*, 2014, **14**, 3996–4004.
92. (a) N. Issa, P. G. Karamertzanis, G. W. A. Welch and S. L. Price, *Cryst. Growth Des.*, 2009, **9**, 442–453; (b) M. Habgood and S. L. Price, *Cryst. Growth Des.*, 2010, **10**, 3263–3272.
93. (a) S. Mohamed, D. A. Tocher and S. L. Price, *Int. J. Pharm.*, 2011, **418**, 187–198; (b) P. G. Karamertzanis and S. L. Price, *J. Phys. Chem. B*, 2005, **109**, 17134–17150.

Subject Index